£31.7

SATELLITE TECHNOLOGY AND ITS APPLICATIONS

P.R.K. CHETTY

Fairchild Space Company
Germantown, Maryland

TAB BOOKS Inc.

Blue Ridge Summit, PA 17294

Note: The author makes no representation that the use of the material described herein will not infringe on existing or future patent rights, nor do the descriptions contained herein imply the granting of licenses to make, use, or sell equipment constructed in accordance therewith. No liability is assumed with respect to the use of the information herein. The views and contents of this book are solely those of the author and not of Fairchild Space Company.

FIRST EDITION
FIRST PRINTING

Copyright © 1988 by TAB BOOKS Inc.
Printed in the United States of America

Library of Congress Cataloging in Publication Data

Chetty, P.R.K.
Satellite technology and its applications / by P.R.K. Chetty.
p. cm.
Bibliography: p.
Includes index.
ISBN 0-8306-2931-9 :
1. Artificial satellites. I. Title.
TL796.C45   1987   87-26234
629.44—dc19   CIP

Questions regarding the content of this book should be addressed to:

Reader Inquiry Branch
Editorial Department
TAB BOOKS Inc.
Blue Ridge Summit, PA 17294

Cover photograph courtesy of NASA.

Contents

Preface

IN THE 1960s, THE DEMANDS OF THE SPACE PROGRAMS LED TO THE DEVELOP-
ment of highly reliable, efficient, and lightweight spacecraft. Despite limited technology, engineers found innovative solutions for manufacturing efficient spacecraft. These helped usher in the space era. In recent years the satellite applications and the number of satellites in practical use and demands for newer satellites are continuously increasing. Better and challenging performance demands await the spacecraft designers/engineers.

The potential of satellite applications is so great that it can change human life. The future will see people receiving mail and doing shopping via satellites. Due to higher costs and time involved with travel, large screen teaching and teleconferencing will become part of our lives. The future office may be spread over small towns and even perhaps neighboring cheap labor countries. The airline industry is already using cheap clerical labor from neighboring countries for processing airline tickets through satellites, and some telephone answering and paging services can be from such places where labor is cheap.

Thus, satellite facilities change telephone and data networks, word processing and mail, travel budgets, training, and human communications as satellite communications become cheaper in cost. There will be many new busi-ness opportunities using the new satellites that can be accessed by small rooftop and parking lot earth stations.

Space research is neither fancy nor prestigious. Many people wonder—
"Why explore space when we have so many grave problems—poverty, hunger,

disease, unemployment and the like before us?" Space research is no direct source of freedom from these grave problems. However, it is a sincere endeavor to put space technology to earth-bound applications and to make this world a happier place. Space research makes valuable contributions directly in the domains of agriculture, communications, education, medicine, mining, material science, etc., besides being a stimulant for the rapid advance in the basic sciences. It has been estimated that even a modest system of earth satellites working with a program for agricultural improvement will increase annual crop yields by an equivalent of many thousands of thousands of dollars, especially in the third world countries. Such satellites can screen wide areas in a short time and radio back the information pertaining to crops, soil, droughts, pests, weather, monsoons, etc. All these countries without such satellites are at the mercy of monsoons. Weather satellites predict the nature of the monsoons well in advance and farmers can plan and take precautions. Earth resource satellites provide information about the unexploited natural resources hidden underneath the earth that are inaccessible using conventional means.

Communications are most economical via satellite. Education through TV can radically change the general outlook of millions of uneducated people in the developing countries. Years of space flight have been rich in discovery, so rich that not until the latter portions of the decade have we been able to start to exploit these discoveries or even to appreciate all of their implications. In brief, space flight has led to growth of knowledge in the following areas: Earth: exosphere, magnetosphere, meteorology, geodesy; Solar System: solar wind, imbedded magnetic field, interplanetary dust; Venus: temperatures; Mars: atmosphere, cratering; Moon: topography, basalt crust, surface properties; and Applications: weather forecasting, communications, navigation, reconnaissance, etc.

Acknowledgments

THIS BOOK WAS DEVELOPED FROM MY 15 YEARS OF WORKING EXPERIENCE IN the spacecraft industry in India, France, and in the U.S. Many colleagues at Fairchild Space Company, especially Francesco Costanzo and Ruth Cholvibul, and N.V. Sivaprasad of INMARSAT (earlier INTELSAT) have generously contributed time in reviewing and constructively critiquing this manuscript. I am very grateful to them. Francesco Costanzo also coauthored Chapter 2. Special thanks to N.V. Sivaprasad my classmate, colleague, and friend who continuously encouraged me throughout the period of writing this book and reviewed the complete manuscript and made this book a success.

The following companies have provided photos, figures, and descriptive material about satellites and satellite components and have given permission to use and reprint the same: Aeritalia, Aerojet Tech-Systems Co., Arianespace Inc., Ball Aerospace Systems Division, British Aerospace, Canadian Astronautics Ltd., Eagle Picher Industries Inc., Earth Observation Satellite Company, European Space Agency, Fokker Space & Systems B.V., General Dynamics Space Systems Division, General Electric Company, Humphrey Inc., INTELSAT, Ithaco Inc., Jet Propulsion Laboratory, Litton Poly-Scientific, Lockheed Missiles & Space Company, Inc., Magnetics, Marconi Space Systems Ltd., Martin Marietta Denver Aerospace, NASA Goddard Space Flight Center, NASA Lyndon B. Johnson Space Center, Navy Space Systems Activity, RCA, Schaeffer Magnetics Inc., SPAR, Spectrolab, Inc., Sperry, TAB Books Inc., TELDIX, and TRW Space Log. Parts of Sections 4.4.1, 4.5.2, 4.7.1, 4.7.2, and 4.7.3 are adapted with permission from Ithaco data sheets; 4.4.3 from Ball Aerospace data sheets; 4.4.5 from Humphrey Inc.'s data sheets; 4.5.1, and 4.5.5 from TELDIX data sheets; 4.5.3 from Fokker data sheets; 9.5 from Marconi brochures; 11.8 from INTELSAT brochures; 12.5.1 from RCA brochures; 12.5.2 from General Electric brochure; and 12.5.3 from European Space Agency's publication -ESA BR-08: ECS databook. Pages A-1A, A-2B, and A-3 are reprinted with permission from TRW Space Log; and A-26, A-28 from the Magnetics Catalog. I am very grateful to each of the above companies.

I am especially appreciative of the patience, help, and encouragement given by my wife, Sarada, during the years that went into writing this book, for iterative typing of the whole manuscript and my son, Santosh and my daughter, Sruthi, for sacrificing many precious hours which rightfully belonged to them.

Introduction

THIS BOOK CONTAINS TWO PARTS. PART I DEALS WITH SATELLITE BUILDING blocks, and Part II deals with the various applications. Part I details the state of the art technology of various building blocks that go into making the satellite after an introduction to satellites, and Part II discusses the various applications of satellites like telecommunications, broadcasting, weather forecasting (meteorology), predicting crop yields, mineral exploration, land survey, status of forestry, growth of ultra-clean crystals for faster computers, and growth of products for making new and cheaper medicines. It also discusses the financial aspects and changes in our daily lives. Also included are some of the important design details, calculations, and examples, etc.

Most of the books in this field largely confine themselves to satellite communications and mainly concentrate on the communications aspects including communications technology. Some books are quite old and some may not be relevant. There are also some books about satellites for juveniles. Thus most of the books present satellite communications and are not about satellites and the design of the satellites. This book, perhaps, is the first attempt to concentrate on defining a satellite and its building blocks, describing each building block, and presenting some applications in detail for today and tomorrow. Also, this book illustrates the various design examples and provides formulas and data about satellite design.

This is a practical and ready to use reference book for satellite design engineers, satellite systems engineers, and managers. This book can be used by engineers of various disciplines—electrical, electronics, electromechanical, and systems engineers working in the field of satellites. This book may also be recommended as an introductory textbook at the BS level.

In Loving Memory of My Father

Part I
Satellite Technology— Building Blocks

Chapter 1

Satellites and Rockets

1.1 SATELLITE AND ROCKET OVERVIEW

THE EARTH IS WRAPPED AROUND BY A LAYER OF GASES KNOWN AS THE ATmosphere and its density decreases as the distance from the earth increases. Above about 100 miles, there is no air and the atmosphere disappears. Thus, space begins above about 100 miles.

The Earth's gravity keeps us, oceans, and other objects, small and large, tied to it. It keeps the atmosphere in place. All the heavenly bodies like the sun, moon, planets, and stars have gravity. Anything thrown up falls down due to gravity and any object traveling in the atmosphere will be slowed down by the atmospheric drag, the lower the density of the atmosphere, the lower the effect of slowing. Thus, an object traveling in space, continues to travel until it meets the atmosphere due to the earth's gravity. But to send an object into space, first it has to overcome gravity and then it has to travel at least at a particular minimum speed to stay in space. Thus, an object traveling at about 5 miles per second can circle around the earth and become a satellite. Such a velocity or thrust can only be generated by rockets at present.

Burning fuel in the combustion chamber of the rocket turns into hot gases. The gases expand and push hard against the front and sides but can't get out. They escape out of the exhaust nozzle. The difference between the pressure of the hot, expanding gases inside the rocket and the lower pressure outside pushes the rocket ahead upward away from the earth, as they stand with nozzles downward.

To obtain maximum advantage, instead of making one big rocket, two or more are stacked one over the other so that when the fuel in one rocket completely burns out, it can be dropped (which burns up in the atmosphere), thereby net weight decreases. Without the weight of the first stage to slow it up the speed of the rocket increases. This process increases the overall system efficiency. Thus any object mounted on the top of the rocket, at the end is thrown into space and becomes a satellite. Figures 1-1 and 1-2 show the Landsat-D satellite and FLTSATCOM satellite in orbit.

Spacecraft carry electronic equipment and solar arrays that convert the sunlight into electrical energy and power the equipment. Spacecraft send information to the ground. They are like broadcasting stations in space. They can also be used as repeaters or relay stations in space.

1.1.1 A Satellite

A satellite is defined as an unattended body revolving about a larger one. Though, it was used to denote a planet's moon, since 1957 it also means a man-made body put into orbit around another large body. Man-made satellites are sometimes called artificial satellites.

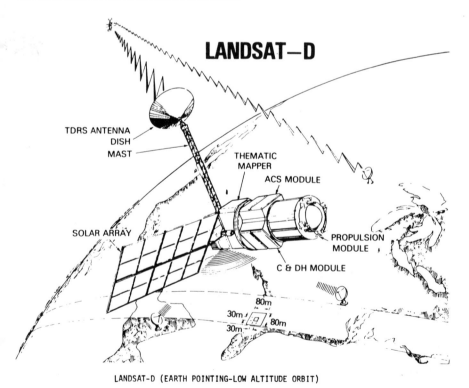

LANDSAT-D (EARTH POINTING-LOW ALTITUDE ORBIT)

Fig. 1-1. Landsat-D satellite in orbit (courtesy of NASA Goddard Space Flight Center).

Fig. 1-2. FLTSATCOM communication satellite in orbit (courtesy of TRW).

An enormous amount of energy is necessary to put a satellite into orbit, and is realized by using a rocket. *Rocket* is defined as an apparatus consisting of a case containing a propellant, i.e., fuel and reagents, by the combustion of which it is projected up into the atmosphere or space. As the payload is carried on the top of the rocket, the rocket is usually separated and dropped after burn-out. Thus, a rocket brings a payload up to a required velocity and leaves it in the orbit.

Sometimes a rocket is also known as a *booster* as a rocket starts with a low velocity, attains some height where air drag decreases, and it attains a high velocity.

1.1.2 A Rocket

The rocket ejects the products of the combustion through a nozzle (Fig. 1-3) and thereby moves in the opposite direction by the principle of recoil and conservation of momentum. Thus, the rocket does not need to push against air or any other matter. Its forward gain in momentum equals the ejected combustion products backward gain in momentum. Thus the rocket is the only vehicle capable of powered flights through a vacuum. Figure 1-4 shows the internal details of the Space Shuttle solid rocket booster.

The lifting capability of the rocket depends on available thrust, structural combination, utilization, and size. It is also dependent on the launch trajectory,

Fig. 1-3. Titan III lift-off (courtesy of Martin Marietta).

the relative sizes of the different stages, the altitude of the final orbit and the accuracy of the final orbit.

The product of the average thrust, which is produced by high temperature gasses at high pressure (generated by burning the fuel with the oxidizer and making it escape through a nozzle), and the operating time is called the impulse. Specific impulse is the amount of total impulse provided by a pound of fuel. If a rocket has to leave the ground, it must have a thrust (T) greater than its weight (W). Then the acceleration, a, in feet per second^2 is given by

$$a = (T\text{-}W)/(W/g)$$

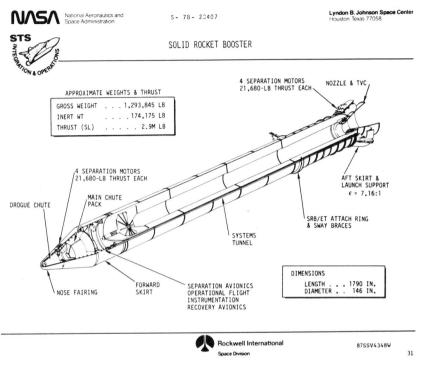

Fig. 1-4. Internal details of Space Shuttle solid rocket booster (courtesy of NASA Lyndon B. Johnson Space Center).

where g is the acceleration due to gravity. As the rocket continuously burns fuel, its weight gradually decreases and with an increase in altitude the value of g decreases. The average weight (W_{av}) is equal to the fuel weight plus the empty weight divided by two. The velocity is equal to the product of the average acceleration and the burning time and is related as

$$V/U = \text{Log}_e(M/m)$$

where U equals exhaust velocity relative to the rocket.
V equals rocket velocity at any instant.
M equals rocket mass at launching.
m equals rocket mass at the instant when the velocity is V.

Multistages. As the empty stages can be discarded as soon as the fuel in them is burned out, the weight of the rocket decreases and this results in an increase of the total impulse. Thus, it is clear why it is better to have a multistage rocket where the stages ignite in sequence. Also staging uses the available thrust effectively. Since there is a definite relationship between orbital altitude and velocity, it may be advantageous for a spacecraft to reach a given altitude at a low vertical speed, i.e., coast up, then accelerate to reach the required horizontal velocity. Staging facilitates this. Figure 1-5 shows the Titan IV characteristics.

Titan IV Characteristics

	Solid Rocket Motors (Two)	Stage One	Stage Two	Centaur Upper Stage
Length	112.9 ft	85.5 ft	32.6 ft	29.3 ft
Diameter	10.0 ft	10.0 ft	10.0 ft	14.2 ft
Thrust	1.6-Million lb per Motor	546,000 lb	104,000 lb	33,000 lb
Propellants	Solid	Storable Liquid	Storable Liquid	Cryogenic

Guidance: Inertial with Digital Computer
Payload Fairing: 200-in. Diameter, 86-ft Length, Tri-Sector Design, Isogrid Construction

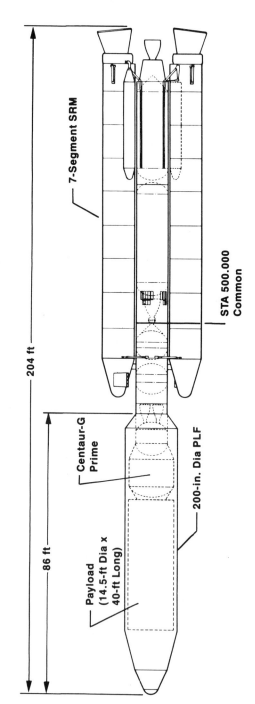

7-Segment SRM

STA 500.000
Common

204 ft

Centaur-G
Prime

86 ft

200-in. Dia PLF

Payload
(14.5-ft Dia x
40-ft Long)

Fig. 1-5. Titan IV characteristics (courtesy of Martin Marietta).

The following illustration shows how effectively the staging influences launching a satellite. Let us suppose that a satellite must be accelerated to 5 miles per second and the weight of the structure and the motor on any stage to be 5% of the total weight of the stage.

☐ **Case-1.** First consider a single stage rocket as shown in Fig. 1-6(A) weighing about 100,000 lbs. The final weight in the orbit after the rocket completely burns its fuel is 5% of the total weight or about 5000 lbs. Let us assume that the payload weight is about 10 lbs.] Thus the mass ratio equals [100,000/5010] or about 20. Referring to Fig. 1-7 (velocity versus mass ratio) the rocket acquires 3.5 miles per second. Thus even a tiny 10 lbs payload fails to become a satellite.

☐ **Case-2.** Now consider a two stage rocket as shown in Fig. 1-6(B), with a total weight of about 100,000 lbs, the first stage weighs about 85,000 lbs and the second stage weighs about 15,000 lbs. The final weight in the orbit after the first stage rocket completely burns its fuel is 5% of the first stage weight or about 4250 lbs plus the second stage weight of 15,000 lbs and the payload weight. Let us assume that the payload weight is about 10 lbs. Thus the mass ratio at the end of first stage burnout equals [100,000 /(15,000 + 4260)] or about 5. Referring again to Fig. 1-7, the rocket acquires 1.6 miles per second.

Thus when the first stage is discarded, the second stage is already traveling at 1.6 miles per second and needs another 3.4 miles per second to bring the rocket to the orbital speed of 5 miles per second.

From Fig. 1-7, the velocity of 3.4 miles per second corresponds to a mass ratio of 18. As the structure of the first stage is thrown off, the initial weight for the second stage plus the payload is 15010 lbs. Hence, at the end of fuel

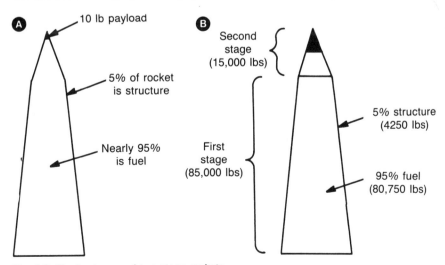

Fig. 1-6. Single stage and two stage rockets.

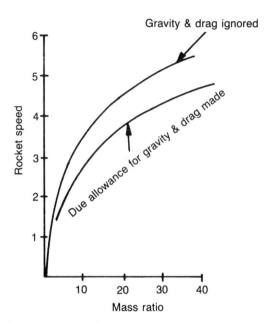

Fig. 1-7. Velocity versus mass ratio.

burnout, the weight should be [15010/18] or 830 lbs. Thus, the final weight of 830 lbs. at the end of second stage burnout, brings the payload to 5 miles per second. However, only 750 lbs is the structure of the second stage rocket and hence the payload can weigh up to [830 – 750] or 80 lbs.

Thus a single-stage rocket cannot even put a 10 lb payload into orbit while a two-stage rocket of the same weight can put up a 80 lb payload. Two-stage rockets offer significant performance improvement over single stage, even at modest velocities. Three-stage rockets provide an improvement over two stage, but not the same amount as two stage over single stage. Four stages provide an even smaller improvement and it saturates. The diminishing improvement in velocity capabilities indicates that there is little to be gained beyond three or four stages. Therefore most of the rockets have only three or four stages. Figure 1-8 shows a typical low-earth-orbit mission profile of launching a space-craft using a two stage rocket.

Start off Methods. Horizontal starting is more advantageous than vertical starting. Consider that a circular orbit is required at a height of 180 miles where the appropriate orbital speed is 5 miles per second. Now consider two methods as shown in Fig. 1-9, (A) a near horizontal takeoff, accelerating to a little over orbital speed, say to 5 miles per second followed by coasting up to the specified orbital speed of 4.9 miles per second. During coasting phase very little speed is lost because it is not climbing steeply. The extra thrust required is only 0.1 mile per second and the total demand for speed is thus 5.1 miles per second; and (B) consider a rocket launched straight up for a vertical climb, comes to a halt at the top and then starts off horizontally. To

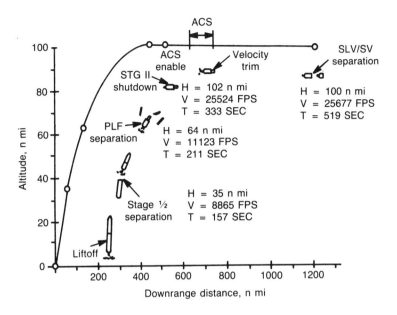

ACS

100

ACS
enable

STG II
shutdown

Velocity
trim

SLV/SV
separation

H = 102 n mi
V = 25524 FPS
T = 333 SEC

H = 100 n mi
V = 25677 FPS
T = 519 SEC

80

PLF
separation

H = 64 n mi
V = 11123 FPS
T = 211 SEC

60

40

H = 35 n mi
V = 8865 FPS
T = 157 SEC

Stage ½
separation

20

Liftoff

0

200   400   600   800   1000   1200

Downrange distance, n mi

Altitude, n mi

LEO mission sequence of events (with ACS)	
Time (sec)	Event
– 5.0	GO INERTIAL
– 2.0	STAGE I IGNITION SIGNAL
0.0	LIFTOFF
9.0	ROLL TO FLIGHT AZIMUTH
19.0	END STAGE I ROLL MANEUVER
20.0	START PITCH RATE 1
25.0	DRIVE ANGLE OF ATTACK TO ZERO (PITCH RATE 2)
30.0	BEGIN ZERO LIFT;
110.0	END ZERO LIFT; START PITCH RATE 3
135.0	START PITCH RATE 4
156.9	STAGE I END STEADY STATE
157.7	STAGE ½ SEPARATION
161.0	BEGIN LINEAR SINE STEERING
210.9	PAYLOAD FAIRING SEPARATION
332.8	STAGE II SHUTDOWN
349.0	ENABLE ACS
378.0	CONVERGE TO COMMANDED ATTITUDE, BEGIN VELOCITY TRIM
406.0	END VELOCITY TRIM (NOMINAL)
434.0	START MANEUVER TO SV SEPARATION ATTITUDE
519.0	SV SEPARATION

Fig. 1-8. Typical low earth orbit mission profile (courtesy of Martin Marietta).

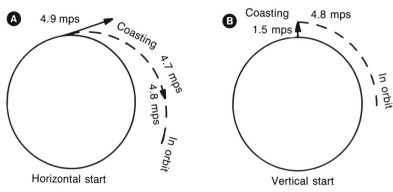

Fig. 1-9. Satellite launching—horizontal and vertical start.

reach a height of 180 miles demands an initial speed of 1.5 miles per second. Then another 5 miles per second is needed at the end of the climb giving a total of 6.5 miles per second to be provided whereas the first method required only 5.1 miles per second. Figures 1-10 and 1-11 show the horizontal start off method used by the Space Shuttle.

A rocket always suffers a loss in speed of about 0.2 miles per second through the atmosphere. However, the earth helps to cancel this loss if used advantageously. Since the earth rotates from West to East at a speed of 0.29 miles per second at the equator, a bonus in speed of this amount can be obtained by launching the satellite from West to East.

1.1.3 Satellite Building Blocks

A spacecraft typically comprises a payload and a number of subsystems, together with their consumable supplies. Partitioning of the spacecraft into building blocks is necessary for engineering convenience and ensures that design and manufacture is managed sensibly in an organized manner. Figure 1-12 shows the exploded view of EXOSAT spacecraft and its building blocks.

The subsystems play a vital role from initial launch through to the end of its operational life and make sure that the mission payloads or instruments are sustained environmentally and are provided with the adequate power and other services necessary to meet the mission objectives successfully. Various building blocks are described below:

Structure. The spacecraft structure is designed to provide a stable and strong platform for payload instruments and subsystems. It offers rigidity and is capable of withstanding the intense mechanical stresses imposed during various phases of the mission. Structures are manufactured from very lightweight materials while maintaining the strength. Although various materials have been used, the aluminum honeycomb combined with glass or carbon fibre reinforced plastic is becoming increasingly common.

Power. The primary power source on-board most of the spacecraft is the solar cell arrays. The solar arrays are secured to the outer casing on spinners, whereas they are deployed in accordion fashion, in case of three axis stabilized satellites, once the spacecraft is in orbit. They are then oriented continuously

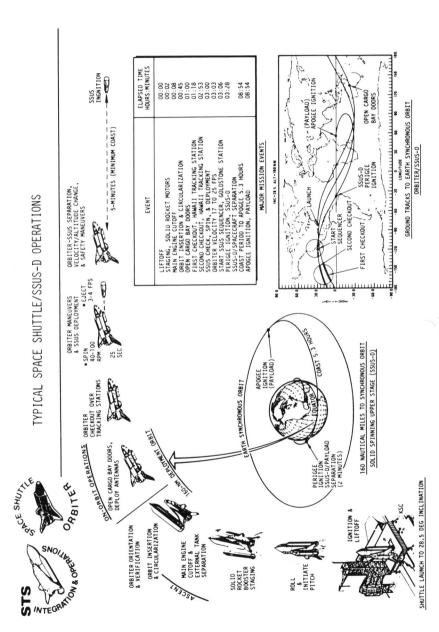

Fig. 1-10. Typical Space Shuttle/SSUS-D operations (courtesy of NASA-JSC).

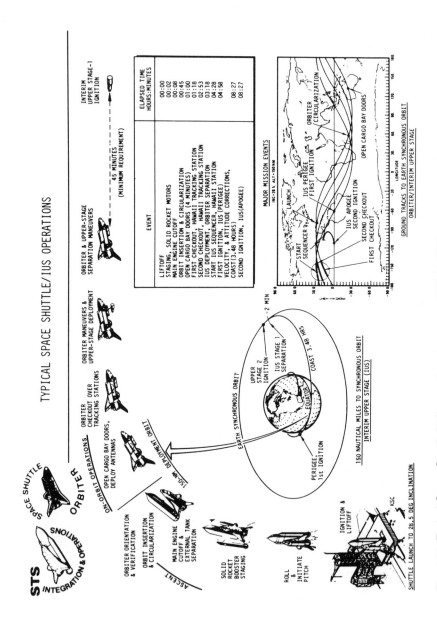

Fig. 1-11. Typical Space Shuttle/IUS operations (courtesy of NASA-JSC).

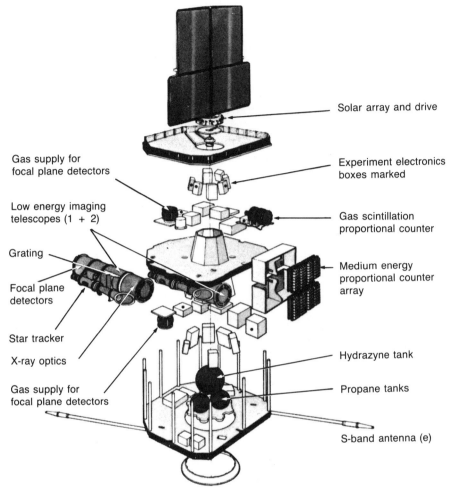

Solar array and drive

Gas supply for
focal plane detectors

Experiment electronics
boxes marked

Low energy imaging
telescopes (1 + 2)

Gas scintillation
proportional counter

Grating

Focal plane
detectors

Medium energy
proportional counter
array

Star tracker

X-ray optics

Hydrazyne tank

Gas supply for
focal plane detectors

Propane tanks

S-band antenna (e)

Fig. 1-12. Exploded view of EXOSAT spacecraft (Marconi).

such that they are perpendicular to the sun. During eclipse periods or when the load power requirements exceed the solar array capability, that power is provided by on-board storage batteries. Power conditioning, switching, and distribution units are employed to match and distribute power to all electronic systems.

Attitude and Orbit Control. The attitude and orbit control subsystem checks that a spacecraft is placed in its precise orbital position, and maintains, thereafter, the required attitude throughout its mission. Control is achieved by employing momentum wheels (which produce gyroscopic torques) combined with an auxiliary reaction control gas thruster system. Various sensors are employed to sense attitude errors. The sun is also used, generally as a reference for initial orientation purposes. Depending upon the precision requirements, simple to sophisticated equipment and techniques are employed. For example, to achieve accuracies of a few seconds of arc, advanced techniques involving gyroscope and star sensor measurements, and on-board computers are employed.

Telemetry, Tracking, Command, and Communication. The telemetry, tracking, command, and communication equipment enables data to be sent continuously to earth and received from earth and allows ground control stations to track the spacecraft and to monitor the health of the spacecraft and to send commands to carry out various tasks like switching the transponders in and out of service, and switching between redundant units, etc.

Thermal Control. Thermal control of a satellite is necessary to achieve temperature balance and proper performance of all subsystems. Thermal stress results from high temperature effects from the sun and from low temperatures occurring during eclipse. Heat shields, blankets, heaters and thermal devices are employed to protect on-board equipment. The thermal control system ensures temperature regulation for optimum efficiency and satellite performance.

Propulsion/Reaction Control System. The propulsion subsystem facilitates for large orbit change, station keeping, and spacecraft attitude pointing control maneuvers. Usually, orbit change is achieved by the apogee kick/boost motor. Auxiliary propulsion uses a low-thrust cold gas or liquid mono or bi-propellant system for station keeping and attitude control. If a bi-propellant is selected, then the apogee boost and auxiliary propulsion systems can be combined into one system. This approach results in fuel economy and in extended operational lifetime. When only station keeping and attitude control is needed, this is achieved by using cold gas or liquid propellant system. Such a system is known as a *reaction control system.*

Mechanisms. Payload instruments and subsystems use many mechanical devices, like deployment hinge and latching mechanisms, rotating drives, rotational and linear pointing mechanisms. Usually, motor drives like the solar array drive are controlled throughout the duration of the mission. Some mechanisms, such as the deployment of antennas and solar arrays operate only once and are achieved by means of spring loaded or pneumatically operated actuators, which are released by firing pyros.

Payloads. Payloads can be classified depending upon the nature of their function and they can be radio frequency experiments (including radio astronomy, radio propagation, etc.), audio frequency experiments, magnetic field measurement experiments, plasma measurement experiments, light frequency experiments (including airglow and aurora photometry, and solar emissions, etc.), particle radiation experiments (including low energy photons, galactic and solar cosmic rays, x-rays, neutral particles, etc.), mass spectrometry experiments (including neutral and ion mass spectrometry), meteorite detection experiments, biological detection experiments, mineral detection experiments, earth resources instruments, meteorological instruments, and communications equipment for various applications.

Usually electrical power is provided to the payloads with the nominal spacecraft bus voltage, leaving any required conversion, inversion, or additional regulation to be performed within the payload(s). This practice simplifies the definition of interfaces and expedites the overall program. However, the space-craft bus voltage is selected as a compromise among all subsystems and payload(s) in view of overall efficiency. Also standard telemetry and telecommand interface is provided using an interface unit, which is located close to the payload(s). It processes the signals or information gathered by the payload(s) including housekeeping of payload(s) and sends to on-board telemetry system which is directly inserted into the data being modulated and transmitted down to the ground stations. Similarly, the command information received by the on-board receiver, after processing through the telecommand system reaches payload(s) via the interface unit. The interface unit decodes and gives appropriate command signals to turn on/off relays, switches, etc.

Spacecraft Harness. The spacecraft Harness provides electrical connections for both signal and power between all subsystems and the instruments or payloads. It also connects the solar array power through slip rings via a power subsystem to the various spacecraft subsystems. The satellite harness includes all interconnecting cables that interface with each of the spacecraft subsystems, i.e., power, telemetry, telecommand, communications, attitude control system, reaction control system, solar arrays, drive mechanisms, and payload instruments or experiments, etc. The harness also includes the umbilical wiring needed for ground checkout and launcher interface, separation switches, terminal blocks, grounding connectors, structural ground bonding, and the battery conditioning resistors.

Continuous harness lacing is avoided, and instead spot ties are made. The harness insulation is usually designed to withstand a 500 V breakdown. Cable routing isolation methods and shielding are employed to minimize electromagnetic interference (EMI). Wire separation between sensitive and EMI generating devices is carefully maintained. Pyrotechnic cables are twisted and shielded, and cables are routed separately from other parts of the harness.

Shielded wires are used in the harness to minimize electromagnetic interference. In addition, twisted pairs of wires are used to minimize magnetic fields due to current flow in wires. Redundant twisted pairs are used for all primary power and power return lines with separate pins at both ends and are bundled and shielded. Primary power, secondary power, and chassis grounds are isolated from one another by at least 10 megohms within each piece of equipment and are connected together only at the common central ground point. Separate bundling and double shielding is employed for pyrotechnic lines in addition to regular twisting and shielding. Twisted, shielded pairs are also used for secondary power, signal, and their return lines. Wires carrying high frequency signals are shielded separately. Individual shields are terminated to chassis ground via connector pins or directly.

The reliability of the harness is very important and environmental conditions like outgassing, radiation, temperature effects are considered in selecting the suitable qualified connectors, wires, terminations, etc.

Earth Station. One of the important basic elements in a satellite link is an earth station. An earth station mainly consists of an antenna, a power amplifier, a low-noise receiver, and ground communications equipment, and iipped with power supplies, control, test and monitoring facilities or ᴛᴇ.ᴇ.. try, tracking, and command systems. Smaller stations usually do not have a tracking facility because of the large beam width of the antenna. Earth stations can be a fixed, ground mobile, maritime, or aeronautical terminal.

For transmission of mainly commands, there may be one or many transmit chains, depending on the number of separate carrier frequencies and satellites with which the station operates simultaneously. For receiving the data, more than one receiver down converter chain is used depending on the number of separate frequencies and satellites to be received, and various operating modes. Usually one antenna serves for both transmission and reception. The antenna usually comprises a reflector and feed, separate feed systems to permit automatic tracking, and a duplex and multiplex arrangement to permit the simultaneous connection of many transmit and receive chains to the same antenna.

For uninterrupted operation of the earth station, it houses in addition to the commercial power source, a locally generated and or battery supplied power source. Also some test equipment are housed in the earth station for routine checking of earth station and terrestrial interfaces, and possible monitoring of satellite characteristics. Each building block will be dealt in detail in the following chapters.

1.2 ORBIT

The orbit of bodies in space is governed by the well established laws of celestial mechanics and the laws of planetary motion, which were set forth by Johannes Kepler in the 17th century.

When a spacecraft is released at a particular altitude with a particular velocity in a particular direction then the spacecraft stays at that altitude traveling with the same velocity, if the centripetal and centrifugal forces are balanced. Otherwise, it may be slowed down by air drag to a lower altitude, in that process the spacecraft attains higher velocity and results in a balance of two forces and attains a new orbit. The velocity increase is equivalent to the loss in altitude, which is given by $V^2 = 2 \times g \times h$, where V is the velocity, g is the acceleration due to gravity and h is the orbital altitude. A satellite gains 1% in speed at the expense of a 2% loss in height, measured from the center of the earth. At the same time, the distance to be traveled by the satellite is also reduced, so that a 2% loss in height decreases the orbital path length by 2% and the orbital time by 3%. However, if this type of new orbit is not achieved, then the spacecraft falls due to centrifugal force (earth's gravity) and might burn due to the heat produced by the atmospheric friction. Sometimes, if the mass of the spacecraft is large enough, some portions might fall back on to the earth.

1.2.1 Injection into Orbit

Thus, as mentioned above, there is a specific velocity required for attaining a specific orbit at a given altitude. Due to aerodynamic considerations, normally payloads are injected into an initial orbit with an altitude of greater than 100 nm and up to about 160 nm altitude orbits, usually the payload is directly injected into the proper orbit. Just like multi-stages are employed to obtain maximum benefit from the rockets, for higher circular orbits, say 400 nm and higher altitude orbits, it would be efficient and beneficial to perform orbit injection in more than one step. Thus, in injecting a payload into orbit, the most economical and standard procedure is to first establish a low altitude parking orbit (Fig. 1-13) (usually circular) and subsequently change/depart into a second orbit, which intersects the desired final orbit. Thus, from circular orbit, the satellite is first injected into the perigee of a highly eccentric ellipse (Fig. 1-13). When it reaches the apogee, it is at the desired altitude for the circular orbit. However, its velocity is still not enough to hold it in a circular orbit. The additional velocity is now imparted by firing the apogee kick motor. This approach requires multi restart capability to perform two burns one at perigee and the other at apogee. Propulsion requirements are minimized in following this multi-step approach.

The original eccentric orbit as shown in Fig. 1-13, is called a *transfer orbit,* or *Hohmann ellipse* and is defined as that ellipse which is simultaneously tangential to two circular orbits. Hence, the perigee of the Hohmann lies on the smaller circle and the apogee on the larger. This method is also applicable to the case where two orbits are circular but do not lie in the same plane.

Spin Stabilization. In space, the spin axis of a rotating body remains pointed in the same direction for durations of several hours to several days or longer before any substantial spin axis drift builds up. Whenever solid rocket perigee or apogee kick motors are used, some sort of stabilization system is always employed. This is because thrust misalignment of only a fraction of a degree will result in tumbling and consequently in an improper orbit injection.

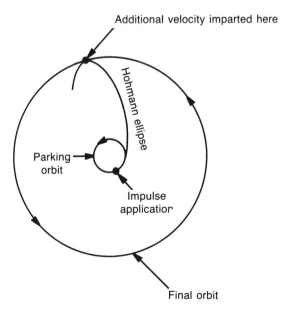

Fig. 1-13. Hohmann ellipse for transfer between coplanar circular orbits.

Usually, the last stage of the launch vehicle orients the payload before the payload is separated so that its spin axis will be aligned in a direction parallel to the required direction of the perigee/apogee thrust vector.

Satellite Coverage. Satellite coverage as illustrated in Fig. 1-14, depends upon its altitude and it increases as its altitude increases. For continuous coverage of an area, the number of satellites required depends upon their orbital altitude. The higher the altitude, the lower the number of satellites required. A high altitude satellite, on the other hand, requires additional launch vehicle capability to place the satellite in the desired higher altitude orbit. Also, the satellite and/or the ground station have to be more powerful to communicate back and forth with each other.

Sun Effects. The sun exerts solar radiation pressure and the effect increases as the spacecraft area-to-mass ratio increases. At lower altitudes (below 300 miles above the earth), aerodynamic drag exerts forces on spacecraft that perturb the spacecraft orbit. Spacecraft orbits about the earth are also perturbed by the non-spherical mass distribution of the earth and by electromagnetic forces (both due to interactions between the magnetic field of the earth, the electromagnetic fields produced by current loops on the spacecraft, and the electrostatic charging of the spacecraft in the plasma) in addition to the gravitational forces exerted by the sun and the moon.

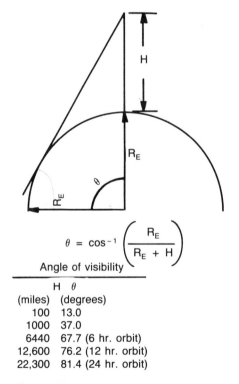

$$\theta = \cos^{-1}\left(\frac{R_E}{R_E + H}\right)$$

Angle of visibility

H (miles)	θ (degrees)
100	13.0
1000	37.0
6440	67.7 (6 hr. orbit)
12,600	76.2 (12 hr. orbit)
22,300	81.4 (24 hr. orbit)

Fig. 1-14. Satellite earth coverage.

1.2.2 Mathematical Relationships

After launch, a spacecraft is released/separated from the launch vehicle at a particular altitude with a particular velocity in a particular direction and the spacecraft possesses kinetic energy, E_k, and potential energy, E_p, given by

$$E = E_k + E_p = (m \times v^2/2) - (\mu \times m/r)$$

and $\mu = G \times m_e$

where m = spacecraft mass
   v = spacecraft velocity
   r = distance between the spacecraft and the center of the earth
   μ = gravitational parameter, and is equal to $G \times m_e$
   G = universal gravitational constant
   m_e = mass of the earth

If the spacecraft is left like that without any disturbance, then the energy of the spacecraft will be conserved and will remain constant. The moving spacecraft also possesses momentum and is conserved throughout the lifetime of the s' ecraft. The angular momentum of a spacecraft moving in an elliptical orbit (as shown in Fig. 1-15) is given by

$$H = m \times r \times v \times \cos \theta$$

where v = tangential velocity of the satellite
   θ = angle between the tangential velocity vector (direction of velocity) and the normal to r, also known as the local vertical. m, and r are defined above.

The spacecraft energy E and angular momentum H determine the orbit altitude as a function of time. The orbital relationship is defined by Kepler's first law, which, when applied to spacecraft, states that if two objects in space interact gravitationally, each will describe an orbit that is a conic section with the center of mass at one focus. If the bodies are permanently associated, their orbits will be circles or ellipses; if they are not permanently associated, their orbits will be hyperbolas. Thus, Kepler's first law is mathematically written as

$$V^2 = \mu(\frac{2}{r} - \frac{1}{a}) \qquad \textbf{Equation (A)}$$

where V = velocity of the spacecraft
   a = semimajor axis

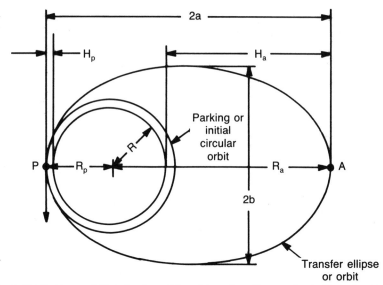

Fig. 1-15. Parking or initial circular orbit and transfer ellipse or final elliptical orbit.

Elliptical orbit. Figure 1-16 shows the Swedish Viking scientific satellite in elliptical polar orbit around the Earth and orbital parameters of such an orbit (Fig. 1-15) are related to spacecraft energy and angular momentum as follows:

$$a = -\mu/2E$$
$$b^2/a = H^2/\mu$$

where a = semimajor axis
       b = semiminor axis

Eccentricity is an indication of the non-circularity of an orbit and eccentricities of elliptical orbits range from 0 to 1. The eccentricity of the ellipse, e, is geometrically related to a and b as follows:

$$e^2 = 1 - (b^2/a^2)$$

Ballistic Trajectories. Ballistic trajectories are in principle the simplest space vehicle flight paths. After ascent through the atmosphere, the rocket places the spacecraft in an elliptical path as shown in Fig. 1-17, whose perigee is inside the earth. Usually when the spacecraft reaches its highest point above the earth (apogee), additional velocity is imparted by firing a rocket to raise the perigee altitude above the earth to put the spacecraft into a stable orbit. If no additional velocity is imparted at apogee, then the spacecraft comes back to earth (when it enters the atmosphere, it may burn out if it cannot withstand the atmospheric friction). There is only one chance, if it is not changed into a stable orbit, the spacecraft will be lost. This approach utilizes the minimum

Fig. 1-16. The Swedish Viking scientific satellite in elliptical polar orbit around the earth (courtesy of Canadian Astronautics Ltd.).

amount of fuel to put a spacecraft into orbit compared to other methods of orbit injection.

☐ Altitude in elliptic orbit: The relationship between time and position in orbit is given by Kepler's second law which states that if two objects in space interact gravitationally (whether or not they move in closed elliptical orbits), a line joining them sweeps out equal areas in equal intervals of time. Thus

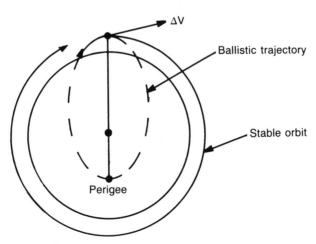

Fig. 1-17. Ballistic trajectory.

an incremental area dA is swept out in an incremental time dt, as given by

$$dA/dt = (r^2/2)(dv/dt) = \text{constant}$$

and

$$r = \frac{a(1 - e^2)}{1 - e \times \cos \gamma}$$

where all symbols are as defined previously and γ = angle between apoapsis and P, known as the true anomaly. This really means that a satellite in elliptical orbit is constantly changing its velocity. The velocity is greatest at the perigee or lowest altitude, and slowest at the apogee, or highest altitude. Thus, a synchronous satellite in an equatorial but elliptical orbit would oscillate east and west relative to the surface of the earth because of variations in its speed.

Some of the other relationships in the case of elliptical orbits are given below:

Eccentricity,

$$e = \frac{H_a - H_p}{2R + H_a + H_p}$$

Semi-major axis,

$$a = R + \frac{H_a + H_p}{2}$$

Semi-minor axis,

$$b = R_p \left[(1 + e)/(1 - e)\right]^{\frac{1}{2}}$$

Radius at Apogee,

$$R_a = 2a - R_p$$

Perigee Velocity,

$$V_p = \left[(2\mu/R_p) - (\mu/a)\right]^{\frac{1}{2}}$$

Apogee Velocity,

$$V_a = (V_p \times R_p)/R_a$$

Circular Orbit. For a circular orbit $r = a$ and equation (A) reduces to

$$V_c^2 = \mu/\text{r}$$

where V_c is known as the circular velocity. Eccentricity for a circular orbit is zero

The gravitational force, defined by Newton, between two bodies of masses m (say, a spacecraft) and m_e (say, the earth) separated by a distance r is given by

$$\text{Gravitational Attraction} = G \left(\frac{m \times m_e}{r^2} \right)$$

It is found for the earth that the force of gravity at a distance r corresponds to a value of G, the universal gravitational constant of 95,700 miles^3 second^{-2}. Thus if r is taken to be 3964 miles, the value of g at sea-level is found to be 32.18 feet/second^2.

Changing Circular to Elliptical. Let us assume that an object is in circular orbit as shown in Fig. 1-15. Say, now, at point P, a rocket motor is fired thrusting in the direction of the orbital motion of the object and assume that the thrust is such that its velocity increases to VP less than the velocity of $\sqrt{2}$ times the circular velocity. This results in the orbit change from circular to elliptical with point P the low point, and point A the high point. These points P and A are known as the perigee (closest point to earth) and the apogee (farthest point to earth), respectively. Thus, a velocity increase at perigee causes an increase in apogee altitude and perigee altitude does not change. If the perigee velocity is decreased, then it causes a decrease in the apogee altitude. A similar situation occurs when a forward thrusting maneuver is performed at apogee.

In elliptical orbit, the velocity with which the object is traveling changes continuously and is maximum at perigee and is minimum at apogee. For an elliptical orbit, the true anomaly of a spacecraft is the angular distance traversed since its last apogee as measured about a principal focus of the ellipse.

Parabola and Hyperbola. When the spacecraft possesses the escape velocity, V_e, the orbit becomes a parabola with a = infinity. Thus, its velocity is given by

$$V_e^2 = 2\mu/r$$

Thus, the escape velocity is 141.4% of the circular orbit velocity, the circular orbit velocity (at the surface of the earth) and the escape velocity 4.91 and 6.95 miles per second respectively. In fact anything leaving the earth at a speed in excess of 6.95 miles per second will never return, irrespective of the direction of the initial aim. This calculation neglects the kick due to stepping off a rotating earth, which has a maximum value of ± 0.29 miles per second.

Thus, in any circular orbit, if the orbital velocity is increased to a value equal to the square root of 2 times the circular orbit velocity for that altitude, then the orbiting object escapes from the body's gravitational field and the escape trajectory will take the shape of a parabola (Fig. 1-18). If the velocity

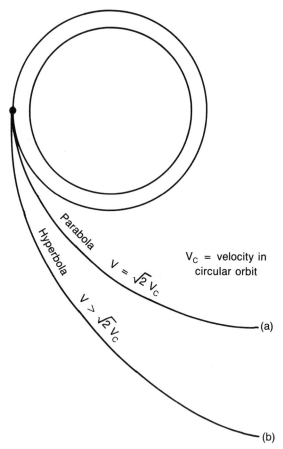

Fig. 1-18. Escape trajectories, (a) parabola, (b) hyperbola.

is greater than $\sqrt{2}$ times the circular orbit velocity, then it escapes in a hyperbolic trajectory (Fig. 1-18). The eccentricity for a parabola is one while it varies between 1 and infinity for a hyperbola.

Orbit Period. The period of a satellite in an orbit is measured by the time between successive passes of a particular point on the orbit. Kepler's third law states that if two objects in space revolve around each other due to their mutual gravitational attraction, the sum of their masses multiplied by the square of their period of mutual revolution is proportional to the cube of the mean distance between them; that is,

$$(m + M)\, T^2 = (4\, \pi^2)\, a^3/G$$
$$\text{or } T^2 = k \times a^3$$

where $k = 4\, \pi^2/\mu = 99.04 \times 10^{-6}$

where T is their mutual period of revolution, a is the mean distance between them, m and M are the two masses, and G is Newton's gravitational constant. Note that the period is independent of eccentricity.

Alternately, the period of revolution in seconds for circular orbit is given by the distance traveled divided by the speed, namely,

$$T = 2 \pi r (\mu r)^{-\frac{1}{2}} = 2 \pi (r^3/\mu)^{\frac{1}{2}} = 2.7644 \times 10^{-6} (r)^{\frac{1}{2}} \text{ hours}$$

This will have its shortest value when r is equal to the radius of the earth, 3964 miles and this value is found to be about 1.4 hours. Figure 1-19 presents speed and period of revolution versus spacecraft altitudes.

Plane Change. Plane change is carried out sometimes due to various reasons. A launch site may launch into a particular inclination, due to safety or other reasons. The plane change as illustrated in Fig. 1-20, is achieved by a vector addition of a velocity component that is necessary to change the direction of the orbital velocity vector from the initial heading through an angle to its final direction. Thus, a change in satellite orbital plane without changing any other orbital characteristics is accomplished by adding an out of orbit plane ΔV. The magnitude of ΔV and the direction in which it is directed depend on the initial orbital velocity and the angle of the desired plane change. This relationship is illustrated in Fig. 1-20 (B). $V1$ is the original velocity vector of the spacecraft at the time the plane change is initiated. $V2$ is the velocity

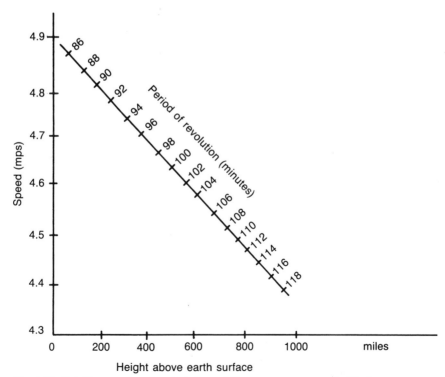

Fig. 1-19. Satellite speed and period of revolution versus spacecraft altitude.

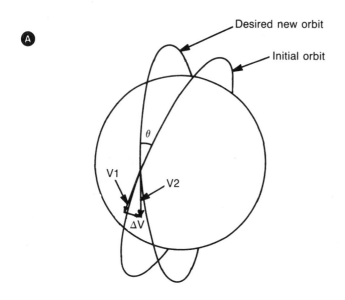

Desired new orbit

Initial orbit

θ

V1

V2

ΔV

B

V1 V2
θ
α α

ΔV

$2\alpha + \theta = 180$

$\therefore \alpha = 90 - \dfrac{\theta}{2}$

$$\dfrac{|\Delta V|}{\sin\theta} = \dfrac{|V1|}{\sin\alpha} = \dfrac{|V1|}{\sin\left(90 - \dfrac{\theta}{2}\right)} = \dfrac{|V1|}{\cos\dfrac{\theta}{2}}$$

$$\therefore |\Delta V| = \dfrac{\sin\theta}{\cos\dfrac{\theta}{2}} \; |V1| = 2|V1|\sin\dfrac{\theta}{2}$$

Fig. 1-20. Plane change relationship.

vector after the change. θ is the angle through which the orbit plane is rotated, and ΔV is the incremental velocity necessary to effect this change.

The required velocity change, ΔV, for plane change angles less than 5 degrees may be approximated as being equal to $2 \times V_i \times \sin(\theta/2)$. For example for an object in a 100 nm circular orbit with $V_i = 25,570$ ft/s, a plane change of only 5 degrees imposes a major penalty of 2,230 ft/s. This can be avoided or minimized by selecting a proper launch site and launching directly into that inclined plane.

Nodal Regression Rate. As the Earth is not perfectly spherical, it causes small, but noticeable, perturbations in a spacecraft's orbit. The nodes are on the line where the plane of the satellite's orbit intersects the plane of the Earth's equator. Any satellite which is in non-equatorial orbit has two nodes, (i) the descending node, when the satellite crosses the equator going north to south, and (ii) the ascending node when it crosses going south to north. The line connecting these nodal points continually shifts with time, due to the Earth's asphericity regressing along the equator at a more or less constant rate. The regression rate depends upon orbital inclination, eccentricity, and semi-major axis.

Apsidal Rotation. In an elliptical orbit, the major axis is also known as the line of apsides. The oblateness of the Earth causes the line of apsides to rotate about an axis normal to the orbit plane and going through the center of the Earth. The rotation is in the direction of motion of the satellite if the orbital inclination is less than 63.4 degrees, and opposite to the direction of motion for inclinations greater than 63.4 degrees. The rotation rate depends upon orbit altitude and inclination.

Illumination of the Orbit Plane. The angle of sunlight incidence on the orbit plane, β, is defined (see Fig. 1-21) as the geocentric angle between the solar vector (the earth-sun line) and the local vertical in the orbit plane when the spacecraft is closest to the sun (orbit noon) and is given by

$$\sin \beta = \sin i_a (\cos \epsilon \times \sin \gamma \times \cos \Omega - \cos \gamma \times \sin \Omega) \\ - (\cos i_a \times \sin \epsilon \times \sin \gamma)$$

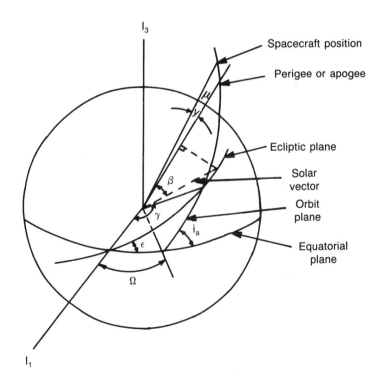

Fig. 1.21. Angle of sunlight incidence definition.

where γ is defined below and β is positive when the sun is seen from the earth to lie above (north of) the orbit plane (Fig. 1-21). If γ is related to a specific angle γ_0 at time t_0 (such as the launch or equinox), γ is given a later time t by

$$\gamma = \gamma_0 + (t - t_0)\, \frac{d\gamma}{dt}$$

and

$$d\gamma/dt = 360/365.24 = 0.98565 \text{ degrees per day}$$

At time t after the launch time, t_0, and referring to Fig. 1-21, Ω is given by

$$\Omega = \Omega_0 + (t - t_0)\, \frac{d\Omega}{dt}$$

For circular orbits

$$\frac{d\Omega}{dt} = \frac{J}{(R} \times \frac{R^2}{} \times \frac{\mu^{\frac12}}{+} \times \frac{\cos}{} \frac{i_a}{H)^{\frac12}}$$

and for elliptic earth orbits

$$\frac{d\Omega}{dt} = \frac{J \times R^2 \times \mu^{\frac12} \times \cos i_a}{(R + H)^{\frac12} \times (1 - e^2)^2}$$

where

$$J = 1.624 \times 10^{-3}$$
$$\mu = 3.986 \times 10^5 \text{ km}^3 \text{ s}^{-2}$$

Solar Eclipses. The solar eclipse time is dependent upon the orbit altitude and the angle of sunlight incidence on the orbit plane, β. For circular orbits, it is given by

$$T_e = \frac{1}{2} + \frac{1}{\pi} \sin^{-1} \left[\frac{[1 - (R/r)^2]^{\frac12}}{\cos \beta} \right]$$

Where　　R = mean radius of the Earth
r = radius of the satellite orbit or the satellite altitude

Synchronous Satellites. A satellite in circular orbit at 35786 km altitude (orbit radius = 42164 km) will circle the earth in exactly 24 hours. Should such a satellite be in a zero degree inclination orbit moving eastward it would stay "fixed" over the same spot on the equator. Figure 1-22 shows a satellite injection into geostationary orbit. The satellite is first placed in an elliptical

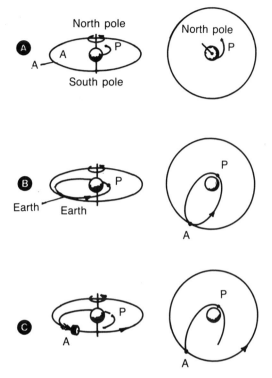

Fig. 1-22. Satellite injection into geostationary orbit. Same scene viewed from two different angles: in profile (left) and plan view (right) (courtesy of Arianespace Inc.).

transfer orbit (perigee 200 km, apogee 35,786 km), making one or more revolutions of the Earth in this orbit (see A and B). When the satellite reaches its apogee A, the apogee motor is ignited, imparting the necessary thrust to initiate a new, circular orbit at an altitude of 35,786 km (see C).

A satellite in an equatorial synchronous orbit experiences acceleration forces due to nonuniform gravitational field of the earth and the influence of the gravitational fields of the moon and the sun. Consequently, over a long period of time, the mass asymmetry of the earth (equatorial bulge) forces the satellite out of exact synchronism and so the period deviates a small amount from 24 hours. While moon and sun also contribute an additional small deviation in the east-west velocity of the satellite, the primary effect of the moon and the sun is to cause north-south deviations of the orbital plane. Usually station keeping is attempted only in the east-west direction, because north-south deviations require more velocity increments or propulsion to correct them.

Synchronous satellites can be transferred from an initial station to a station at another longitude; for example, the initial position might be above the Atlantic Ocean and a second required position over the Indian Ocean. Propulsion requirements for such station changes are not large if adequate time can be allowed. A 100 ft/s decrease or increase in orbital velocity will provide a net easterly movement of 11 degrees per day or westerly transfer movement of 10 degrees per day. Thus, in 10 or 11 days a longitude shift of 110 degrees could be accomplished. Of course, a second propulsion burn of 100 ft/s would l upon arrival at the new station to bring the movement along o a halt.

Earth Orbit Lifetime. The atmosphere produces drag forces that retard a satellite's motion and alter the orbit's shape. A low altitude satellite will eventually be slowed until it spirals into the atmosphere and burns up. The rate of orbital decay depends on atmospheric density, which varies with time and geographic location. The lifetime of a satellite mainly depends upon (a) spacecraft size that determines the effective drag area, (b) spacecraft shape that determines drag coefficient, (c) spacecraft mass, (d) atmospheric density as a function of attitude, (e) initial apogee and (f) initial perigee. The ballistic coefficient is defined as the ratio of spacecraft weight to the product of ballistic constant for the spacecraft and the projected area of the spacecraft along the velocity vector and is expressed as:

$$\text{Ballistic Coefficient} = W/ (C_D \times A)$$

where W = 32.2 m = the spacecraft weight
C_D = drag coefficient
A = projected frontal area

The value of C_D depends on the shape of the vehicle, its attitude with respect to the velocity vector, and whether it is spinning, tumbling, or stabilized. Above 100 nm altitude, the drag coefficient varies from about 2.2 for a sphere to about 3.0 for a cylinder, with other shapes being somewhere in between. Due to the atmospheric density changes that result from the eleven year solar cycle, large variations in orbital lifetime can occur during the course of several years. As solar activity (solar flares and sunspots) increases the effective density, height of the upper atmosphere is increased.

Figure 1-23 gives the lifetime of a satellite in a circular orbit as a function of ballistic coefficient, Fig. 1-24 gives the lifetime of a satellite in an elliptical orbit for various eccentricities, Fig. 1-25 gives apogee decay rate versus perigee altitude for various eccentricities, and Fig. 1-26 gives perigee decay rate versus perigee altitude.

Knowing the spacecraft physical dimensions and other details, the ballistic coefficient is calculated first. Then Figs. 1-23 to 1-26 are used to find out the satellite lifetimes corresponding to the orbits, i.e., 160 nm circular orbit, and geosynchronous transfer orbit. If the lifetime is in the order of hours then two modifications can be considered, (i) reconfigure the spacecraft structure so that it results in a different ballistic coefficient or (ii) insert into an orbit where the satellite lifetime is more than the previous case.

1.3 LAUNCH VEHICLES AND LAUNCHING

Some of the US launch vehicles are Scout, Delta, Atlas, Titan, and Saturn. Upper stages, which can be mated to these launch vehicles include Agena, Burner II, Centaur, Delta, OV-1, and Transtage. Figure 1-27 shows (A) the

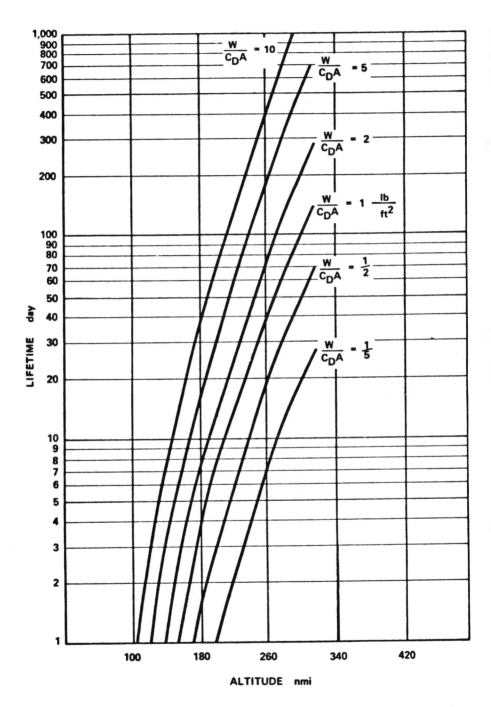

Fig. 1-23. Earth orbit lifetimes—circular orbits (courtesy of Navy Space Systems Activity).

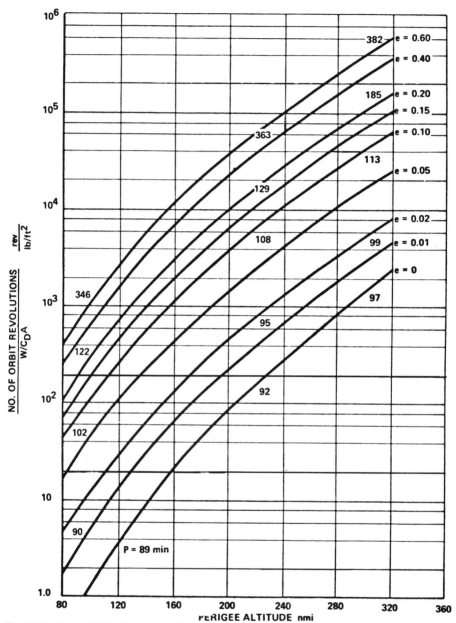

Fig. 1-24. Earth orbit lifetimes—elliptical orbits (courtesy of Navy Space Systems Activity).

more common existing launch vehicle/upper stage combinations, including (B) Titan launch vehicle family and (C) Ariane launch vehicle family. Figures 1-28 and 1-29 show the Payload Capability for these launch vehicles for launches from the Western Test Range (WTR) and the Eastern Test Range (ETR) in USA. In addition, Europe, Japan, China, USSR, India and other countries have launch vehicles.

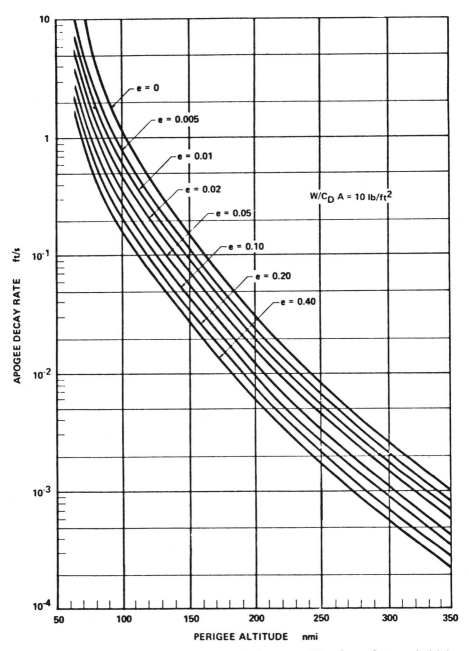

Fig. 1-25. Apogee decay rate versus perigee altitude (courtesy of Navy Space Systems Activity).

Weight lifting capability of existing launch vehicles can be increased by uprating the performance of the lower stages by increased booster engine thrust, by the addition of strap-on solid rocket motors, and by using bigger booster propellant tanks. Velocity capability can be increased by improving

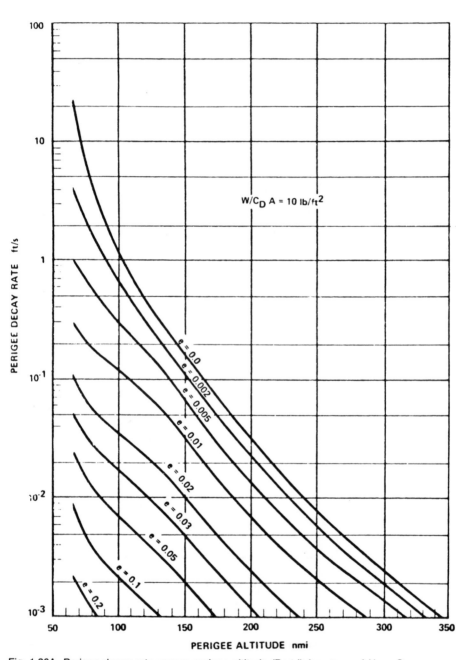

Fig. 1-26A. Perigee decay rate versus perigee altitude (Part-I) (courtesy of Navy Space Systems Activity).

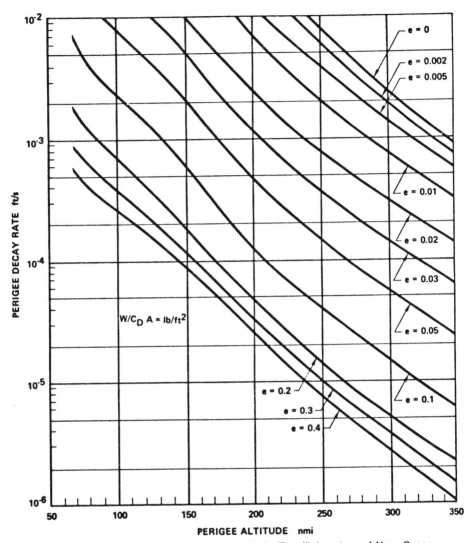

Fig. 1-26B. Perigee decay rate versus perigee altitude (Part-II) (courtesy of Navy Space Systems Activity).

the existing upper stages or by employing additional stages. Sometimes propellant off-loading may be required to tailor the off-the-shelf rocket motor designs to provide the exact velocity increment required for a specific mission.

There are some losses associated with the launching of a payload into orbit. Actual velocity losses in achieving the initial low orbit differ for each specific launch vehicle/upper stage/launch pad combination; the losses typically range from 4,000 ft/s to 6,000 ft/s.

Factors like gravity, aerodynamic drag, rotation of the earth, trajectory shaping, and orbit plane changes influence the velocity demands on a launch

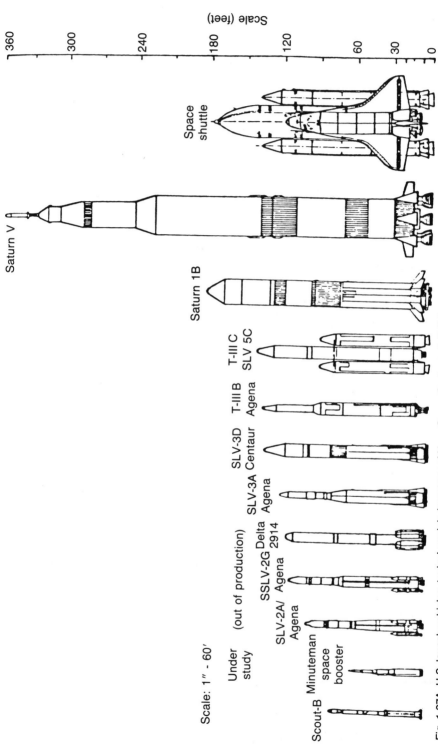

Fig. 1-27A. U.S. launch vehicles, typical models (courtesy of Navy Space Systems Activity).

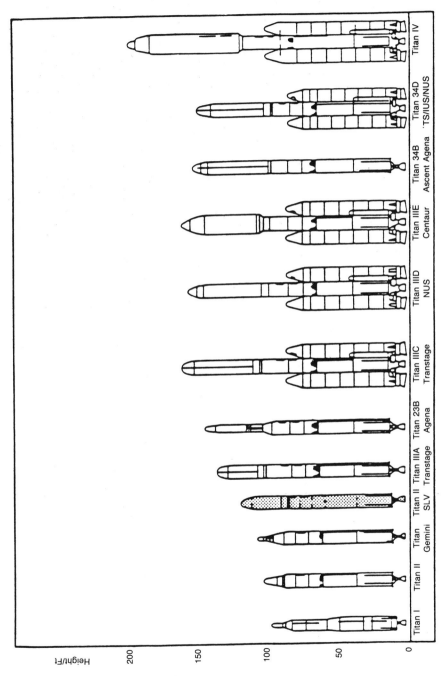

Fig. 1-27B. Titan launch vehicle family (courtesy of Martin Marietta).

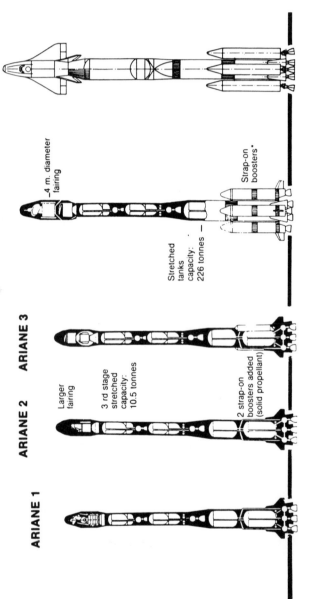

ARIANE 5

ARIANE 4

ARIANE 3

ARIANE 2

ARIANE 1

_4 m. diameter fairing

Stretched tanks capacity: 226 tonnes —

Strap-on boosters*

Larger fairing

3 rd stage stretched capacity: 10.5 tonnes

2 strap-on boosters added (solid propellant)

* To meet the desired performances, Ariane can be launched either without strap-on boosters, or with a combination of 2 or 4 solid or liquid propellant strap-on boosters.

	Height	Mass in G.T.O.
ARIANE 1	47.4 m	1 750 kg
ARIANE 2	49 m	2 175 kg
ARIANE 3	49 m	2 × 1 195 kg
ARIANE 4	58.4 m	1 900 to 4 200 kg
ARIANE 5		4 500 to 5 000 kg

Fig. 1-27C. Ariane launch vehicle family (courtesy of Arianespace Inc.).

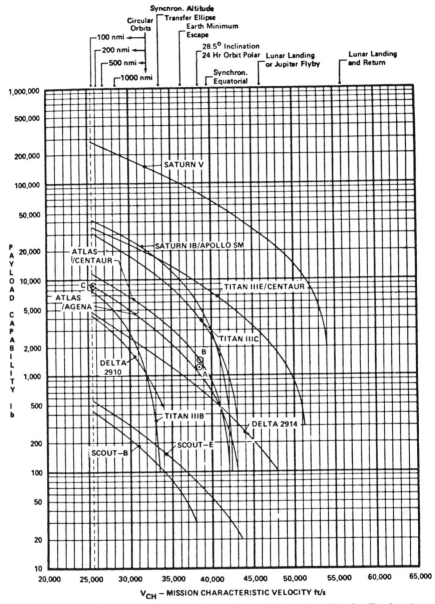

Fig. 1-28. ETR, upper performance limits of operational, U.S launch vehicle families (courtesy of Navy Space Systems Activity).

system during the injection of a payload into the orbit. Because of launch site/range safety considerations specific launch sites have to be considered for particular inclination angles. Thus, payloads requiring a low inclination angle must be launched from the Eastern Test Range (ETR) (Fig. 1-30) in a general easterly direction and payloads requiring high inclination or polar orbit must

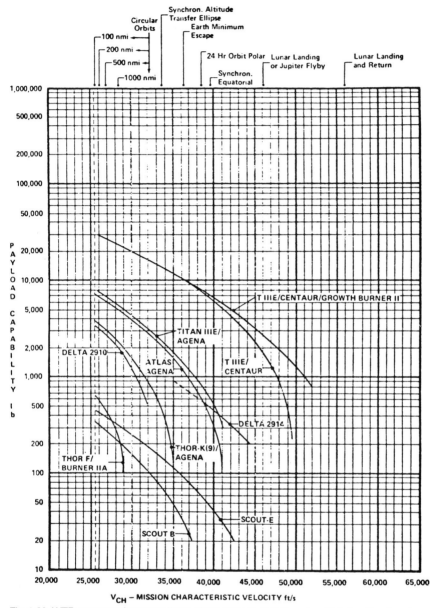

Fig. 1-29. WTR, upper performance limits of operational, U.S. launch vehicle families (courtesy of Navy Space Systems Activity).

be launched from the Western Test Range (WTR) (Fig. 1-31) in a general southerly direction.

Payloads launched in an easterly direction gain an additional velocity of about 0.29 miles/sec to the eastward movement of the earth.

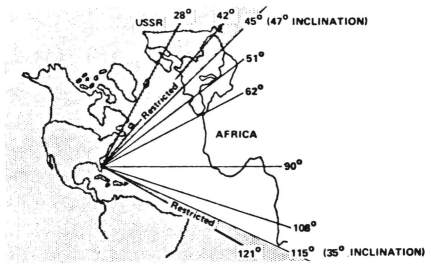

Fig. 1-30. ETR geographic launch constraints (courtesy of Navy Space Systems Activity).

1.3.1 Auxiliary Propulsion

As there are many launch vehicles, one can select more than one launch vehicle for the purpose. Also sometimes, a spacecraft has to be designed with an aim to be launched by a particular launcher due to various constraints. If the aimed launcher cannot directly insert the payload into the desired orbit, then it is accomplished by on-board propulsion or with the aid of an auxiliary upper stage or kick motor. Thus, the following three cases are considered, i.e., (i) hydrazine propulsion system (N_2H_4), (ii) bipropellant system, and (iii) solid apogee kick motor.

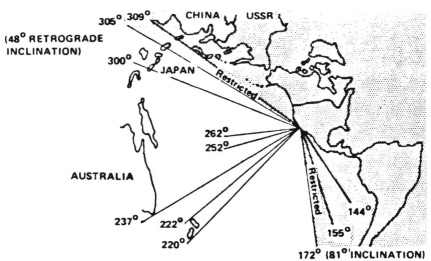

Fig. 1-31. WTR geographic launch constraints (courtesy of Navy Space Systems Activity).

If N_2H_4 or bipropellant is used, then the propulsion system is part of the spacecraft. On the other hand, if a solid motor is used, it is mounted on a separate skirt between the adaptor and the spacecraft.

The spacecraft weight on station at the beginning of the mission shall include the propellant required for orbital maneuvers, reaction control system, and station keeping. The propellant requirements vary depending on the spacecraft design and configuration.

In case of a bipropellant system, using the same propellant weight, a heavier payload can be put into orbit due to the higher specific impulse of the bipropellant. This is clear from the examples worked out.

Sometimes, depending upon the qualified propellant tanks, one may be carrying more propellant than what is needed to put the payload into the orbit, in which case the payload weight margin increases.

If the aimed launch vehicle and on-board propulsion together are not able to put the payload into the desired orbit, then a solid kick stage or motor is employed.

An Example. Let us consider a payload weighing about 1000 lbs to be put into geosynchronous orbit (22,236 miles) with an inclination of 30 degrees. It is assumed that the launcher first places the payload into a 185 nm circular orbit and then the launcher final stage puts the payload into geosynchronous orbit via a Hohmann transfer orbit.

The first step in the preliminary selection of the launch vehicle, is to calculate the characteristic velocity (the sum of all velocity increments attained) necessary to establish the desired orbit. First let us calculate the velocity required to put the payload (with some stages of rocket still attached) into 185 m circular orbit using the following formula:

$$V_{185} = [\mu/(R + H)]^{\frac{1}{2}}$$

where μ is the gravity of the earth, R is the radius of the earth and H is the height of the orbit from the surface of the earth. The velocity required is calculated to be 25,354 ft/sec. Thus, an object moving in an orbit with an altitude of 185 m will circle the earth with a nominal velocity of 25,354 ft/sec.

Once the payload is in a circular orbit a rocket is fired from a point, say, P, with a velocity of ΔV_p to change the circular orbit into an elliptical orbit whose apogee will be at an altitude of 22,236 m. The point P becomes the perigee and apogee occurs exactly 180 degrees opposite to the perigee. The velocity ΔV_p is obtained first calculating the perigee velocity and then subtracting from it the circular velocity calculated above. The perigee velocity is calculated using the formula given below:

$$V_p = \{(\mu) [(2/R_p) - (1/a)]\}^{\frac{1}{2}}$$

where R_p is the distance between the center of the earth and the perigee point, "a" is the semimajor axis. The perigee velocity is calculated to be 33,316 ft/sec and thus the ΔV_p is 7962 ft/sec.

Now the ΔV_a required to change the payload from the elliptical orbit into circular orbit at apogee altitude is obtained first calculating the velocity of the geosynchronous orbit and then subtracting from it the apogee velocity. The apogee velocity of the elliptical orbit is calculated using the formula given below:

$$V_a = \{(\mu)\,[(2/R_a) - (1/a)]\}^{1/2}$$

where R_a is the distance between the center of the earth and the apogee point. The apogee velocity is calculated to be 5274 ft/sec and the geosynchronous velocity is calculated to be 10,088 ft/sec and thus the velocity increment ΔV_a required is 4814 ft/sec. Thus the characteristic velocity is the sum of 25,354 + 7962 + 4814 or 38,130 ft/sec.

The payload is usually mounted on the launch vehicle using a truss structure and it weighs typically 10% to 20% of the payload weight. Assuming 15% for the truss structure, the total weight to be put into the geosynchronous orbit is 1150 lbs. Figures 1-28 and 1-29 present the characteristic velocity versus the payload lifting capability for various launch vehicles, launched from ETR and WTR respectively. Figures 1-30, 1-31 and 1-32 show the possible achievable launch inclinations and launch azimuth requirements in relation to orbit inclinations. As the payload is to be put into 30 degrees inclined orbit, referring to Figs. 1-30 and 1-31, ETR launch is selected. Now referring to Fig. 1-28, the Atlas/Centaur is selected because the Atlas/Agena's capability is just below the requirement. Of course, the Atlas/Agena with an apogee kick motor can be used for the same purpose.

One can carry out a trade off of two or three options considering various factors like cost (see Figs. 1-33 and 1-34), ease of integration, reliability and other aspects and select final launch vehicle.

Alternate Approach. Let us consider a payload weighing about 3000 lbs to be put into geosynchronous orbit (22,236 miles) with an inclination of 30 degrees. It is assumed that the launcher first places the payload into a 185 m circular orbit and the launcher stages are there further to put the payload into geosynchronous orbit via a Hohmann transfer orbit.

The payload is usually mounted on the launch vehicle using a truss structure and it weighs typically 10% to 20% of the payload weight. Assuming 15% for the truss structure, the total weight to be put into the geosynchronous orbit is 3450 lbs. Figures 1-35 through 1-53 present the payload injecting capabilities of various launching vehicles as a function of various orbit inclinations. Referring to these figures, a payload weighing 3450 lbs can be launched by Titan II or Atlas/Centaur into 185 m circular orbit whereas Titan III can put this payload directly into geosynchronous transfer orbit. Thus both approaches can be analyzed to see which one is the most advantageous and efficient approach.

☐ Case-1: Considering a launch vehicle, which can put the payload into 185 m circular orbit, the two velocity increments, (i) one to change the payload into a Hohmann transfer ellipse and (ii) the other to change the elliptical orbit into geosynchronous circular orbit, are calculated.

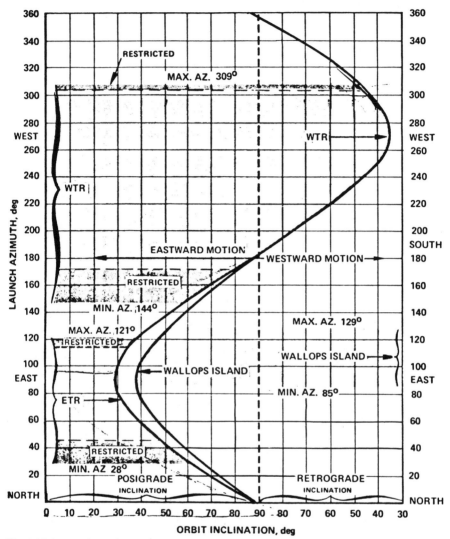

Fig. 1-32. Interactions of inclination and launch azimuth (courtesy of Navy Space Systems Activity).

The velocity V_p is obtained first calculating the perigee velocity and then subtracting from it the circular velocity. The perigee velocity is calculated using the formula given below:

$$V_p = \{(\mu) [(2/R_p) - (1/a)]\}^{1/2}$$

where R_p is the distance between the center of the earth and the perigee point, "a" is the semi-major axis. The perigee velocity is calculated to be 33,316 ft/sec and thus the ΔV_p is 7962 ft/sec.

Now the ΔV_a required to change the payload from the elliptical orbit into

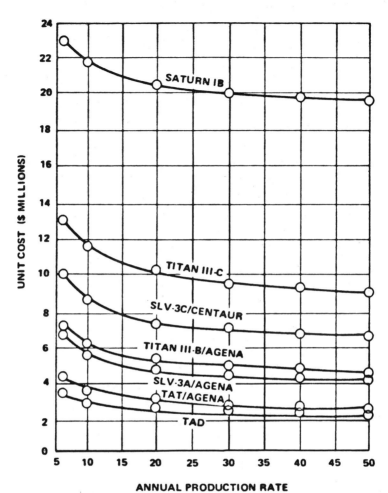

Fig. 1-33. Launch vehicle costs versus production rate circa 1966 (courtesy of Navy Space Systems Activity).

circular orbit at apogee altitude is obtained by first calculating the velocity of the geosynchronous orbit and then subtracting from it the apogee velocity. The apogee velocity of the elliptical orbit is calculated using the formula given below:

$$V_a = \{(\mu)\,[(2/R_a) - (1/a)]\}^{\frac{1}{2}}$$

where R_a is the distance between the center of the earth and the apogee point. The apogee velocity is calculated to be 5274 ft/sec and the geosynchronous velocity is calculated to be 10,088 ft/sec and thus the velocity increment ΔV_a required is 4814 ft/sec. Thus the total velocity required is the sum of 7962 + 4814 or 12,776 ft/sec.

☐ Propellant Requirements: Various propellants are discussed in Chapter 6. Monopropellant (N_2H_4) has a specific impulse of 220 sec whereas bi-

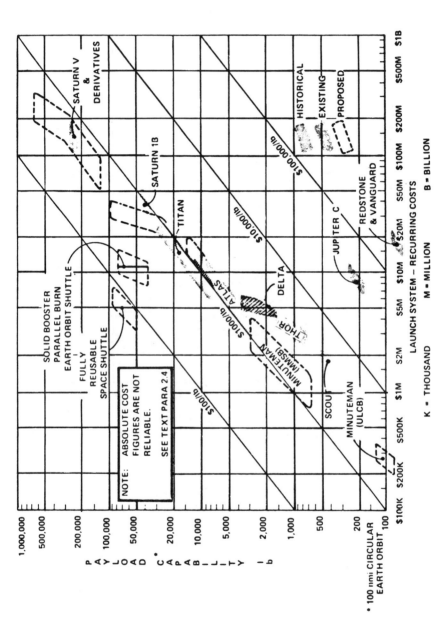

Fig. 1-34. Launch system—recurring costs (courtesy of Navy Space Systems Activity).

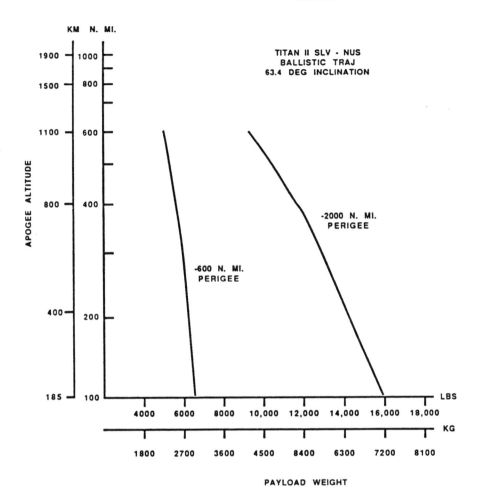

KM   N. MI.

TITAN II SLV - NUS
BALLISTIC TRAJ
63.4 DEG INCLINATION

-2000 N. MI.
PERIGEE

-600 N. MI.
PERIGEE

APOGEE ALTITUDE

LBS

4000   6000   8000   10,000   12,000   14,000   16,000   18,000

KG

1800   2700   3600   4500   8400   6300   7200   8100

PAYLOAD WEIGHT

NOTE: Structural modifications to the Titan II SLV may be required
depending on payload configuration

Fig. 1-35. Ballistic performance capability for Titan-II SLV-NUS for 63.4 degree inclination (courtesy of Martin Marietta).

propellant has a specific impulse of 300 sec. The propellant weight is calculated using the formula given below:

$$W_{prop} = W \{1 - \exp[-\Delta V/(I_{sp} \times g)]\}$$

where g is the earth's gravity and the weight of the monopropellant required is 0.84 W. Thus the weight left for actual payload (W) after allowing for truss and propellant is 54% of the launcher's lifting capability.

☐ Case-2: Considering another launch vehicle, which can put the payload into geosynchronous transfer orbit, the velocity increment required to change the elliptical orbit into geosynchronous circular orbit is calculated first

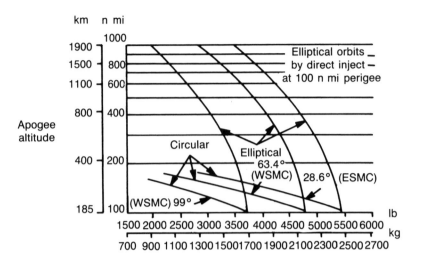

Payload weight

Fig. 1-36. Titan II SLV-NUS performance capability from ESMC (courtesy of Martin Marietta).

calculating the velocity of the geosynchronous orbit and then subtracting from it the apogee velocity. The apogee velocity of the elliptical orbit is calculated using the formula given below:

$$V_a = \{(\mu) [(2/R_a) - (1/a)]\}^{1/2}$$

where R_a is the distance between the center of the earth and the apogee point. The apogee velocity is calculated to be 5274 ft/sec and the geosynchronous

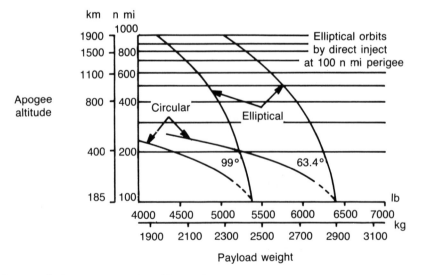

Payload weight

Fig. 1-37. Performance capability of Titan II SLV with second stage propulsion system from WSMC (courtesy of Martin Marietta).

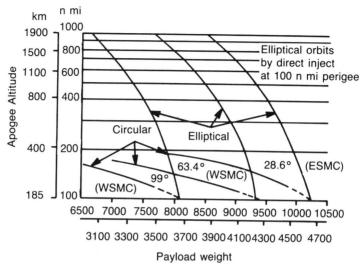

Fig. 1-38. Performance capability of Titan II SLV with four SRM's and Payload Assist Module-D2 from WSMC and ESMC (courtesy of Martin Marietta).

velocity is calculated to be 10,088 ft/sec and thus the velocity increment ΔV_a required is 4814 ft/sec.

☐ Propellant Requirements: The propellant weight is calculated using the formula given below:

$$W_{prop} = W \{1 - \exp[- V/(I_{sp} \times g)]\}$$

where g is the earth's gravity and the weight of the monopropellant required is 0.49 W. Thus, the weight left for actual payload (W) after allowing for truss and propellant is 67% of the launchers lifting capability.

1.3.2 Orbit Maintenance

Most of the time orbit maintenance is required for relatively low earth orbits. Usually above 500 nm altitude circular orbits have relatively longer lifetimes (from Fig. 1-23, more than 4 years for a ballistics coefficient of more than 1) and for lower altitude orbits, however, some kind of maintenance is required. For example, say that the orbit of a satellite shall be maintained for 3 years with a nominal circular orbit altitude of 250 ± 50 nm.

One method is to boost the spacecraft or directly inject the spacecraft into 300 nm orbit and allow it to decay depending upon its ballistic coefficient and other orbit and satellite characteristics. It might take more than 3 years to decay to 200 nm; alternately if it takes, say, only 2 years to decay to 200 nm orbit, then boost it again to about 200 nm-plus orbit such that it takes about a minimum of one year to decay to 200 nm orbit.

With another method, the satellite is initially injected or boosted into a 300 nm orbit and is reboosted at regular (time) intervals, for example once

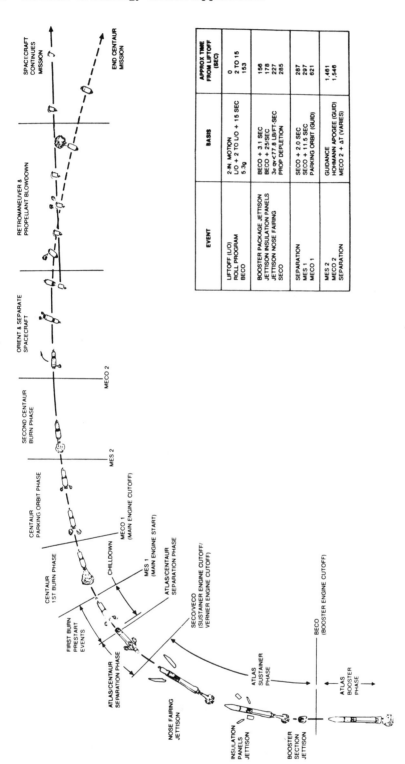

EVENT	BASIS	APPROX TIME FROM LIFTOFF (SEC)
LIFTOFF (L/O) ROLL PROGRAM BECO	2-IN. MOTION L/O + 2 TO L/O + 15 SEC 5.3g	0 2 TO 15 153
BOOSTER PACKAGE JETTISON JETTISON INSULATION PANELS JETTISON NOSE FAIRING SECO	BECO + 3.1 SEC BECO + 25/SEC 3σ qv<77.8 LB/FT-SEC PROP DEPLETION	156 178 227 285
SEPARATION MES 1 MECO 1	SECO + 2.0 SEC SECO + 11.5 SEC PARKING ORBIT (GUID)	287 297 621
MES 2 MECO 2 SEPARATION	GUIDANCE HOHMANN APOGEE (GUID) MECO 2 + ΔT (VARIES)	1,461 1,546

Fig. 1-39. Atlas/Centaur flight profile (courtesy of General Dynamics Space Systems Division).

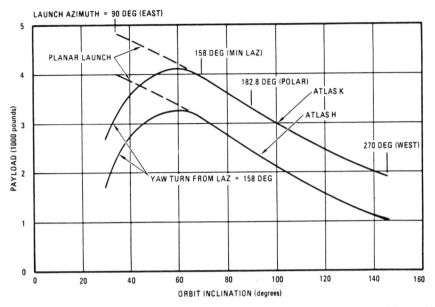

Fig. 1-40. Atlas H/K performance, 100 n.mi circular orbit (WTR Launch) (courtesy of General Dynamics Space Systems Division).

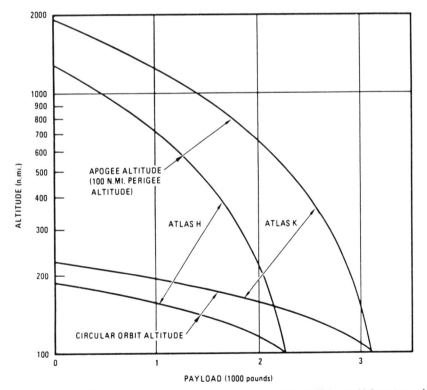

Fig. 1-41. Atlas H/K performance sun-synchronous inclination (WTR Launch) (courtesy of General Dynamics Space Systems Division).

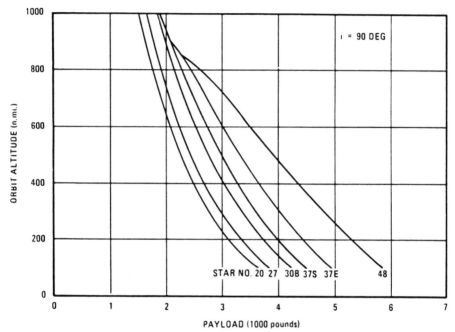

Fig. 1-42. Atlas H/AKM circular orbit performance (90 degrees inclination, seven-foot nose fairing) (courtesy of General Dynamics Space System Division).

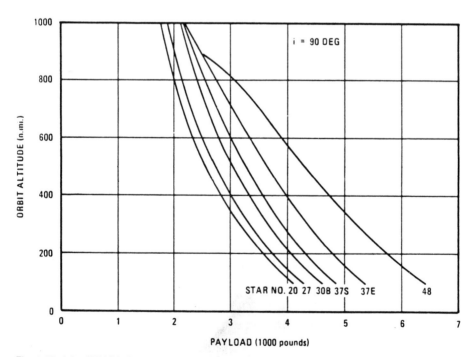

Fig. 1-43. Atlas K/AKM circular orbit performance (90 degrees inclination, ten-foot nose fairing) (courtesy of General Dynamics Space Systems Division).

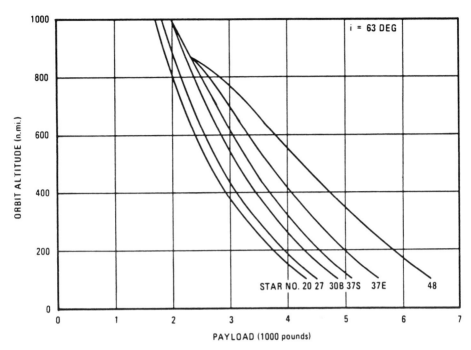

Fig. 1-44. Atlas H/AKM circular orbit performance (63 degrees inclination, seven-foot nose fairing) (courtesy of General Dynamics Space Systems Division).

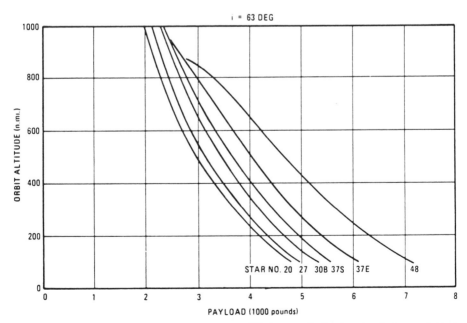

Fig. 1-45. Atlas K/AKM circular orbit performance (63 degrees inclination, ten-foot nose fairing) (courtesy of General Dynamics Space Systems Division).

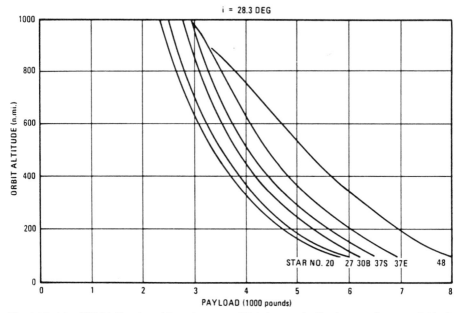

Fig. 1-46. Atlas K/AKM Circular orbit performance (28.3 degrees inclination, ten-foot nose fairing) (courtesy of General Dynamics Space Systems Division).

in a month or so. It is found that this method requires a minimum amount of total velocity increment and propellant requirement. This offers the safest and most practical approach. Either the N_2H_4 or bipropellant system is acceptable. The N_2 system would be prohibitive from a weight standpoint.

1.3.3 Space Shuttles

U.S. Space Shuttles

Dimensions:	Length	System	56.1 m
		Orbiter	37.2 m
	Height	System	23.3 m
		Orbiter	17.3 m
	Wingspan	Orbiter	23.8 m
Weight	Gross at lift off		1,995,840 kg
	Orbiter at landing		84,800 kg
Thrust	Solid-rocket boosters (2)		1,305,000 kg each at sea level
	Orbiter main engines (3)		168,750 kg each at sea level
Cargo bay	18 m long, 4.5 m in diameter		

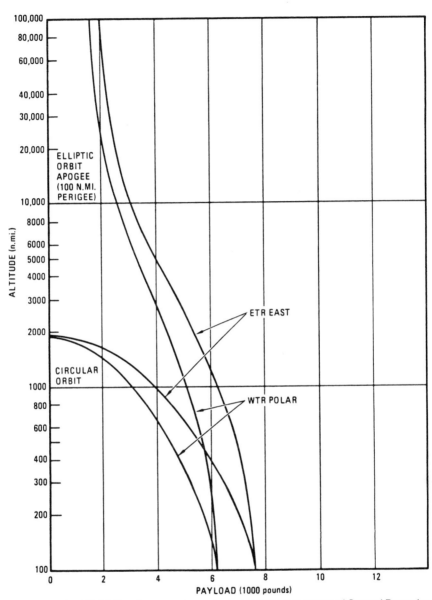

Fig. 1-47. Atlas H/SGS-II performance (seven-foot nose fairing) (courtesy of General Dynamics Space Systems Division).

During launch and ascent, Columbia's main engines and solid rocket boosters (SRBs) fire in parallel, generating more than 6 million pounds of thrust. The SRBs burn for about two minutes, separating at an altitude of 28 miles. They are then lowered by parachute to the ocean, where they are recovered, refurbished and reused. The three main engines, together producing more than a million pounds of thrust, burn for about six more minutes, propelling the

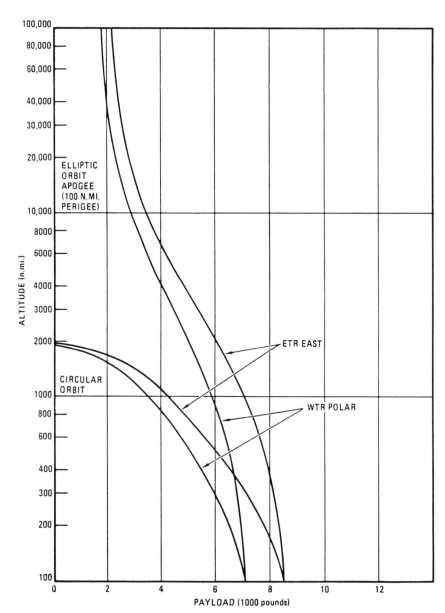

Fig. 1-48. Atlas K/SGS-II performance (courtesy of General Dynamics Space Systems Division).

vehicle to orbital altitude. After the engines are cut off, the external propellant tank is jettisoned into a little used area of the Indian Ocean. At this point, the Orbital Maneuvering System (OMS) is fired. The two OMS engines provide thrust for orbit insertion and for maneuvering in orbit to transfer from one orbit to another, rendezvous with satellites, and deorbit. The first firing of the OMS engines inserts the spacecraft into an elliptical orbit; the second burn circularizes the orbit at an altitude of 130 nm.

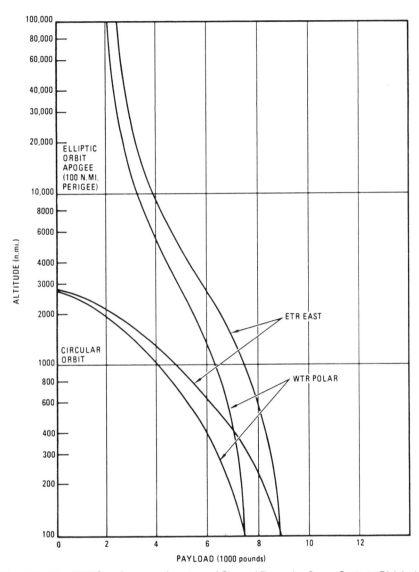

Fig. 1-49. Atlas K/AUS performance (courtesy of General Dynamics Space Systems Division).

Soviet Space Shuttle.

Weight	Gross at lift off	2,000,000 kg
Thrust	at lift off	3,000,000 kg
Maximum Payload Capability	30,000 kg	
Unmanned Cargo Module	110,000 kg	

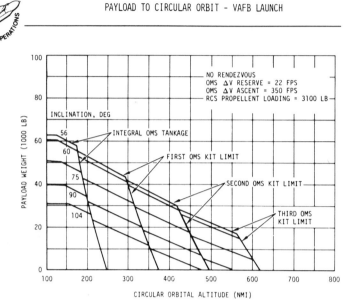

PAYLOAD TO CIRCULAR ORBIT - VAFB LAUNCH

REF: PAGE 3-14 OF JSC 07700 VOL XIV

Fig. 1-50. US Space Shuttle—payload weight versus circular orbital altitude VAFB launch (courtesy of NASA-JSC).

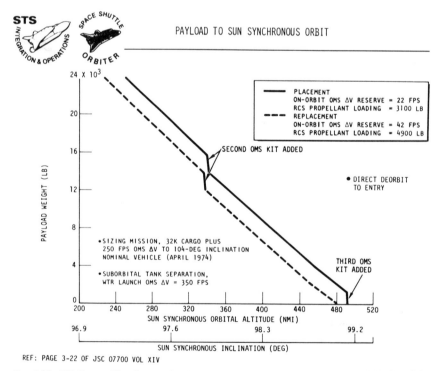

PAYLOAD TO SUN SYNCHRONOUS ORBIT

REF: PAGE 3-22 OF JSC 07700 VOL XIV

Fig. 1-51. US Space Shuttle—payload to sun synchronous orbit (courtesy of NASA-JSC).

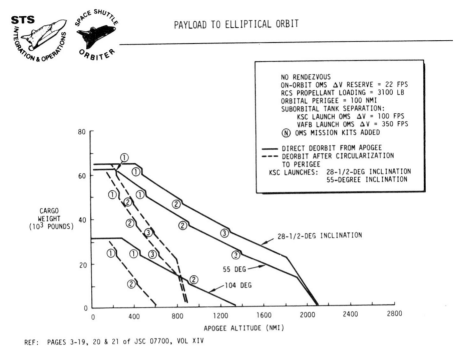

REF: PAGES 3-19, 20 & 21 of JSC 07700, VOL XIV

Fig. 1-52. US Space Shuttle—payload to elliptical orbit (courtesy of NASA-JSC).

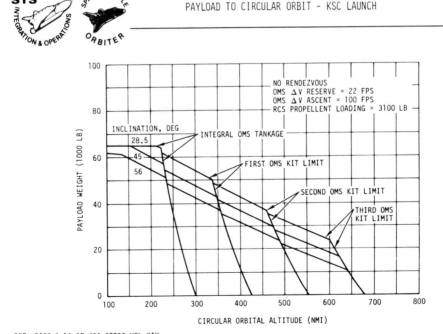

REF: PAGE 3-13 OF JSC 07700 VOL XIV

Fig. 1-53. US Space Shuttle—payload to circular orbit (courtesy of NASA-JSC).

The unmanned cargo module can ride piggyback on the external tank in place of the orbiter.

1.3.4 Performance Details of Launchers

Table 1-1 presents the performances of satellite launchers. Figures 1-35 to 1-38 present Titan II performance capabilities with various upper-stage com-

Table 1-1. Performance of Satellite Launches - Payload Weight (Kg).

Country	Launcher	Low Altitude Orbit	Transfer Orbit	Geostationary Orbit
China	Long March-1	300 (400 km)	—	—
	Long March-1C	600 (400km)	500 (200 by 1500 km)	—
	Long March-2	2200 (400 km)	1800 (200 by 1500 km)	—
	Long March-3	3800 (400 km)	4100 (200 by 1500 km)	1400
ESA	Ariane I	—	1,700	1,006
	Ariane II	4,500	2,000	1,241
	Ariane III	—	2,420	1,325
	Ariane IV	—	3,500	—
	Ariane V	—	5,700	—
Japan	N-I	1,200		130
	N-II	2,000		350
	H-I	3,000		550
USA	STS	29,500 (185-1,110 km) (Inclination 26 degrees)	— 1,973	2,270 1,039 1,180 545
	Titan IIIC	15,000	4,500	1,500
	Atlas/Centaur	6,500	1,850	910-980
	Delta 3914	2,500	900	440
	Delta 2914	—	—	350
U.S.S.R.	STS	30,000	—	—
	A-1	4,750	—	—
	Soyuz A-2-e	7,500	2,400	1,100
	B-1	450	—	—
	Cosmos C-1	700	350	150
	Zonda D-1-e	22,000	5,000	1,600
	G-1	135,000?	—	—

STS : Space Transportation System (Space Shuttle)

binations and for different inclinations and for ballistic trajectory. Figure 1-39 shows the Atlas/Centaur flight profile while Figs. 1-40 to 1-49 present Atlas performance capabilities with various upper-stage combinations and for different inclinations. Figures 1-50 to 1-53 present performance capabilities for different inclinations and for circular and elliptical orbits.

Chapter 2

Spacecraft Structure

THE SPACECRAFT STRUCTURE IS DESIGNED TO MECHANICALLY SUPPORT ALL bus support systems, payloads or instruments and is the backbone of the spacecraft. It also provides precise alignment for attitude control/propulsion systems and isolates equipment on booms, and shields important equipment/instruments from the external environment. Its design closely ties with the surface finish requirements for proper thermal control. In addition, there are some other constraints, i.e., a) inertia ratio within certain limits, b) certain specified stiffness requirements, c) principal inertia axes in preferred directions, d) structure shall sustain loads during various phases of the mission, etc. Thus, meeting all these constraints and requirements, the spacecraft structure can take various shapes i.e., cuboid, cylinder, a long box, octohedral sphere, winged polyhedrons, etc., as shown in Fig. 2-1.

2.1 STRUCTURAL DESIGN REQUIREMENTS

The primary requirements for the structure are:

☐ a) The structure configuration must be such that it meets the functional requirements of all subsystems, providing stable support and maintaining its integrity during all phases of the mission. It shall further provide a compatible interface with the launch vehicle and be within the envelope constraints imposed by the launch vehicle during the launch phase.

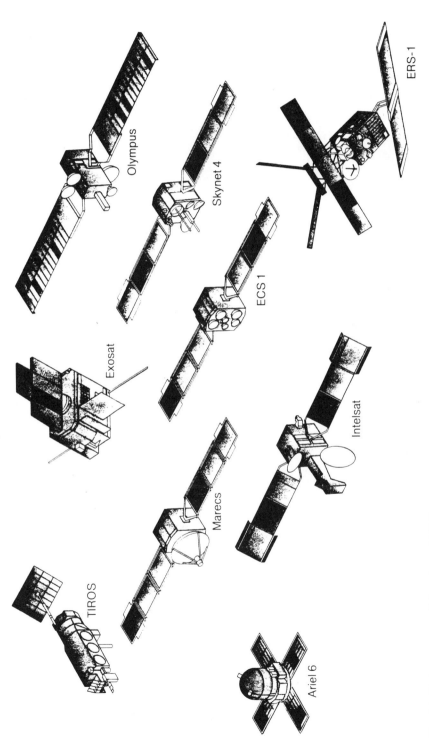

Fig. 2-1. Various spacecraft (courtesy of RCA)

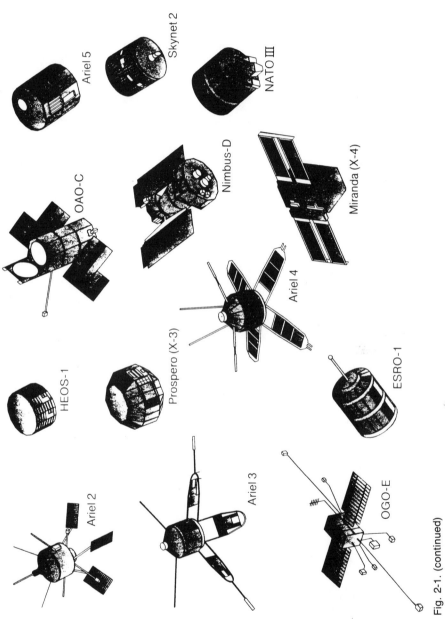

Fig. 2-1. (continued)

The absence of weight and air in orbit gives the flexibility in selecting arbitrary shape, although some missions might dictate a particular shape. Spin stabilized spacecraft shall have symmetry around their spin axes and thus, cylinders and cylindrical polygons are very common.

□ b) The structure shall be designed to withstand launch loads, ground qualification and acceptance test loads and on-orbit loads, shocks, vibration and to a stiffness level such that fundamental resonant frequencies do not occur below certain frequencies, typically below 30 cycles in the longitudinal axis and below 10 cycles in the lateral axes.

The minimum frequency requirements are imposed in order to decouple the spacecraft main resonance from the launch vehicle dynamic excitation, thus maintaining the spacecraft dynamic response to within acceptable design limits and limiting the environment on the equipment. Meeting minimum frequency requirements greatly simplifies the structure design and reduces the overall spacecraft weight in general.

Structural resonances in the vicinity of the launch vehicle resonances shall therefore be avoided. Launch loads include steady state and dynamic accelerations transmitted through the launch vehicle interface and acoustic noise through the fairing. Steady state acceleration developed by the engine thrust is superimposed with the dynamic accelerations due to ignition and cut off transients, wind gusts, etc. The dynamic loads consist of low-frequency transients and random accelerations.

Typical launch limit load factors are given in Tables 2-1, 2-2, 2-3, and 2-4. Load factors vary greatly for different launch vehicles. On-Orbit loads are generally very small compared to the launch environment; an exception is given by deployment loads on appendages and shuttle docking loads. These normally apply to the secondary structure and do not affect the primary structure design.

□ c) Design factors shall be used to account for uncertainties which cannot be thoroughly analyzed. These shall be applied to the limit loads and pressures and to the stresses arising from temperature differences and gradients. Typical design and qualification minimum factors of safety are presented in Table 2-5. The special factors of safety are given in Table 2-6 and are to be used to account for uncertainties in load/stress distributions in fittings and joints and variations in the control of welding and bonding processes.

Table 2-1. Flight Limit Quasi-Static Loads (G's) (Ariane).

Flight Event	Longitudinal Axis	Lateral Axes
Maximum Dynamic Pressure	− 3 g (compression)	+/− 1.5 g
Before Thrust Termination	− 7 g (compression)	+/− 1.0 g
Thrust Lift off	+ 2.5 g (Tension)	+/− 1.0 g

Table 2-2. Sinusoidal Vibration Acceptance Level (Ariane).

Axis	Frequency range (Hz)	Level (0 - peak)
Longitudinal	5 - 100	1.0 g
Lateral	2 - 18	0.8 g
	18 - 100	0.6 g

Table 2-3. Random Vibration Flight Limit Environment (Ariane).

Frequency	Level (PSD)	Overall (rms)
5 - 150	+6 dB/oct to 0.04 g²/Hz	7.3 g
150 - 700	0.04 g²/Hz	—
700 - 2000	-3 dB/oct	—

Table 2-4. Acoustic Vibration Flight Level (Ariane).

Octave Band (Center Frequency, Hz)	Acceptance Level** (Flight)
31.5	114
63	120
125	131
250	136
500	139
1000	133
2000	128
4000	121
8000	120
Overall	142

**Zero dB : (ref. 2×10^{-5} Pascal)

Table 2-5. Typical Minimum Factors of Safety.

Event	Qualification	Yield	Ultimate
Ariane Launch	1.25	1.4	1.6
Shuttle Launch/Landing	1.1	1.2	1.6
Shuttle Emergency Landing	—	—	1.0

Table 2-6. Spacial Factors of Safety.

	Special Factors of Safety	
Item	Yield	Ultimate
Fittings	1.00	1.15
Welded Joints	1.00	1.50
Bonded Joints	1.00	1.50
Fasteners	1.00	1.15

In addition, a positive margin of safety is required for all stresses induced by the above loads. The margin of safety is defined as

$$MS = \frac{\text{Material allowable stress (or load)}}{\text{Applied stress (or applied load)}}$$

For minimum weight design, the margins of safety shall be the smallest practical value equal to or greater than zero.

Pressure vessels (such as propellant tanks, Ni-H$_2$ batteries, etc.) shall be designed to higher safety factors, depending on their size and amount of stored energy.

☐ d) The structure shall minimize distortions and maintain alignments and stability for the sensors consistent with payload alignment and attitude determination and control requirements. This requirement extends over temperature limits ranging typically from zero degrees C to 40 degrees C.

☐ e) As accessibility of components and subsystems is very important, modular approach shall be followed to a maximum extent. Modular bays or drawers or other types of modular arrangements shall be used where it is easy to reach or replace a component or a board, etc.

☐ f) Scientific instruments shall have obstruction-free view angles. Some instruments like magnetometers may have to be isolated from the satellite by employing long booms.

☐ g) Some subsystems like a solar array need surface area, while others need 4π steradians of solid-angle field of view. This may necessitate the use of deployable appendages, like booms, panels, antennas, etc.

☐ h) Spacecraft with appendages folded, may be required to have mass symmetry about spin axis although the spacecraft may not be finally (in-orbit) spin stabilized.

Most upper stages used for orbit transfer are spin stabilized (e.g., PAM-D) at 50/60 rpm. Also if initial spin stabilization is required by the spacecraft, this is normally provided by the launch vehicle before separation.

☐ i) The structure shall be electrically conductive so that it can act as a common ground plane for all equipment.

☐ j) As all the heat generated inside the spacecraft shall be radiated outside of the spacecraft, the structure shall provide low or high resistance heat paths to outside as per thermal requirements.

2.2 DESCRIPTION

For better understanding, consider Fig. 2-2 which shows a typical geosynchronous telecommunications spacecraft. This is primarily a four sided box and all six faces are flat and are designated as north, south, east, west, aft, and forward faces or platforms as equipment are mounted on all sides.

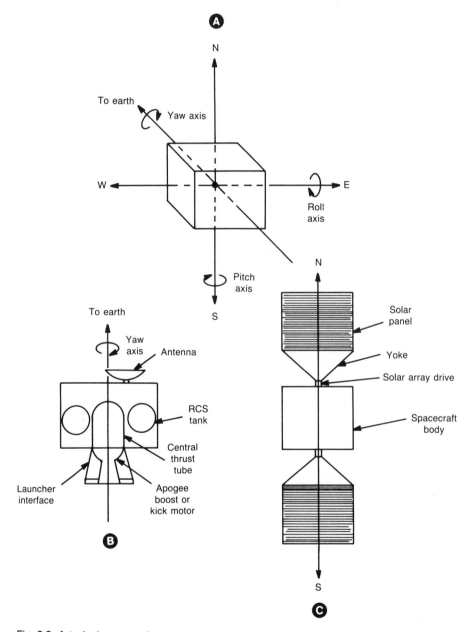

Fig. 2-2. A typical spacecraft structure.

Figure 2-3 shows the exploded view of the SATCOM Ku-Band spacecraft detailing various structural components as described below as well as spacecraft subsystems. Most geosynchronous spacecraft carry a solid Apogee Kick Motor (AKM) for transferring the spacecraft from the initial elliptical parking orbit to the geosynchronous altitude. The AKM is located normally in the center part of the structure, inside a central tube. Thus, the primary structural elements are the central thrust tube, the aft and forward platforms, the equipment panels and the support struts. The remaining elements form the peripheral structure, which is supported by the primary structure. The peripheral structure consists of the trusses and bracketry connecting these platforms and panels to the primary structure.

The thrust cylinder/cone combination acts as the main central load-bearing column and consists of a magnesium or aluminum sheet-metal skin stiffened with a certain number of rings made from aluminum alloy forgings and machined to fit. Graphite fiber reinforced plastic central tubes are also common, if additional weight saving is needed. The lower ring provides the mechanical interface with the launch vehicle adaptor. The other rings are located to interface with the antenna and base panel and with the AKM. The sheet metal skin is normally riveted to the rings. On each of the north and south sides, two honeycomb sandwich bulkheads are attached to the central cylinder. These support the solar array drives to the central cylinder and an edge of the north and south equipment panels. The east and west bulkheads consist of an aluminum honeycomb sandwich and provide support for compounds of the reaction control subsystem. A honeycomb sandwich is used for all bulkheads to provide high strength to weight capability. Thrust loads from the box structure are transmitted to the cylinder via the six bulkheads and from there to the conical section to the launch vehicle adaptor. If struts are used to support the outer edge of the base panels, these are made of thin-walled tubing and loaded in compression or tension only. Struts are pinned to the platform edge and lower interface ring at their extremities.

Aluminum honeycomb panels are extensively used in spacecraft structure for their high strength-to-weight and stiffness capability. The aluminum honeycomb panel is composed of two thin sheets (typically 0.5 to 1 mm) bonded on either side of an aluminum honeycomb core. Each sheet could also be made of so-called composite materials (such as graphite epoxy) to further reduce weight. Heavily loaded areas, such as strut attachment points and support points of heavy equipment are often locally filled with denser core and reinforced with thicker face sheet to withstand the high concentrated loads. Boxes are fastened to the honeycomb panels through inserts. These consist basically of an aluminum cylinder threaded inside and bonded within the honeycomb, are in correspondence with attachment points.

2.3 ANALYSIS

In the spacecraft structure design process, the structure analysis is performed in order to verify that the structural design and material selection meets the imposed launch environment requirements with the application of the

Satcom Ku-Band Spacecraft Exploded View

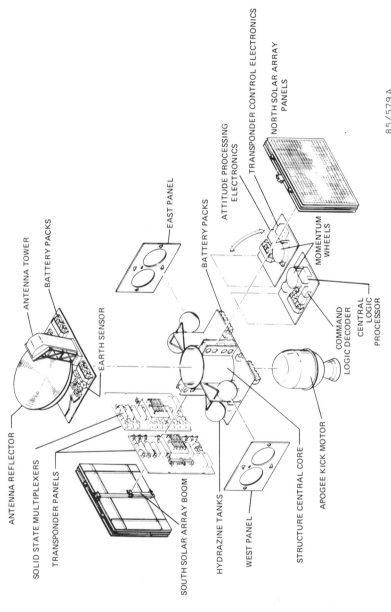

ANTENNA TOWER

BATTERY PACKS

EARTH SENSOR

EAST PANEL

BATTERY PACKS

ATTITUDE PROCESSING ELECTRONICS

TRANSPONDER CONTROL ELECTRONICS

NORTH SOLAR ARRAY PANELS

MOMENTUM WHEELS

COMMAND LOGIC DECODER

CENTRAL LOGIC PROCESSOR

85/579 A

ANTENNA REFLECTOR

SOLID STATE MULTIPLEXERS

TRANSPONDER PANELS

SOUTH SOLAR ARRAY BOOM

HYDRAZINE TANKS

WEST PANEL

STRUCTURE CENTRAL CORE

APOGEE KICK MOTOR

Fig. 2-3. SATCOM Ku-band spacecraft exploded view (courtesy of RCA).

minimum factors of safety and appropriate margins. A standard technique consists in generating a finite element mathematical model of the structure, using a general purpose finite element analysis program, such as NASTRAN. In the model, mass distribution, material properties, geometry and boundary conditions are defined and static and dynamic behavior of the structure are therefore formulated.

Upon application of the external loads, the load distribution in all structural elements is determined. Resonant frequencies and mode shapes are also calculated; these are used for transient response analysis and for coupled loads analysis with the launch vehicle dynamic model. The coupled loads analysis is performed by combining the spacecraft and launch vehicle dynamic model and subjecting it to the launch forcing functions. The dynamic response of the spacecraft determines loads, accelerations, and displacements within the spacecraft. Several analytical techniques are available, whose description goes beyond the scope of this text.

Results of the coupled loads analyses are also used to determine the design loads for the equipment and to define environmental test requirements. Once the static and dynamic load distribution within the spacecraft is characterized, detailed stress analysis of all structural components is then carried out to determine the adequacy of the stress margins.

Thermal distortions can also be calculated using the same finite element models and the predicted temperature distribution from the thermal analysis. The process, which has been briefly described, is iterative and at the end of each cycle, the structural design is evaluated and modified, if necessary, until stress margins requirements and stiffness (minimum frequency) requirements are met.

For shuttle compatible payloads, additional analysis may have to be performed to determine a crack propagation under repeated loads during the lifetime of the structure for fracture critical items in order to meet the safe-life or fail-safe criteria. This type of analysis is called *Fracture Mechanics analysis* and is of particular relevance for pressurized components. Computer techniques are available to define the initial crack size and evaluate its growth as a result of the applied load spectrum.

2.4 WEIGHTS

Table 2-7 gives a typical list of spacecraft equipment. Table 2-8 shows the items considered to arrive at the lift off weight, that the launcher is capable of placing into a transfer orbit. The apogee of this transfer orbit is so chosen that the required altitude/orbit can be achieved by firing the apogee motor with the fuel already accounted for. The contingency weight is that amount that can be used to cover design weight changes, or can be used for additional payloads. Structure weight represents 5% to 20% of the spacecraft launch weight.

2.5 LAUNCHER INTERFACE

Payload Adapter. The adapter section forms the structural connection

Table 2-7. Spacecraft Subsystems/Equipment.

Attitude Control System
  Electronics
  Horizon Sensor
  Inertial Reference Unit
  Star Sensor

Telemetry, Telecommand & Communications
  Transponder, directional (On-orbit)
  Transponder, omni (transfer & on-orbit)
  Telemetry
  Command Decoders
  Antennas, Omni and Directional

Electrical Power System
  Solar Array
  Solar Array Drive
  Batteries
  Electronics

Initial Acquisition/Transfer Orbit Control
  Injection Attitude Control Propellant
  Thrusters - Large
  Thrusters - Small
  Plumbing, valves
  Injection Motor

Vernier Propulsion
  Impulse Propulsion
  Tankage
  Plumbing, Transducers, Valves
  Pressurization gas
  Thrusters

Station Keeping/Attitude Control Propulsion
  Thrusters
  Tankage/Regulator
  Plumbing, Transducers

Wiring Harnesses

Thermal Control

Structure

Adapter

Payloads (for example COMMUNICATION PAYLOAD contains Phased Array, Transponders, Antenna reflector, Antenna feed, etc.)

between the spacecraft and the launcher forward bulkhead to provide an acceptable load distribution at the launcher payload adapter ring. Typically, the payload adapter includes attachments for payload separation mechanisms, an inflight electrical disconnect mechanism, and a range-safety destruct unit.

Depending upon the launcher and spacecraft structures, the adapter can take different shapes. An adapter is shown in Fig. 2-2(B), which is a truncated

Table 2-8. Weight at Lift-off.

Item	Weight
On-orbit spacecraft weight	= A
Contingency	= B
Injection Error/Station Acquisition Propellant	= C
Weight after apogee firing	= D = A + B + C
Apogee kick motor and attitude control propellant	= E
Weight at apogee firing	= F = D + E
This is the maximum weight launcher puts into transfer orbit.	
Performance loss due to inclination change	= G
Adapter weight	= H
Spacecraft weight at lift-off	= L = F + G + H

aluminum skin and ring conical structure. It is mounted on a ring on the launcher forward bulkhead. The separation system for the spacecraft consists of a pyrotechnic-actuated separation latch (Mormon clamp) and a set of compressed helical springs. Upon separation, the springs located close to the latches impart a longitudinal motion of typically about one to two feet per second. Some launchers also impart an angular rotation rate when required by the spacecraft.

Electrical Interface. Typically an electrical interface connector is located between the adapter and the spacecraft. A wiring harness is provided leading from the electrical connector to terminal points on the launcher. The terminal points include the electrical ground connect umbilical, the telemetering system, and flight control system. The electrical interface permits transmission of payload environmental data and engineering data through the launcher to the ground station. The same electrical interface is used for prelaunch checkout of the spacecraft. An inflight umbilical disconnect is normally used. The two halves of the umbilical connector are demated usually by the pulling action of the spacecraft being separated, aided by a connector spring located in correspondence of the connector on the adaptor side.

Payload Fairing. The spacecraft is enveloped by a shroud during the first few minutes of launch. Its function is to protect the spacecraft and the launch vehicle against aerodynamic loading during the ascent phase through the denser part of the atmosphere. The shroud is normally released when the aerodynamic pressure is reduced to a negligible level so as not to damage the spacecraft.

Table 2-9. Mechanical Properties of Spacecraft Materials.

Materials	Density Kg/m^3	Transverse ultimate tensile strength N/mm^2	Longit. ultimate tensile strength N/mm^2	Longit. tensile yield N/mm^2	Fracture Toughness (N/m$^{3/2}$ × 10^6)
Aluminum Alloy					
Sheet (2024-T36)	2770	—	480	410	36
Sheet (7075-T6)	2800	—	520	450	30
Beryllium					
Extrusion	1850	—	620	410	—
Lockalloy					
(Be-38% A1)	2100	—	427	430	—
Graphite/Epoxy, V$_f$ 55%					
[0] HTS	1490	67	1340	—	—
[0$_1$ +/−45]HTS	1490	290	640	—	—
Invar 36					
Annealed	8080	—	490	280	—
Magnesium					
Sheet (AZ31B-H24)	1770	280	270	200	—
Steel					
PH15-7 (RH1050)	7670	—	1310	1200	—
Ti6AL-4V					
Sheet	4430	—	1100	1000	46

Materials	Shear Modulus N/mm^2	Young's Modulus N/mm^2	Specific Strength (Km/s)^2	Specific Stiffness (Km/s)^2
Aluminum Alloy				
Sheet (2024-T36)	28000	72000	0.17	26
Sheet (7075-T6)	27000	71000	0.19	25
Beryllium				
Extrusion	140000	290000	0.33	158
Lockalloy (Be-38% A1)	—	190000	0.20	88
Graphite/Epoxy, Vf 55%				
[0] HTS	5900	150000	0.89	100
[0$_1$ +/−45]HTS	—	83000	0.43	56
Invar 36	56000	145000	0.06	18
Magnesium				
Sheet (AZ31B-H24)	16000	45000	0.15	25
Steel				
PH15-7 (RH1050)	76000	200000	0.17	26
Ti6AL-4V				
Sheet	43000	110000	0.25	25

The shroud is usually a lengthened cylindrical section and it contains access doors to the payload adapter and launcher equipment. It has a split line along the launcher longitudinal axis to allow fairing jettisoning during flight. A gas, typically nitrogen, stored in the bottles near the forward end of the conical section and preloaded springs provide the jettison force. The gas is released from the bottles by igniting an explosive, which operates a pin-puller mechanism; the resultant reactive force and the acceleration of the launcher together rotate the fairing halves clear of the launcher.

2.6 MATERIALS

The materials used for the spacecraft structure shall be of lightweight material that will retain their strength and shape throughout the spacecraft mission in orbit. Aluminum is most commonly used for its good strength/weight capability, easy workability in various shapes and forms, and its availability. For lighter weight structures, magnesium is often used. To meet even more stringent requirements for light weight, high stiffness and minimum thermal distortion, advanced materials such as beryllium and advanced composites (carbon fiber reinforced plastics) are also used. However, higher cost and, in the case of beryllium, toxicity of its dust particles limit the extent of their application. Advanced composite materials find excellent application in load bearing structures, highly stable structures (e.g., optical benches) and

Table 2-10. Thermal Properties of Spacecraft Materials.

Materials	Type	Specific Heat J/kg K	Thermal Expansion 10^{-6}/K	Thermal Conductivity w/m K
Aluminum Alloy				
Sheet	2024-T36	880	22.5	120
Sheet	7075-T6	840	22.9	140
Beryllium				
Extrusion		1860	11.5	180
Lockalloy	Be-38% AL	—	16.9	210
Graphite/Epoxy, V_f 55%				
[0]	HTS	—	− 0.4	—
[0]	UHM	—	− 1.04	—
Invar 36	Annealed	510	1.3	13.5
Magnesium				
Sheet	AZ31B-H24	1050	25.2	97
Steel				
PH15-7	RH1050	—	11.0	15
Ti6AL-4V				
Sheet		500	8.8	7.4

lightweight large appendages (solar array). For other applications, such as pressure vessels, titanium and stainless steel are most commonly used.

Mechanical properties of spacecraft materials, i.e., material type, density, longitudinal tensile strength and yield, transverse ultimate tensile strength, Young modulus, shear modulus, specific strength, and specific stiffness are given in Table 2-9. Thermal properties of spacecraft materials, i.e., material type, specific heat, thermal expansion, and thermal conductivity are presented in Table 2-10.

Chapter 3

Power Systems

THE SUCCESSFUL FULFILLMENT OF A SPACE MISSION IS DEPENDENT ON THE proper and reliable functioning of the power system of the spacecraft in orbit. The stringent demands on performance, weight, volume, reliability, and cost make the design of the spacecraft power system a truly challenging exercise.

Significant advances have been made in this area resulting in the development of reliable and lightweight power systems for long duration missions, typically more than five years. Since a space mission is inherently expensive, the necessity of optimization and built-in reliability becomes a rule rather than exception for all the on-board systems. Therefore, continuous efforts are being made to realize better performance of power systems. The different elements of the power system include energy sources, energy converters, energy storage, power conditioning, and control systems.

3.1 ENERGY SOURCES

The amount of electrical power required on-board a spacecraft is dictated by the mission goals, viz., the nature and operational requirements of the payloads, the antenna characteristics, the data rate, the spacecraft orbit, etc. In the particular case of communication satellites, the power requirements range from 500 W to 2000 W depending upon the channel capacity. Further, the power requirements are to be met uninterrupted for durations typically in excess of five to seven years. On the other hand, the unmanned scientific probes vary from a few months to three to four years and manned space stations for space laboratories like Skylab, Space Stations require 2 to 100 kW of power depending upon the load and the specific nature of the mission.

General Constraints. The generation of electrical power on-board a spacecraft generally involves four basic elements i.e.,

☐ A primary source of energy such as direct solar radiation or nuclear power generators, chemical batteries, etc.

☐ A device for converting the primary energy into electrical energy.

☐ A device for storing the electrical energy to meet peak and/or eclipse demands.

☐ A system for conditioning, charging, discharging, regulating and distributing the generated electrical energy at the specified voltage levels.

Foremost among the sources of primary power for use in spacecraft is the solar radiation that impinges in the vicinity of our planet at a level of 135.3 mW/cm^2. Nearly all the spacecraft use solar radiation as the primary source of power. However, use of solar radiation would need a supplementary source that can store the electrical energy. Chemical sources such as rechargeable storage batteries serve such a purpose. These batteries employing electrochemical processes have typical efficiency of 75%. As an alternate to solar energy, radioactive isotope generators have also been used especially for outer planetary missions because of the distance effects resulting in a low level of solar radiation. For example, the solar radiation reduces to about 58 mW/cm^2 in the Mars orbit and to about 5 mW/cm^2 in Jupiter orbit. So it becomes necessary to use other primary sources of energy for spacecraft on missions to Jupiter and beyond.

The most suitable electrical energy sources for different power output levels and mission durations are shown in Fig. 3-1. It is clear from the figure that the exact demarcation with regard to the choice of a power source for a specific mission is difficult. Other factors such as the continuous power requirements and eclipse conditions, are also major factors in the final choice. Sometimes a combination of two or three energy sources yields favorable results. However, since most of the spacecraft uses solar radiation as the primary source, much of the discussion that follows is with reference to this.

The basic configuration of a spacecraft power system based on a solar energy source is shown in Fig. 3-2. It consists of (a) solar cell array, (b) rechargeable secondary storage batteries for energy storage and (c) the power conditioning and control system (PCCS) which transfers power from the solar cell array directly and/or indirectly through the battery to the different loads.

3.2 ENERGY CONVERSION—
SOLAR CELLS AND ARRAYS

The solar cell basically works on the principle of photovoltaic effect and converts incident radiation into electrical energy usually represented by the so-called Johnson curve. Solar cells are made using different materials. But,

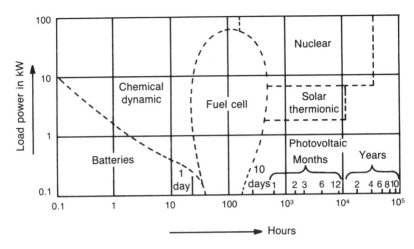

Fig. 3-1. Most suitable electrical energy sources for different power output levels and mission durations.

in spite of the excellent developmental efforts in various materials, such as Si, Ga, As, Inp, CdTe, Cds, etc., since 1955, all significant practical applications of solar cells utilize silicon devices, since none of the other materials provide higher efficiencies than silicon for production-type cells. It appears that this is primarily due to the excellent technology associated with silicon devices.

Recently GaAs solar cells have exhibited higher efficiencies. These, also offer lower radiation degradation and less temperature dependence compared to silicon solar cells. These cells have been flown on LIPS spacecraft, launched in February 1983. Figure 3-3 shows the I-V Characteristic of GaAs solar cell and it is clear that it offers about 18.7% efficiency. An additional advantage with these cells is that these cells have an open circuit voltage of about 1 volt and hence less series cells are required (compared to silicon solar cells) to obtain a particular voltage. This means lower production costs. Figure 3-4 shows efficiency versus temperature for GaAs cells and silicon solar cells and this

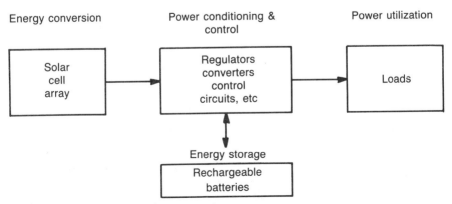

Fig. 3-2. Basic configuration of a spacecraft power system.

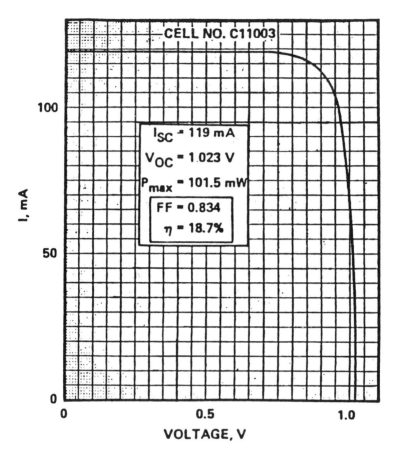

Fig. 3-3. I-V characteristic of a GaAs solar cell (reproduced with permission from Spectrolab "Products/Capability Overview" catalog).

shows that the GaAs solar cells are less temperature dependent. However, due to their higher densities (2.2 times compared to silicon cells) and higher cost (2 to 5 times compared to silicon cells), it may be a few more years before one can effectively use the GaAs cells.

3.2.1 State of the Art of Silicon Solar Cells

The silicon solar cells were developed first at the Bell Laboratory in 1953. They were used for the first time in space application in 1958 on-board the VANGUARD-1 Spacecraft for the generation of electrical power. Since then, there have been continuous improvements in the performance characteristics of solar cells. And to date, these cells continue to be the primary means of generating electrical power in earth orbiting spacecraft. Major developments in the area of solar cell technology for use in space applications during the last two decades are summarized in Table 3-1. As can be seen from this table, the efficiency of conversion has steadily improved from about 4% in 1953 to

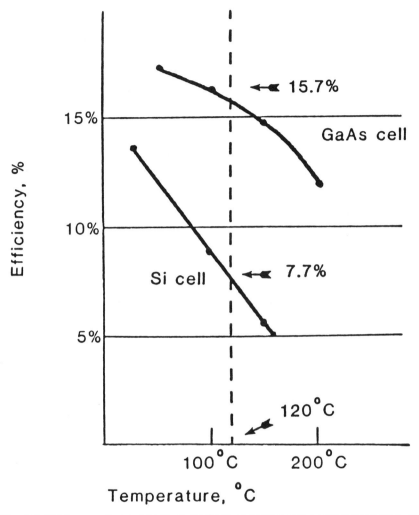

Fig. 3-4. Efficiency versus temperature for GaAs and silicon solar cells (reproduced with permission from Spectrolab "Products/Capability Overview" catalog).

Table 3-1. Major Developments in Silicon Solar Cells for Space Applications.

Year	Description	Efficiency
1953	Lithium diffused p-type n on p cell	4%
1954	Boron doped large area p on n cell	6%
1955	Theoretical understanding of solar cells and analytical treatment	

Table 3-1. Major Developments in Silicon
Solar Cells for Space Applications. (Continued From Page 83.)

Year	Description	Efficiency
1957	Wrap-around type solar cell so that both the contacts are on the back surface and optimum thickness to reduce resistive losses.	10%
	Both the contacts on the front surface resulting in consistent efficiency	>10%
	Finger-like contacts for redundancy, reducing resistive losses	>10%
1963	Improved n on p cells for better radiation resistance to result in reduced damage from radiation.	>10%
	"Blue" sensitive solar cells with very shallow diffused n layers having good surface properties and anti-reflection coating to radiation hardness	>10%
1973	$\frac{1}{4}$ micron diffused layers and $Ta_2 O_5$ anti-reflection coatings to result in improvements in short wavelength collection efficiency and overall efficiency	
	Back surface reflector placed on the rear side of the solar cell would reflect some of the long wavelength infrared energy back out the front face of the cell.	12.9%
1974-80	Zero reflectivity cells (black cells) leading to further improvements in efficiency	
	Selective etching to form numerous pyramidal shaped structures, which trap more of the incident light by multiple reflection. Addition of a back surface reflector and a P+ field adjacent to the back contact.	15.0% (approx.)
	Introduction of a layer of palladium in between silver and titanium contacts to eliminate moisture contamination.	
##	Gallium Arsenide solar cells developed recently (since 1981) exhibit about 18.7% efficiency and are flown on LIPS spacecraft. However, due to their higher density (2.2 times that of silicon cells) and higher cost (2 to 5 times that of silicon cells), it will be a few more years before one can effectively use these cells to take advantage of their higher efficiency.	

Table 3-2. Comparison of Characteristics of a Commercial and a COMSAT Cell.

Parameter	Commercial Cell	COMSAT Cell
Collection efficiency	71%	79%
Reflection losses	9.5%	4.9%
Open circuit voltage	0.580 V	0.595 V
Series resistance	0.17 ohm	0.05 ohm
Conversion efficiency	10.4%	14%

around 15% in present day solar cells suitable for space applications. To appreciate the trends in the latest technology, the COMSAT cell is compared with conventional technology in Table 3-2. It is clear from this table that reflection losses and series resistance of solar cells are considerably lower for COMSAT cells (0.05 ohm as against 0.17 ohm for a conventional cell) compared to conventional cells, whereas open circuit voltage and efficiency are considerably higher.

3.2.2 Description of Solar Cell

Figure 3-5 shows the structure of a typical solar cell. The cell consists of a small wafer of single crystal silicon (p-type) in which the junction (shallow) is formed by diffusion of phosphorous (n-type) through one surface. The nominal resistivity of silicon is about 10 ohm-cm. The junction depth is usually a fraction of a micron, such that photon energy is enough to create electron-hole pairs near the junction. The freed electrons traverse the junction and diffuse in the n-region. Similarly, the holes diffuse in the p-region. This exchange of electrical charge carriers through the junction reduces the electrical resistance and causes a current to flow in the external load. This current is proportional to the magnitude of incident luminous flux and the angle of its incidence. Some of the electrons and holes liberated away from the junction recombine before they reach the junction. Such electrons and holes do not contribute to electrical

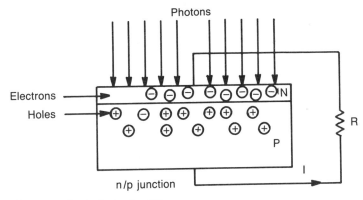

Fig. 3-5. Structure of a typical solar cell.

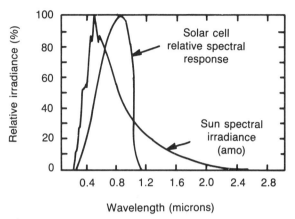

Wavelength (microns)

Fig. 3-6. Spectral response curve of a typical solar cell in relation to the solar radiation.

current. The spectral response curve of a typical silicon solar cell in relation to the solar radiation is shown in Fig. 3-6. The cell response generally extends over the wavelength 0.35 microns to 1.10 microns. For obtaining a maximum output, it is desirable to have the peak spectral response of the cell at the maximum energetic response of the solar spectrum, i.e., in the neighborhood of 0.5 microns. In fact, for technological reasons, the maximum of this response is located around 0.8 microns. The short wavelength response (blue response) can be increased by making junctions closer to the surface, while long wavelength response can be improved by making the junctions deep below the surface. The solar cells on-board the spacecraft undergo a large number of thermal cycles (from > + 60 degrees C to < – 170 degrees C in the case of geosynchronous orbits). Figure 3-7 shows how the current-voltage (I-V) characteristic of a typical cell varies due to different temperatures.

The change in cell open circuit voltage due to temperature increase is typically in the range of – 0.3% per degree C to – 0.4% per degree C of the

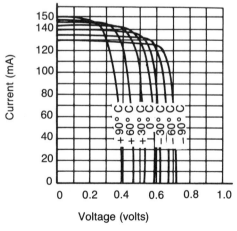

Voltage (volts)

Fig. 3-7. I-V characteristic of a typical solar cell for various temperatures.

30 degrees C value and results from changes in the diffusion length and the attendant forward and reverse current changes. The change in the cell short circuit current due to temperature increase is typically in the range of + 0.025% per degree C to + 0.075% per degree C of the 30 degrees C value. These result in a decrease of overall efficiency by about 0.5% per degree C of 30 degrees C value as the temperature increases. To get good power output from a cell, it would be desirable to have the I-V characteristic as nearly a rectangle as possible, so that the maximum power-point voltage is near to open circuit voltage of the cell.

Figure 3-8 gives the equivalent circuit of the solar cell and its mathematical model. The series resistance (R_s) imposes a limitation on the maximum obtainable conversion efficiency. It is easy to see that the shunt resistance (R_{sh}) even as low as 100 ohms does not appreciably decrease the power output, whereas R_s of only 0.5 ohm can reduce the available power to less than 90% of the optimum power (i.e., at R_s = 0 ohm). The R_s of a cell depends on the junction depth, impurity concentrations of p-and n-type regions and the arrangement of the front surface contacts. Typically R_s is 0.10 ohm and R_{sh} is about 250 ohm for a 2 × 2 cm^2 silicon solar cell.

Cover Glass: Cover glass is usually made of quartz or sapphire or cerium doped silica and with a 150 to 300 microns thick cover slip attached to the cell with UV resistant adhesive for the purpose of thermal control and protection from radiation and micrometeorites. The cover slip incorporates an antireflection coating on the front surface optimized at 0.6 microns and a multi-layer

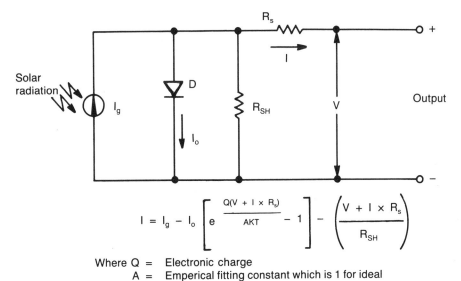

$$I = I_g - I_o \left[e^{\frac{Q(V + I \times R_s)}{AKT}} - 1 \right] - \left(\frac{V + I \times R_s}{R_{SH}} \right)$$

Where Q = Electronic charge
A = Emperical fitting constant which is 1 for ideal junction
K = Boltzmann constant
T = Absolute temperature °K
I_o = Diode saturation current
I_g = Generated current due to incident light energy

Fig. 3-8. Solar cell equivalent circuit and its matnematical model.

UV rejection filter at the back surface. The cover glass will have 98% trans-mittance over the spectral range of 0.35 microns to 1.10 microns.

Temperature of the cells has to be kept low to obtain higher output power and hence the ratio of solar absorptivity to the black body emissivity (a/e) of the cells should be made a minimum. The cover glasses, which acts as a filter cut down the total energy absorbed by the cell, achieve this to a good extent. A solar cell complete with its cover glass exhibits an a/e of around 0.94.

3.2.3 Radiation Damage

The efficiency of the solar cell decreases as a function of time in space due to its susceptibility to the particle radiation. A large flux of protons and electrons arise in the spacecraft environment because of the Van Allen Radiation Belts (trapped radiation) and solar flares. Most of the solar cell degradation effects occur because of the solar flare protons and trapped electrons. The energy of these particles varies from a few KeV to several MeV in the case of electrons and is in the range of several hundred MeV for protons.

The bombardment of these high energy particles produces crystalline defects in the cells, which then become recombination centers. Low energy particles create damage close to the junction and therefore, raise the dark current and lower the open circuit voltage. High energy particles penetrate far into the base and lower the life-time of electron hole pairs, thereby decreasing the short circuit current. This results in a reduced cell power output. Figure 3-9 shows the effects of successive doses of 1 MeV electrons in the I-V characteristics of solar cells.

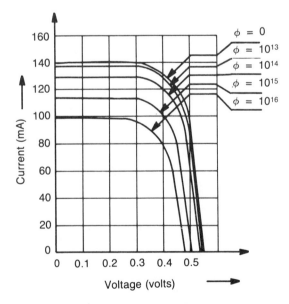

Fig. 3-9. I-V Characteristic of a typical solar cell subjected to successive doses of 1 MeV electrons.

These effects have been primarily minimized by changing the configuration of silicon solar cells to n/p from earlier used p/n. The cells of n/p type are considerably more radiation tolerant than the p/n type. This is due to (a) the higher mobility of electrons, which are the electrical charge carriers in the n-region, which overcome some of the recombination centers to form the current flow, and (b) the radiation incident energy required to damage the crystal lattice on n-type is more than on the p-type.

Further, the radiation effects have been minimized by protecting the cells using cover glasses (Corning microsheet 7940 or OCL cover glass). The amount of radiation falling on the cell depends upon the thickness of the cover glass employed. Normally, these cover glasses are also coated with anti-reflection coating (e.g., silicon monoxide) to minimize reflections, which further helps to enhance the overall efficiency of the cell.

3.2.4 Some Recent Developments in Solar Cells

The solar cell technology has considerably advanced in step with the advances in solid-state technology. Some of the recent developments in this area that hold promise are:

☐ Lithium doping of silicon cells so that the radiation damaged sites on the cells are annealed in orbit itself, resulting in an increase of the life of the cells by more than 50% for any mission. However, this technology is yet to be proven in an actual space mission.

☐ Thermo-compression bonding of interconnecting tabs to solar cells so that the cell modules can withstand high temperatures (>300 degrees C).

☐ Fabrication of integral cover on solar cells so that handling problems are avoided.

☐ Ion implantation of doping impurities in silicon solar cells so that their sensitivity at short wavelengths is improved.

☐ Fabrication of multiple junction edge illuminated silicon solar cells to yield arrays with high voltage, low current capability.

☐ Use of 0.1 ohm-cm p-type material to optimize conversion efficiency, the theoretical limit of this being 24.5%.

☐ Development of tandem cells consisting of two p-n junctions placed back to back and connected suitably to have an additive effect, so that efficiencies as high as 40% are possible.

☐ GaAs single junction large area cells have been manufactured using liquid phase epitaxy and metal-organic chemical vapor deposition processes which exhibited efficiencies of 20% AMO.

☐ Since the maximum solar incident energy is 135.3 mW/cm², the multi-junction approach utilizes the solar spectrum more efficiently by stacking several band gap cells (example: a thin GaAs cell stacked on top of a silicon cell) in series such that successive junctions convert different frequency ranges of sunlight.

☐ Development of 2 mil thick GaAs cells to obtain lower density and more radiation resistance.

It appears from the present trends that further improvements in solar cell technology have to necessarily come from the development of materials technology that could lead to increased open circuit voltage and improvement in efficiency.

3.2.5 Solar Array

A solar array normally is comprised of suitable series-parallel combinations of solar cells connected as shown in Fig. 3-10, to yield suitable voltage, current, and power to meet a particular mission requirement. The cell mounting can be either (a) shingled type, where the cells are attached in a staircase manner as shown in Fig. 3-11, in which case, series connection between adjacent cells is automatically achieved, or (b) flat type, where the cells are placed side by side as shown in Fig. 3-12, so that both the contact terminals are used for interconnection. Table 3-3 presents a comparison of the characteristics of

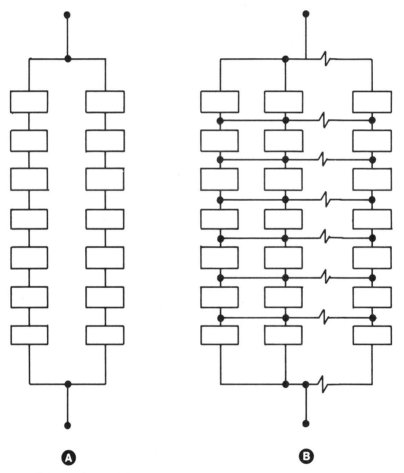

Fig. 3-10. Solar array comprising series-parallel (A) and parallel-series connected solar cells (B).

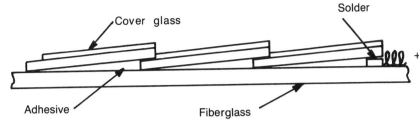

Fig. 3-11. Shingle type mounting.

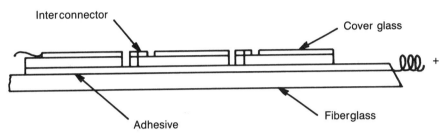

Fig. 3-12. Flat type mounting.

Table 3-3. Comparison of Characteristics of Flat and Shingle Mounted Cells.

Parameter	Flat Type	Shingle Type
Replacement of broken cells	Easy	Difficult; A group of cells has to be removed to replace any broken cell.
Series-parallel interconnection of cells	More freedom	Less freedom
Number of physical connections per array	More	Moderate
Thermal properties	Better	Good
Bonding of cells to the substrate	Stronger	Strong
Active area occupied by solar cells per unit of projected area	Relatively less, as the cell bus bars are exposed	High
Packing factor	High	Moderate

flat and shingle-mounted cells in a solar array. The loss of exposed area in both cases is about 5-10% of the total cell area. The electrical connections are generally made by soldering fingered connections for redundancy. Use of wrap-around contact cells, simplifies cell interconnections. Gold plated kovar bus bars are used to connect the top and bottom ohmic contacts of the cells. Invar interconnects are used due to their low thermal expansion, which increases the life of the solder joints. Recently developed parallel-gap resistance welding proved highly reliable and also it reduces production time and costs.

In any array, if some cells are shadowed, then they act as load for the other cells instead of serving as energy converters. Also if a shadowed cell is made to carry through it a current more than 3 or 4 times a single cell current in the reverse direction, the shadowed cell might go into zener breakdown and dissipate more power, resulting in a hot spot. This is more so common if the series string has a 3- or 4-cell width. Diodes are normally connected across a single or a group of cells to minimize such shadow effects, as shown in Fig. 3-13. Alternately, solar cells with integral diode protection can be used to meet this requirement. Use of series strings with only one cell width, eliminates the hot spot problem and in this case diodes are not needed. However, if a cell in the string is shadowed, the power from the complete string is lost.

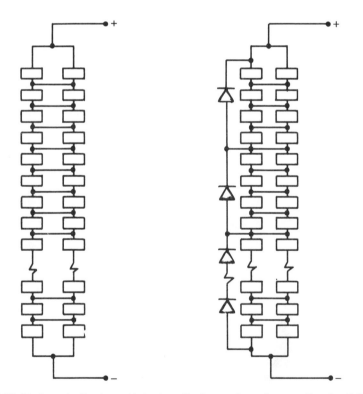

Fig. 3-13. Diode protection to avoid shadow effects on solar cell arrays. For simplicity a series string with a two cell width is shown, however this problem becomes severe if the string has a three or four cell width.

The larger the area of the solar array, the strength and stiffness of the array becomes more crucial due to the launch loads. To decouple the solar array from the spacecraft resonance frequency effects, the solar array is designed to be structurally lightweight using high modulus materials. Such arrays have natural frequencies above 25 Hz. Some less stiff and lightweight arrays exhibited natural frequencies as low as 10 Hz.

During on-orbit phase, the natural frequencies of the deployed solar array is important as it interacts with the attitude control system and the attitude control system must accommodate the low natural frequencies of the deployed solar arrays.

3.2.6 Array Types

Solar arrays are usually fabricated by mounting the cells either on honeycomb or fiberglass substrates, the former leading to 5-10% weight saving as compared to the latter. The other consideration is that these substrates should be able to withstand the initial launch vibration and acceleration stresses (typically 10-15 g up to 2500 Hz) as well as thermal stresses from $> +60$ to < -170 degrees C in the case of geosynchronous orbits. Further, depending on the power requirements, mission, etc., the solar arrays, as shown in Fig. 3-14, are fixed (a) either on the skin of the spacecraft and are known as body mounted arrays or (b) on extendable panels and are known as deployable arrays.

Body Mounted Arrays. This is a very simple approach and has been employed in many of the spin-stabilized spacecraft. One of the important characteristics of this array is that it facilitates easy temperature control. The instantaneous power output of this array is proportional to the projected area, and there is usually a definite maximum power output for a chosen configuration. This can be seen from the calculations of maximum projected power output for two shapes of spacecraft, viz., cylindrical and spherical corresponding to the shroud dimension of different launchers given in Table 3-4. It is easy to see from the table that the cylindrical shape gives more power, but requires a particular orientation as compared to the spherical shape, i.e., spin axis of the cylindrical spacecraft has to be perpendicular to the sun's rays.

Deployable Arrays. There are two types of deployable arrays, viz., those having (i) the deployed orientation remaining constant with respect to the spacecraft axis, and (ii) the deployed orientation continuously controlled (using solar array drive assembly) with respect to the spacecraft axis so that they remain pointing towards the sun as the spacecraft moves in its orbit. Some deployable solar arrays in deployed configuration are shown in Figs. 3-15 to 3-18. The deployable arrays can be grouped into rigid, semi-rigid and flexible types according to the type of substrate on which the solar cells are mounted. A comparison of characteristics of different deployable arrays based on the available information is given in Table 3-5.

When deployable arrays are stowed, they are folded with protective pads, restraining hardware and mounting and tensioning members are sandwiched between two honeycomb vanes, which hold tightly against the spacecraft body during launch by several post and slot latch restraining mechanism pads keep

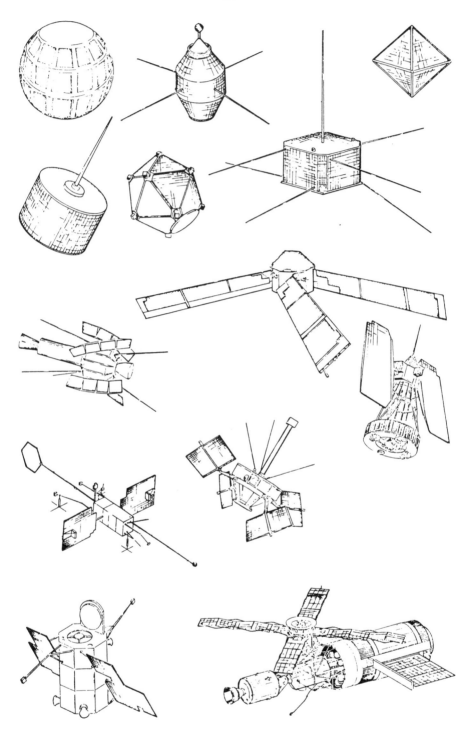

Fig. 3-14. Evolution of solar cell array configurations (courtesy of Jet Propulsion Laboratory).

Table 3-4. Maximum Body Mounted Array Power Restrictions.

Launcher	Diameter (m) (2R)	Cylindrical**		Spherical	
		Maximum Projected Area (m²)	Maximum* Power (Watts)	Maximum+ Projected Area (m²)	Maximum* Power (Watts)
Diamont B	0.763	0.504	75	0.457	70
Scout (Bulbous)	0.965	0.806	120	0.731	110
Black Arrow	1.280	1.42	210	1.287	190
Thor Delta – 2914	2.180	4.12	610	3.733	550
Titan IIIC or Titan III/ Agena	2.640	6.04	895	5.474	810
Centaur Intelsat Shroud	2.690	6.27	930	5.683	840
Ariane	3.000	7.79	1155	7.070	1045
Centaur Standard Shroud	3.730	12.05	1785	10.927	1620
Space Transportation System (STS) + +	4.260	15.71	2325	14.250	2110

NOTE: * Computed on the basis of 74 mw/cell of 2 by 2 Cm size at AMO (25 degrees C) with an array packing factor of 80%.

+ The spherical surface area of the spacecraft is, normally, modified from a continuous curved to some flat panels to enable simpler mounting of the modules. This adds some error in maximum power calculations which is not taken into account.

+ + STS has got a total height of 19.81 meters and it is planned to fly 5 spacecraft of 426 cm diameter and 366 cm height.

H Height of cylindrical spacecraft and is equal to 1.732R assuming uniform mass distribution.

** End circular surfaces are not considered as the spin axis (along height) of spacecraft is perpendicular to sun.

Fig. 3-15. The OLYMPUS spacecraft. The electrical power generated by its solar arrays will be 3.5 kW (courtesy of British Aerospace).

opposing cover glasses from touching during storage. After deployment the pads are stowed underneath the lower honeycomb vane.

When deployable arrays are employed and if they are not exposed when they are folded or stowed, then the outer surface of the spacecraft is covered (body mounted) with some solar cells. Although the power from these cells will not be used during operational phases, these give some insurance (redundancy) against difficulties and these body mounted cells provide power during transfer orbit operation with deployable solar arrays folded.

Comparison of Arrays. Table 3-6 gives a comparison of characteristics of typical body-mounted and deployable arrays. Trade-off studies have shown that it will be advantageous to use deployable arrays for end-of-life power of about 500 W or more, though the orientation and power transfer are added complexities. The choice of an array type depends on the relative weightages given to the various criteria, viz., achievable reliability, design simplicity and compatibility with structure, thermal design, stabilization, antenna orientation, cost, weight, ease of handling, and serviceability.

3.3 SOLAR ARRAY DRIVE UNIT

For other than body-mounted solar arrays, some kind of deployment system is required, which normally consists of a pulley-cable-actuator system or an extendable boom. As mentioned above, a solar array drive is employed to control the solar array continuously such that it is perpendicular to the sun all the time. Depending upon the attitude control system and other payload requirements, solar array drives can be single axis or double axis type. Figure 3-19 shows the Schaeffer Magnetics two axis solar array drive system with power transfer slip rings. As the array is rotating with respect to the spacecraft

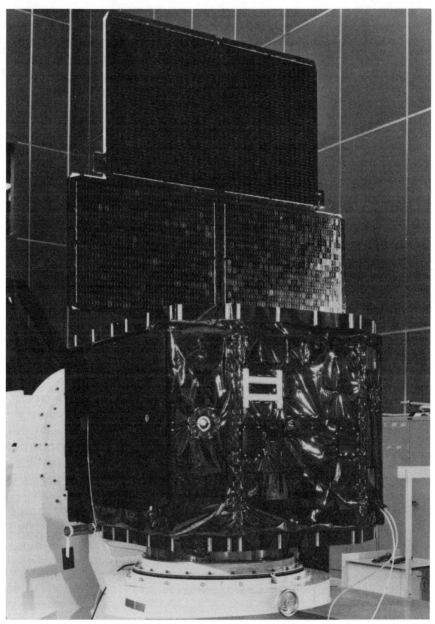

Fig. 3-16. EXOSAT spacecraft with its solar array in its deployed configuration (courtesy of Marconi).

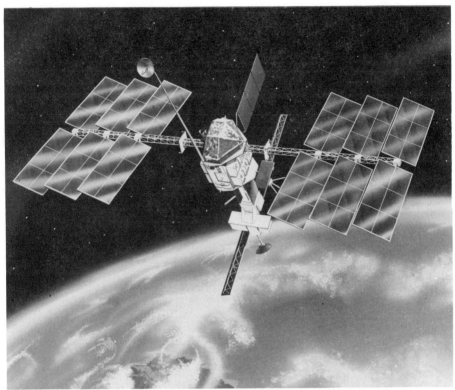

Fig. 3-17. Columbus polar platform—in sunlight the solar array produces 15 kW of power (courtesy of British Aerospace).

body, the power from the solar array is transferred to the spacecraft body using what is known as the power transfer assembly. Hence the name *Solar Array Drive and Power Transfer Assembly* (SADAPTA). Structurally, it also provides the mechanical interface and structural support for the solar arrays and houses various components.

The SADAPTA consists of (a) drive motor (dc or stepper motor), (b) reduction gear assembly to transmit the drive motor power to the solar array, (c) suitable slip ring-brush assembly to transfer power and signals via the drive shaft to the spacecraft body, (d) shaft encoder to indicate relative position between the solar array and spacecraft body.

The solar array drive mechanism performs the two basic tasks of acquiring and tracking the sun. Subsequent to initial deployment and following an eclipse of the sun or loss of track, a nominal fast slew mode is required for sun acquisition. After acquiring the sun, the control system will track the sun at 1 revolution per orbit, maintaining the solar array normal to the sun.

In a closed loop control approach, SADAPTA uses the signal from the sun sensor mounted on the solar array, attains the sun and causes the array to rotate such that it is perpendicular to the sun. In semi-closed loop control approach, SADAPTA uses the signal from the sun sensor, attains the sun and then follows a particular rate of rotation. This approach is more common in the geosynchronous communications spacecraft where the rotation rate is one rotation per day or 24 hours. The Solar Array Drive typically uses a dc brush/brushless motor or a stepper motor. Some drives have provision for bidirectional rotation.

The material used for the slip rings and their sizes depend upon the ratings of the current and voltage. Electrical insulation is very important for high voltage systems. Lubricant selection depends upon the amount of heat produced

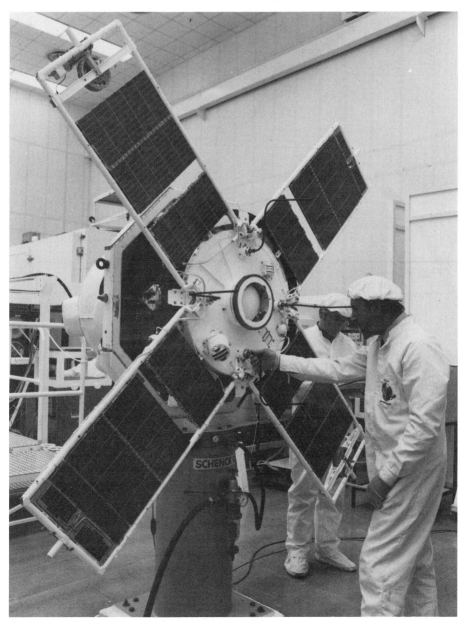

Fig. 3-18. Ariel-6 satellite solar arrays (courtesy of Marconi).

Table 3-5. Comparison of Different Deployable Solar Arrays.

			Deployable Solar Array Types			
Order of Merit	Rigid	Semi-Rigid	Flexible Roll-up	Flexible Fold-up (Extendable boom)	Flexible Fold-up (Pantagraph)	Flexible Fold-up (Telescope)
Power to weight Ratio	6	5	4	3	2	1
Stowed Volume	4	4	3	1	1	2
Stiffness	1	1	3	3	2	2
Adaptability	3	3	1	1	1	2
Development Potential	3	2	1	1	1	1
Cost	5	4	3	2	2	1
Power Transfer Mechanism	1	1	2	1	2	2
Power in Transfer Orbit	Some Power	Some Power	No Power	No Power	No Power	No Power
Examples	11 KW Apollo Telescope Mount on Skylab; Symponie	Boeing 46 KW Array	1.5 KW Hughes Solar Array; AEG Solar Array	1 KW Solar Array for CTS	Developed by SNIAS, France	RAE Proto-Type Solar Array

Note:— Usually the Solar arrays are deployed in the parking or transfer orbit whenever flexible solar arrays are carried, in the absence of any small body mounted array for generating transfer orbit power.

due to the contact resistance in the brushes and the slip ring conductors and the life of the mission. Tribological research carried out at Poly-Scientific has led to various alternative approaches to the selection of contact material for use in space (hard vacuum) environment and are solid film lubricated composite brushes, single element oil lubricated gold wires (with a replenishing reservoir) and fiber brush operated without lubrication. Figure 3-20 shows the photo of slip rings being assembled (manufactured by Poly-Scientific) and Fig. 3-21 shows the assembled slip rings.

3.4 ENERGY STORAGE—STORAGE CELLS AND BATTERIES

In any spacecraft power system that uses solar radiation, the storage battery is the main source of continuous power, as it is called upon to respond to peak and eclipse demands of power depending upon the spacecraft orbit. The eclipse seasons in a geosynchronous orbit occur twice per year, viz., in Spring and Autumn. Each season has 45 eclipses. The battery is charged during the sunlit portion of the orbit and discharged during the eclipse. In the case of low-earth-orbit (LEO) spacecraft, the number of eclipses increases as the altitude decreases. Typically, for a 550 km orbit, there will be about 15 eclipses per day or about 5500 per year. Several times in a year the spacecraft is in continuous sunlight for long periods in the case of high inclination orbits during which the daily average solar array power exceeds the average power demand. Also when the spacecraft comes out of eclipse, the power output of solar array is much higher (as the array is cool and its temperature is very low) than the steady-state power output (when the array attains steady-state temperature).

Table 3-6. Comparison of Characteristics of Body Mounted and Deployable Arrays.

Parameter	Body-Mounted Arrays	Deployable Arrays
Free room inside the nose cone of the launcher	Less	Moderate
Free windows on the spacecraft structure	Difficult	Easy
Handling and integration with spacecraft	Little difficult	Easy
Temperature Control	Easy	Not so easy
Design with projected shadows	Easy	Not so easy
Projected array area per solar array area (sun-oriented)	Less (Usually 25-35% of the total array area)	All the array
Maximum power for a chosen launcher	*Limited	Not limited
Adaptability for design changes	Less	More
Weight per unit power	More (3 to 4 times)	Less
Cost per unit power	More (2 to 3 times)	Less

*Double drum approach has changed this limit and recently launched was a spacecraft with two drums one inside the other, practically doubling the capability.

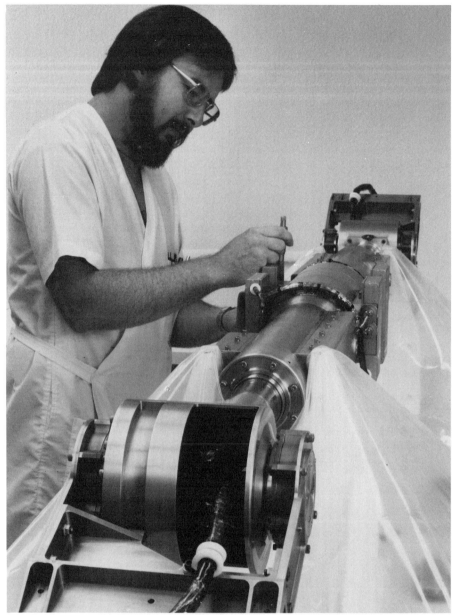

Fig. 3-19. Two axis solar array drive system with power transfer slip rings (courtesy of Schaeffer Magnetics).

This extra power can be optimally utilized only if the battery is capable of being charged at high rates.

Thus for spacecraft applications a storage cell shall have high capacity per unit of weight. Chemical effects shall not cause deterioration or loss of stored energy. The transformation of electrical into chemical energy as in charging, and of chemical into electrical energy as in discharging, should proceed nearly

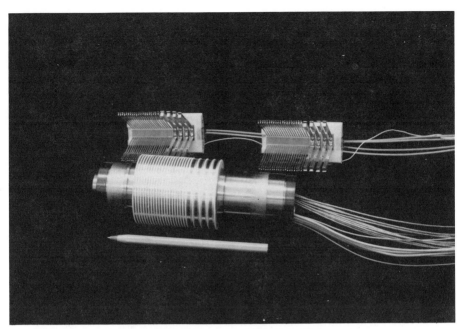

Fig. 3-20. Slip rings being assembled (courtesy of Poly-Scientific).

Fig. 3-21. Assembled slip rings (courtesy of Poly-Scientific).

reversibly. An ideal storage cell should have low impedance, have simplicity and strength of construction, be durable and be producible at low cost.

3.4.1 Storage Cells

A difference in potential is set up when two unlike metals are dipped into an electrolyte. When a resistor is connected externally between these two metals, the potential across the resistor causes a current to flow in the resistor and this is known as a cell. Thus a cell is an electrochemical device that stores energy in the chemical form and this chemical energy is converted into electrical form during discharge. Chemical reactions taking place inside the cells, produce electrical energy and its magnitude depends upon various cell characteristics, i.e., cell potential or voltage, efficiency of the electrochemical reactions, and size or the cell, etc.

Energy cells are classified into two types, (a) *primary cells* and (b) *secondary cells*. A primary cell basically converts chemical energy into electrical energy and it cannot convert electrical energy back into chemical energy. Primary cells once used, cannot be recharged to use it again and hence they are one-shot devices. On the contrary, a secondary cell converts chemical energy into electrical on discharge and back to chemical on charge and this process can be repeated many times. Thus, secondary cells are reversible to a high degree, in that the chemical conditions may be restored by causing current to flow into the cell on charge and hence are known as storage cells. A storage cell is a single device and a storage battery is an arrangement of more storage cells, usually connected in series to supply the necessary voltage. If a single battery is not sufficient, more than one battery is employed to meet a particular mission requirement. Some of the important storage cells are nickel-cadmium, nickel-hydrogen, silver-zinc, silver-cadmium, silver-hydrogen, lead-acid, lithium, and sodium cells. Important space-worthy energy storage cells only are described in the following sections.

3.4.2 Energy Storage Cell Requirements

As mentioned previously, energy storage cells are the continuous source of power on-board a spacecraft and they play an important role in keeping the spacecraft alive since its launch. Proper selection of the storage cells and their ratings depend upon various factors, i.e., spacecraft orbital altitude, inclination, eclipse power requirements, peak power requirements during daytime, operating temperature, mission life, etc.

Orbital Altitude and Inclination. Orbital altitude and inclination play an important role in specifying the energy storage requirements. The eclipse seasons in a geosynchronous orbit occur as shown in Fig. 3-22, twice per year, namely, in Spring and Autumn. Each eclipse lasts for about 45 days with a maximum shadow time of 1.2 hours per day. Thus, the charge-discharge cycles for any storage battery on-board a spacecraft in the geosynchronous orbit will be about 90 per year, as the battery is charged during the orbital day and

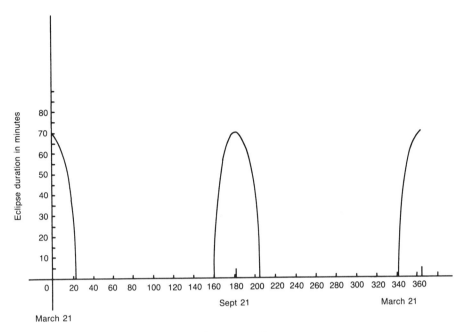

Fig. 3-22. Eclipse seasons in a geosynchronous orbit.

discharged during the orbital night. However, in the case of low earth orbit spacecraft, the number of eclipses increases as the altitude is low. Typically, for a 550 km orbit, there will be about 15 eclipses per day with a maximum shadow duration of 36 minutes for every orbital period of about 96 minutes. Thus, the charge-discharge cycles in this case will be about 5500 per year. Depending upon the inclination of the orbit, several times in a year the spacecraft may be in continuous sunlight for long periods.

Temperature Effects. Whenever the spacecraft comes out of eclipse, the solar array is cool and hence produces more power and as the solar array gets hot, it produces relatively less power. However, this extra power which is available when the spacecraft comes out of eclipse, can be optimally utilized only if the battery is capable of being charged at high and variable rates.

Requirements of an Energy Storage Cell. Thus from the above discussions, the primary requirements of a storage cell for spacecraft applications are, a) capability of accepting and delivering unscheduled power at high rates, b) large number of charge-discharge cycles or long charge-discharge cycle life under a wide range of conditions, c) high recharge efficiency, d) good hermetic seal to prevent loss of electrolyte and corrosion throughout thousands of electrical cycles involving pressure and thermal changes, e) possibility of operation in all positions, f) availability of cells with well matched characteristics, g) capability to withstand launch and space environments, h) stable long-term overcharge characteristics, i) maximum usable energy per unit weight or low weight, j) low volume, k) low cost and l) high and proven reliability.

3.4.3 Storage Cells

There are many storage cells available, however only some are worth considering and suitable for space applications. Important characteristics of such storage cells are presented in Table 3-7.

Ni-Cd Cells. Ni-Cd batteries have been used extensively and have demonstrated their performance in space. Figure 3-23 shows the storage cell volume and weight versus storage cell capacity for (i) standard cells and (ii) low profile cells. The weight penalty that results from their low energy density has prompted the study, design, and development of more efficient cells. Naval labs at Crane, Indiana conducted charge-discharge life cycles on the NASA Standard 50 AH Ni-Cd cells and have successfully undergone more than 27000 cycles at a depth of discharge of about 25%, at a temperature of 20 degrees centigrade.

Ni-H$_2$ Cells. Ni-H$_2$ cells appear to be an improvement over the Ni-Cd in applications requiring longer lifetime and reduced weight. The Ni-H$_2$ cell consists of a catalytic gas negative electrode coupled with the nickel-positive electrode borrowed from the nickel-cadmium system. Electrochemically, the reactions at the positive electrode are the same as those occurring in the parent system. At the negative electrode, hydrogen is displaced from water by electrolysis during charge; during the discharge cycle, hydrogen is consumed by oxidation reaction at the negative electrode, producing electrical power. Figure 3-24 shows typical nickel-hydrogen charge/discharge cycles. Figure 3-25(A) shows a nickel-hydrogen battery being assembled in an environmentally controlled area and Fig. 3-25(B) shows another integrated battery. A detailed comparison of Ni-Cd cells versus Ni-H$_2$ cells is presented in Table 3-8.

To improve the specific energy, various packaging techniques are incorporated, known as the *Common Pressure Vessel* (CPV) cell and the *Bipolar cell* compared to a simple single cell having its own container. A single cell is known as an *Individual Pressure Vessel* (IPV) cell. These are briefly described below:

☐ Individual Pressure Vessel Cell: An Individual Pressure Vessel (IPV) Cell is only one cell in one mechanical enclosure or package. If more power is needed such IPV cells are connected in series and parallel to achieve voltage and current respectively.

☐ Common Pressure Vessel Cell: A Common Pressure Vessel (CPV) Cell encloses a stack of electrodes connected in series where each stack is parallel connected inside, i.e., parallel connection inside each individual stack, and series connected from one stack to the next. This is equivalent to housing many series and parallel connected bare cells inside a common pressure vessel.

☐ Bipolar Cell: A Bipolar cell is made up of electrodes stacked in series inside the stack assembly; several of those stacks are put on top of each other using a bipolar plate. This plate connects electrically one electro-pair to the next. It also isolates the electrolyte in each electro-pair area.

Ni-H$_2$ cells have been manufactured in different packages like IPVs, CPVs, and Bipolars.

Table 3-7. Characteristics of Different Storage Cells.

Type of Cells	Electrolyte	Nominal Voltage/Cell (Volts)	Energy Density (WHr/kg)	Temperature (Deg C)	Cycle Life at Different Depth of Discharge Levels			Whether Space Qualified
					25%	50%	75%	
Ni-Cd	Diluted Potassium Hydroxide (KOH) solution	1.25	25-30	−10 − 40	21000	3000	800	Yes
Ni-H$_2$	KOH solution	1.30	50-80	−10 − 40	>15000	>10000	>4000	Yes*
Ag-Cd	KOH solution	1.10	60-70	0 - 40	3500	750	100	Yes
Ag-Zn	KOH solution	1.50	120-130	10 - 40	2000	400	75	Yes
Ag-H$_2$	KOH solution	1.15	80-100	10 - 40	>18000	—	—	No
Pb-Acid	Diluted sulfuric acid	2.10	30-35	10 - 40	1000	700	250	—

*Ni-H$_2$ cells are employed on-board the Navigational Technology Satellite (NTS-2) and other geosynchronous satellites. However, these cells have not been used on any low earth orbit satellites.

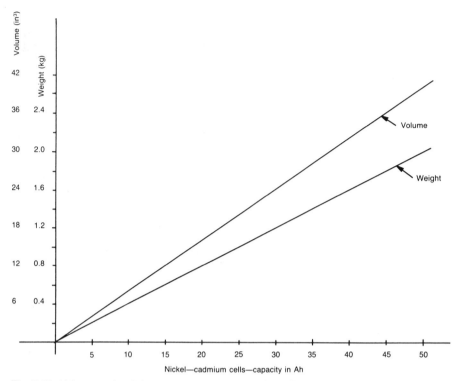

Fig. 3-23. Volume and weight versus capacity for Ni-Cd cells.

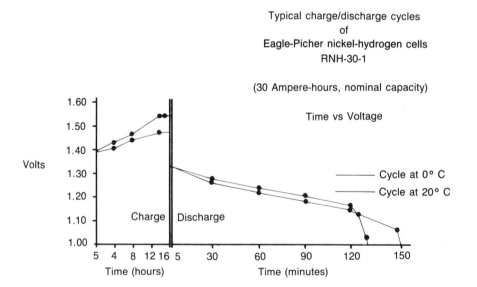

Typical nickel-hydrogen charge/discharge cycles

Fig. 3-24. Nickel-hydrogen storage cell typical charge/discharge cycles (courtesy of Eagle-Picher).

Fig. 3-25A. Nickel-hydrogen battery assembled in an environmentally controlled area (courtesy of Canadian Astronautics Limited, Canada).

Fig. 3-25B. An integrated nickel-hydrogen battery (courtesy of Eagle-Picher).

Table 3-8. Comparison of Ni-Cd Cells Versus Ni-H$_2$ Cells.

Parameter	Ni-Cd	Ni-H$_2$
Change in Electrolyte Concentration	15% (approx.)	10% (approx.)
Electrolyte Quantity	2 cc/AH	2.4 - 3 cc/AH
K$_2$CO$_3$ build up in Electrolyte	1% - 2% per year	0
Redistribution of Electrolyte	Swelling of Electrode	Entrainment loss
Separator degradation	Nylon* is soluble in KOH	None
Negative Electrode	Cadmium migration	No deterioration
Electrolyte Reservoir accommodation	No #	Yes
Overcharge	Limited	Unlimited (relatively)
Cycle life	see Table 3-7	see Table 3-7
Abuse Tolerance	medium	very good
Over-discharge	Less tolerant	More tolerant
Positive Electrode Expansion	Vacuum Deposition Maximum Expansion	Electrochemical Deposition, Minimum Expansion
Expansion Design Accommodation	No	Yes
Temperature Control	Hotter electrode stack	Cooler electrode stack

*A new improved separator known as "pellon" is being qualified.
In case of Ni-Cd, starved electrode condition would result in electrolyte wicking and flooding of negative electrode.

Silver-Hydrogen Cell. Development of a more efficient cell, the silver-hydrogen has been carried out in parallel with the development of the nickel-hydrogen cells. The design of Ag-H$_2$ cells is similar to the existing Ni-H$_2$ types where the electrolyte is a potassium solution. The silver-hydrogen cell is attractive because of its high specific energy in the range of 75 to 80 Wh/kg and an energy density in the range of 50 to 60 kWh/cu m. Accelerated charge-discharge cycle tests simulating ten seasons of geosynchronous orbit conditions, have been conducted on silver-hydrogen cells without any failure. The state of charge can be found out by measuring the pressure of the cell.

Other Cells. All new energy storage cells are usually targeted to geosynchronous orbit applications due to the limited number of eclipses per year in that orbit, i.e., about 90 per year compared to about 5500 per year in the case of a low earth orbit application.

Lithium cells show promise as advanced energy storage devices for future space application. These cells are of great interest due to their high energy density, greater than 100 Wh/kg and long life, greater than 10 years (for geosynchronous orbit applications). Lithium metal sulfide cells presently offer a specific energy of 110 Wh/kg. Improvements to 160 to 200 Wh/kg are possible with chemistry change. Typical cycle life is limited to 300 to 500 cycles because of internal shorts due to volume changes. Operating temperature is 450 degrees centigrade. Of all the advanced battery systems, the sodium/beta-alumina/sulfide battery has received the most development attention due to its theoretical specific energy of over 700 Wh/kg. Sodium/beta-alumina/sulphur cells presently offer 110 Wh/kg and improvements are being made to increase to 160 to 200 Wh/kg with configuration change. Typical cycle life is 500 to 1000 cycles and limited by capacity degradation and ceramic failure. Operating temperature is 350 degrees centigrade. However, lithium and sodium cells are yet to reach perfection and space qualification.

3.4.4 Comparison

From the discussion in the previous section, and the characteristics of storage cells, and considering the storage cell requirements, only Ni-Cd and Ni-H_2 cells are considered for further comparison as others are yet to reach space qualification levels or do not meet some of the other requirements. Compared to Ni-Cd cells, in Ni-H_2 cells one of the two opposing metal electrodes is replaced with hydrogen gas. This minimized the potential for metal-to-metal shorting and the lack of a "wear-out" mechanism for a gas reaction greatly improved system charge-discharge cycle life.

Following are some of the failure mechanisms associated with Ni-Cd cells: a) loss of overcharge protection, pressure buildup, and bulging, b) drying out of the separator due to electrolyte redistribution, and c) shorting due to cadmium migration.

Electrolyte Concentration. An important difference between Ni-H_2 and Ni-Cd cells is that the electrodes and separator function at constant electrolyte loading in Ni-H_2 cells in contrast to the Ni-Cd cells wherein dilution/concentration of the electrolyte successively occurs during charge/discharge cycles. There is no net change in the total amount of water or the KOH concentration while charging or discharging the Ni-H_2 cell. However, in the Ni-Cd cell, one mole of water per Faraday is produced on charge and consumed on discharge. This is a significant advantage for the Ni-H_2 cell in terms of cycle life.

Over Charge. On overcharge of Ni-H_2, oxygen is evolved at the positive nickel oxide electrode. This oxide reacts with hydrogen which is being produced at equivalent Faradic rates to form water, and there is no net change in the water content. Thus, Ni-H_2 cell has an almost unlimited overcharge capability at charging rates at least four times greater than those of the Ni-Cd cells. For practical purposes, the overcharge of Ni-H$_2$ cells will be limited by thermal dissipation capabilities of the system, rather than any inherent cell limitation.

State of Charge Indicator. The state of charge can be easily found out by measuring the hydrogen pressure in the case of Ni-H$_2$ cells whereas, it is not possible in the case of Ni-Cd cells.

Cell Reversal/Over Discharge. Cell reversal protection is achieved by introducing a hydrogen precharge, i.e., a predetermined quantity of hydrogen gas into the cell while it is in the discharge state. Under these conditions, the cell becomes positive-limited on discharge. When all the trivalent nickel hydroxide has been reduced in the discharge part of the cycle, continuation of the discharge will cause hydrogen to be evolved at the nickel electrode surface. The evolved hydrogen is compensated for by an equivalent amount of hydrogen that is oxidized at the hydrogen electrode. Thus, the cell can be continuously operated in reversal without a net change in the amount of water in the electrolyte and with no pressure buildup.

In the case of Ni-Cd, the cell generates gas in reversal, hydrogen if the cell is positive limited (as is usually the case) and oxygen if it is negative limited. In either case, a rapid buildup of pressure results due to the small free volume of the cell, and this buildup will eventually result in catastrophic failure.

Durability. Although Ni-H$_2$ cells presently cost more and do not exhibit as high an energy density as some of the more exotic systems like silver-hydrogen, it does offer an inherent durability or abuse tolerance which will assure years of continuous, dependable service under the most severe conditions.

Space Qualification. The Ni-H$_2$ batteries have been used on-board the following spacecraft: NTS-2 (Navy), INTELSAT V (Ford), SAT COM K (RCA Space Communications), American Sat (American Satellite Corporation), SPACENET (Southern Pacific), Air Force Flight Experiment, G-STAR (GTE), and L-SAT/Olympus (European Space Agency).

NTS-2 is launched into a highly elliptic 12 hour orbit. INTELSAT V, SATCOM K, AMERICANSAT, SPACENET, etc., are geosynchronous orbit satellites. AIRFORCE Spacecraft orbit details are not available. Although the Ni-H$_2$ batteries have been used and are being planned for non-LEO spacecraft, they have not been used on-board a LEO spacecraft to-date.

3.4.5 Battery Charging, Discharging, and Protection

A battery usually consists of series connected storage cells of the required ampere-hour capacity to yield a particular voltage. The battery is charged either at a constant current rate which depends on the operating temperature of the battery or with varying current. Typical charging rates are C/2 to 2C at 40 degrees centigrade, C/10 to C/2 at 20 degrees centigrade and C/30 to C/10 at 0 degrees centigrade. (C = Capacity in ampere hours). When the battery reaches its end-of-charge (EOC) level, it is normally trickle-charged at C/100 rate. The charge rates and operating temperature are mutually related and are selected as a function of various factors including charge-discharge cycle life, temperature control system limitations, and time available for replenishing the battery completely.

As the solar array will be cool when it enters orbital day from eclipse, it generates relatively more power than when it heats up. All this extra power can be used to charge the battery if the variable charge mode is employed. Thus assuming that the load requirement is almost constant, the charging power/current decreases as the array gets heated up/attains higher temperatures. This mode continues until the battery voltage reaches the preselected value as a function of battery operating temperature.

Now the battery is taper charged (decreasing current) while maintaining the preselected voltage level. The end of charge voltage depends upon type of the battery, its aging and its operating temperature. For example, in the case of MMS mission eight temperature dependent end-of-charge voltage levels (V/T curves) are followed. One of the eight levels is selected initially and depending upon the battery performance and operating conditions other levels are commanded.

The battery is usually discharged at rates varying between C/10 and C/2. When the battery reaches its end of discharge (EOD) level, any further discharge of the battery, failure mode arises due to capacity degradation which leads to a capacity imbalance in the battery assembly (since it does not appear simultaneously in all the cells).

This, in turn, results in the possible risk of voltage reversal in weak cells at high *depth-of-discharge* (DOD). In order to keep down the battery weight, normally high DOD is preferred. Therefore, proper monitoring of each cell for its EOC and EOD levels, as well as protection against cell reversal and cell bypass by the use of an appropriate diode network ensures the overall reliability of the battery package.

Charge Control and Battery Protection. Battery charge control (modes and end of charge and discharge voltage levels) is achieved by time-integration of the battery charge/discharge currents (ampere-hours), computation of battery efficiency, computation of battery depth-of-discharge (DOD) and measurement of battery voltage, compared against computed temperature compensated voltage limits.

Battery Under-voltage and Over-temperature Protection. This protection is implemented by monitoring the battery voltage and voltages of each cell. Whenever under-voltage is detected, appropriate commands are issued to remove individual nonessential spacecraft loads. Sometimes, battery under-voltage protection is accomplished by sensing total battery voltage and the half battery analog differential voltages. If the battery voltage is less than preselected end of discharge voltage or a differential voltage of 0.5 or more volts between the two cell groups occurs, then nonessential loads are removed.

If battery temperature rises, indicating increased heat dissipation due to excessive load or overcharge, either further discharge or charge is stopped respectively.

3.5 POWER CONDITIONING
AND CONTROL SYSTEMS (PCCS)

In any spacecraft the outputs of solar array and storage battery are to be conditioned so as to match with the requirements of the various subsystems. The battery has to be charged from the output of the solar array during the orbital day. The storage battery supplies power to the loads during the orbital night or when the load demand exceeds the solar array capability. All these functions are carried out by power conditioning and control systems (PCCS). The PCCS can be grouped into two main systems on the basis of their working principle, i.e., (a) dissipative systems (Fig. 3-26), (b) nondissipative systems (Fig. 3-27). As the name indicates, the dissipative systems are those that do not extract maximum power from the solar array and hence dissipate the unused power, whereas the nondissipative systems extract the maximum power from the solar array and hence dissipate less power intrinsically.

Dissipative systems, further, can be classified into two types depending upon the nature of bus voltage, i.e., (i) regulated bus systems (ii) unregulated bus systems. Figure 3-26(A) shows a simplified schematic of dissipative type regulated bus power system. The shunt regulator maintains bus at a constant voltage. Usually this voltage corresponds to maximum power of degraded solar array (end of life) at the highest temperature. Most of the power at this point is shared between tne load and battery charging. The extra power at cold temperatures and at beginning of life is absorbed by the shunt regulator. During

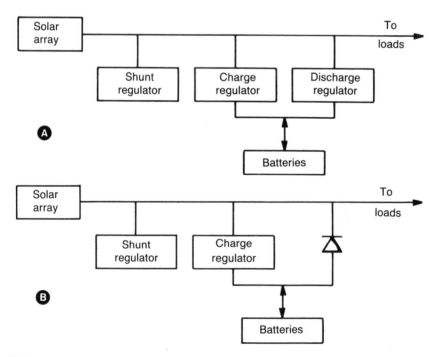

Fig. 3-26. Block schematic of dissipative power systems (A) regulated bus (B) unregulated bus.

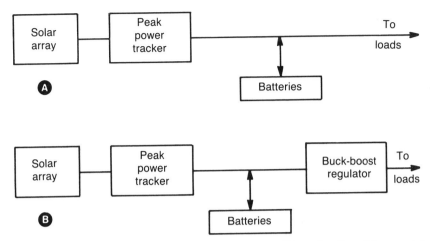

Fig. 3-27. Block schematic of nondissipative power system (A) regulated bus (B) unregulated bus.

eclipse or when power demand exceeds the solar array capability, the discharge regulator supplements from the battery at the bus voltage. This is also known as *direct energy transfer* (DET) power system as the power to the load goes directly from the solar array. Table 3-9 presents briefly the advantages and disadvantages of regulated bus dissipative power systems. Figure 3-26(B) shows a simplified schematic of the dissipative unregulated bus power system. The array is loaded at battery charging voltage. The shunt regulator comes into action at the end of the charge voltage of the battery to protect the battery from overcharge. Table 3-10 presents the advantages and the disadvantages of unregulated dissipative power systems.

Figure 3-27(A) shows a simplified schematic of nondissipative regulated bus power system. The array is loaded near or at the maximum power point controlled by the maximum power transfer control circuit. This circuit can be a tracking type or nontracking type. The different techniques employed in non-dissipative systems, to extract maximum power from the solar array are dealt with in the following sections.

Thus, the bus voltage can be regulated or unregulated, irrespective of whether the PCCS uses dissipative systems or nondissipative systems. But the spacecraft subsystems require different positive and negative voltages at varying regulation requirements and different from the bus voltage. Therefore the bus voltage is further regulated, levelled up, levelled down and/or inverted using regulators and dc-dc converters. From Figs. 3-26 and 3-27, it is clear that the dissipative systems employ shunt, charge and discharge type regulators whereas nondissipative systems employ charge and discharge type regulators. Thus, different types of regulators, dc-dc or dc-ac converters and control circuits become the important building blocks of PCCS, which are described in the following sections.

Table 3-9. Advantages and Disadvantages of Regulated Bus Power Systems.

Advantages	Disadvantages
Permits lighter load regulator/converter units.	Buffering of units from bus is limited.
Some loads may run directly from bus.	Dissipation of excess solar array power in shunt regulator.
Low bus impedance	Three types of regulators are required.
Solar array working point is fixed.	

3.5.1 Regulators

At the unit level, regulators themselves can be classified into dissipative (or linear) and nondissipative (or switch-mode) types. In the dissipative type, the dissipation is a function of input voltage, output voltage and load fluctuations and these regulators have low EMI and less ripple characteristics. The dissipative types of regulators can be classed under series and shunt types.

The nondissipative type regulators operate in the switching mode resulting in high efficiencies. These regulators can be built with low weight and size. Moreover, their output voltage can be greater than, equal to, or less than the input voltage. The regulators can be divided into three types, viz., (i) buck type, (ii) boost type and (iii) buck-boost type. Figure 3-28 shows a simplified schematic of a buck-type switch-mode regulator. The output voltage is

Table 3-10. Advantages and Disadvantages of Unregulated Bus Power Systems.

Advantages	Disadvantages
Easy to avoid single point failure.	Load regulator/converter units complex.
Simple interface	Significant weight penalty particularly with input filter if unit must work
Units are buffered from noise on bus.	over a wide bus voltage range.
	Unit switch-on surge currents may prevent operation of solar array at maximum power point.

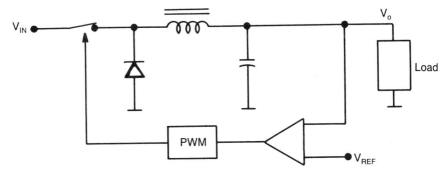

Fig. 3-28. Simplified schematic of buck-type switch-mode regulator.

compared with a reference voltage and the difference signal is used to generate a pulse-width modulated (PWM) waveform that controls the switch ON/OFF periods. When the switch is turned OFF, the stored energy in the inductor at this stage reverses its polarity and takes the path through the diode and sends current into the load while the capacitor maintains a constant output voltage. When all the stored energy in the inductor is used up, the capacitor discharges and V_o decreases. At this stage the switch is turned ON, so that the output voltage is regulated.

Figure 3-29 shows a simplified schematic of a boost-type switch-mode regulator. The output voltage is always greater than the input voltage in this case. When the switch is ON, a current, I, flows through the inductor so that energy is stored. When the switch is OFF, the stored energy tries to collapse and reverses its polarity. Thus, it adds to the input voltage. The output voltage is boosted as a result of this additional charge. Figure 3-30 shows a simplified schematic of the buck-boost type switch-mode regulator. When the switch is ON, inductor charges and stores energy. When the switch is OFF, the stored energy tries to collapse reversing its polarity, thus sending current, I, into capacitor and R_L. This takes the output voltage to an opposite polarity to V_{in}. The charge regulator referred to earlier, used for charging the battery, is an example of the buck-type regulator. Similarly, the discharge regulator is an example of the boost-type of regulator.

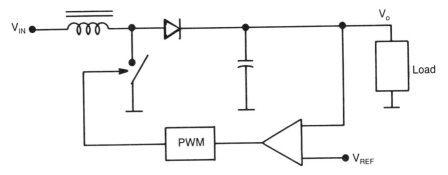

Fig. 3-29. Simplified schematic of boost-type switch-mode regulator.

The switching regulators can be of the self-oscillating type or driven type. As explained before, the output voltage is compared with a reference voltage and the difference signal is used to generate a PWM signal, which is used for the regulator operation. This PWM signal can have one of the following characteristics: (i) Fixed ON period and variable OFF period; in this, circuit complexity is minimum, (ii) variable ON and OFF periods with fixed frequency; it uses simple circuitry and has high efficiency over a wide range of input voltages and output loads, (iii) fixed OFF period and variable ON period; it is simple, cheap, and reliable, and the frequency range over which it operates is lower than the above two types.

To enhance the power capability, the above regulators are usually connected in parallel and are operated in the phase shift (multiphase) mode. This reduces the EMI and problems of electromagnetic screening. To improve the performance characteristics such as regulation, transient response, line interference rejection, etc., the feedback control loop of the above regulators is modified to sense the ac changes besides the dc changes on the bus.

3.5.2 Dc-Dc Converters

Figure 3-31 shows a self-oscillating dc-dc converter in one of its simplest configurations. It is a free-running multivibrator that utilizes the saturating property of a transformer core and the switching properties of the transistors to generate a square waveform. It works efficiently (with efficiencies of greater than 70%) at high frequencies (greater than 100 kHz) and is lightweight. The square voltage waveform can be transformed up or down or inverted and rectified to produce pure dc and in practice little filtering is required at the output of this stage.

3.5.3 Shunt Regulators

The shunt regulator, as shown in Fig. 3-32, maintains the bus voltage at a predetermined fixed value irrespective of the changes in load current and/or solar array output by shunting that portion of the solar array current that is in excess of load current and battery charging current. Depending upon the implementation, these shunt regulators can be divided into three types, i.e.,

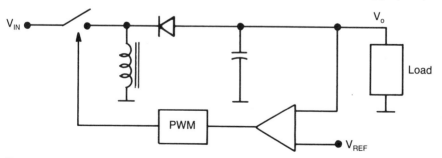

Fig. 3-30. Simplified schematic of buck-boost type switch-mode regulator.

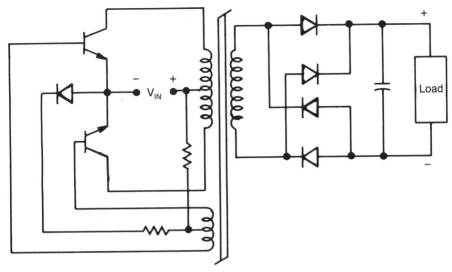

Fig. 3-31. Schematic of a self-oscillating dc-dc converter.

analog/linear, pulse-width modulated (PWM) and digital-shunt regulators and within each type, there are some variations.

Linear Shunt. Figure 3-32 shows a simplified schematic of a linear shunt regulator connected to a solar array. Scaled down main bus voltage is compared with a reference voltage and the difference signal is used to control the current

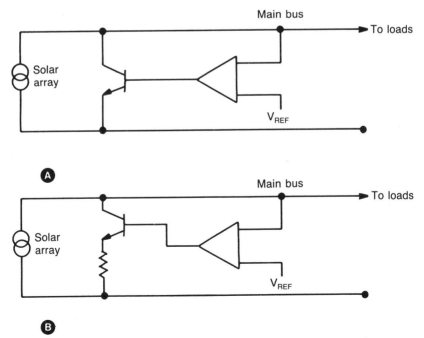

Fig. 3-32. Schematic of shunt regulator, (A) without a resistor (B) with a resistor.

through the shunt transistor such that the bus voltage is regulated. As the excess power (solar array power minus the load requirement at bus voltage) is dissipated by the shunt transistor, it has to be a power device with good heat-sinking capability. Sometimes, a resistor is used in series with the power transistor to share the power dissipation as shown in Fig. 3-32(B).

This type of shunt regulator is relatively simple and weighs more, occupies more volume, and offers lower efficiency. However, it offers large control bandwidth, thereby it responds very quickly for output load or input voltage changes.

Sequential Linear Shunt. A simplified sequential linear shunt regulator shown in Fig. 3-33 is an extension or a variation of linear shunt regulator for handling large power. Just like the linear shunt regulator, the scaled down main bus voltage is compared with a reference voltage and the difference signal is used to control the current through all the sequential shunt stages such that the bus voltage is regulated. The stages are turned ON or OFF sequentially. Only one power transistor is in the active region or in conduction at any time, and the other stage power transistors are either cutoff or saturated; thus dissipating negligible power in the power devices. Solar array excess power is dissipated in the power transistors and resistors.

Pulse-Width Modulated (PWM) Shunt. Figure 3-34 shows a simplified Pulse-Width Modulated (PWM) shunt regulator connected to a solar array. Scaled down main bus voltage is compared with a reference voltage and the difference signal is used to generate a pulse-width modulated control signal. This PWM signal is used to control the current through the shunt transistor such that the bus voltage is regulated. Here the shunt transistor goes ON/OFF in response to the PWM signal. When the shunt transistor is ON, the solar array is shorted to ground and no power goes to output, however, the bus voltage is maintained by the filter. The lost energy is replenished into the filter when the transistor is OFF in addition to supplying the power to the bus. As the power transistor operates in ON/OFF states, it dissipates less power and thereby the regulator runs relatively cooler. However, it needs a bigger filter to maintain the regulated bus voltage.

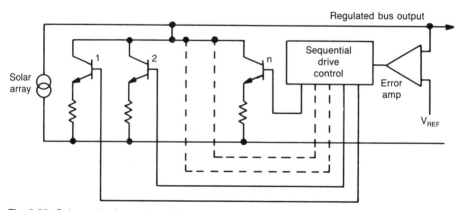

Fig. 3-33. Schematic of a sequential linear shunt regulator.

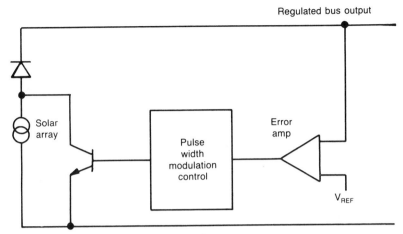

Fig. 3-34. Schematic of a PWM shunt regulator.

This type of shunt regulator (compared to analog/linear shunt) weighs relatively less, occupies less volume, and offers high efficiency. However, it offers medium control bandwidth, thereby it responds reasonably well for output load or input voltage changes.

Multi-stage PWM Shunt Regulator. Figure 3-35 shows a multi-stage PWM shunt regulator which is a natural extension of single stage PWM shunt, for handling large power. The solar array is divided into N sections. Each section is shunted by a transistor switch. The main bus voltage is compared with stable reference voltage and the difference signal is used to generate pulse-width modulated waveforms to control the current through these switches, such that the bus voltage is maintained. Thus the power handling capability of this type of shunt regulator can be increased to very high power levels.

Multi-phase PWM Shunt Regulator. In the above configuration, as all the switches operate in synchronism, large amounts of currents are switched simultaneously and it results in large EMI generation. This can be reduced by operating the switches in phase-shifted mode as shown in Fig. 3-36. Thus

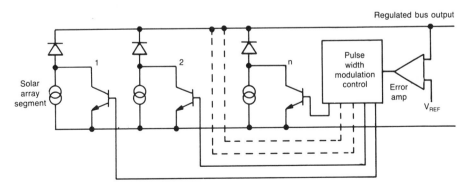

Fig. 3-35. Schematic of a multi-stage PWM shunt regulator.

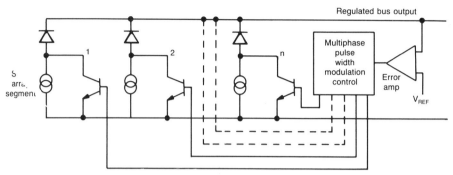

Fig. 3-36. Schematic of a multi-phase PWM shunt regulator.

the phase-shifted PWM waveforms are used to drive the switches. This type of shunt regulator is known as *multi-phase PWM shunt*.

Digital Shunt. Figure 3-37 shows the block diagram of a *digital shunt regulator* (DSR). It regulates the output of the solar array to the needs of the load. The solar array is divided into N sections, one section of which is permanently connected to the bus and all other sections are connected through switches. The DSR contains a small dissipative analog shunt that is designed to regulate one section of the array. When the spacecraft comes out of eclipse the solar array starts generating power, and the dissipative analog shunt comes on immediately. As the voltage crosses the bus voltage, the bus voltage is maintained by controlling the current in the shunt such that the sum of the load current, battery charge current, and the shunt current is equal to the current of the solar array section at the bus voltage. If the load current increases, the shunt current decreases and vice versa. If the load current continues to increase such that the shunt current reduces below the $I_{(min)}$ value, then the $I_{(min)}$ comparator output activates the controller to switch ON one solar array section onto the bus. Now the shunt current rises to a level between $I_{(min)}$ and $I_{(max)}$. Again assume that the load current increases further such that the shunt current reduces below $I_{(min)}$. Then the controller will

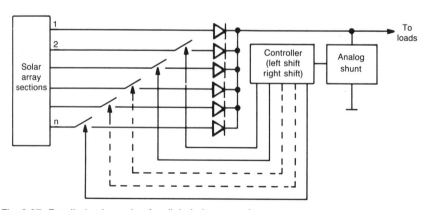

Fig. 3-37. Detailed schematic of a digital shunt regulator.

switch ON one more section. This process continues until all solar array sections are switched ON. Now the load current can increase no more as the system is designed only to this maximum value of load current. Now, say the load current decreases. Then the shunt current increases to maintain the bus voltage. But assume that the load current decreases such that shunt current increases above the $I_{(max)}$ value. Then the $I_{(max)}$ comparator output is used by the controller to remove one solar array section from supplying power to the bus. The shunt current will then decrease. Assume now that the load current decreases again such that the shunt current increases to give an $I_{(max)}$ output. One more solar array section will be switched OFF of the bus. If the load current is decreased to zero, this process continues until all switchable solar array sections are switched OFF and disconnected from the bus. As one solar array section is connected permanently without a switch, linear shunt is designed to handle one section full power and it maintains the bus voltage even if the load is completely disconnected. Thus the DSR maintains the bus voltage at fixed level from no load to full load.

This type of regulator has the advantages of the above regulator types described earlier and is devoid of demerits. This regulator weighs relatively less, occupies less volume, and offers high efficiency. It also offers large control bandwidth, therefore it responds very quickly for output load or input voltage changes. This large control bandwidth is achieved as it employs an analog shunt that responds for (relatively) small changes and the digital portion of the regulator responds for large changes. The DSR can be employed advantageously for high power satellites as its weight and volume do not increase in proportion to the power rating of the DSR. In this DSR, the linear shunt which handles l/N of the solar array power can be single stage shunt or multistage sequential shunt or PWM shunt, etc.

The switches employed in the above digital shunt regulator are in series with the solar array sections. However, these switches can be used in shunt with the solar array sections. Solar Array arrangement for such a DSR is shown in Fig. 3-38.

Figure 3-39 shows a solar array arrangement for another type of DSR. Here the solar array is divided into N sections. N-1 sections are connected to N-1 shunt switches whereas the Nth section is again divided into M subsections. Each subsection is connected to a shunt switch. Each of these subsection switches are operated in OFF/Active/ON mode. For example, a particular subsection switch which was OFF before, becomes active as load demand decreases. The degree of conduction gradually increases as the load demand gradually decreases, eventually it completely saturates. In this process, current through the switch gradually increases, finally reaching saturation limit. As the switch went through the conduction finally into saturation, the dissipation in the switch also reaches a peak before it reduces to a low value. However, when a switch saturates it cannot regulate the bus and hence to have a continuity and smooth regulation control, the operation of these switches is overlapped. Thus, even before a switch saturates, the following switch is in an active state regulating the bus. This arrangement is sometimes called a *staggered-*

stage linear shunt and it reduces linear shunt dissipation by about a factor of M-2. The dissipation of the staggered stage and conventional linear shunt regulators are compared in Fig. 3-40. The main difference is that in staggered-stage linear shunt, separate taps are provided from each solar array subsection whereas all subsections are combined and supplied as single tap in the case of the conventional linear shunt regulator. Separate taps facilitate current drawing from each tap separately, thereby enabling lower dissipation operation. This system increases the solar array harness, necessitates more isolation diodes, and more slip rings. However, slip ring requirement does not apply if there is no relative motion or movement of the solar array with respect to the spacecraft.

Series switches can be used for the array section instead of shunt switches. This permits use of parallel redundancy instead of series redundancy for the switches, with possible deletion of the solar array blocking diodes.

Partial Shunts. The most efficient method of control is to place a tap on the array string of cells as shown in Fig. 3-41 and shunt the current from the tap to the return. In this way, each cell in the tapped section tends to operate on the constant current side of its peak power point, while the untapped cells operate in their constant voltage region. A power mismatch is thus developed so that the array capability falls until it exactly matches the load requirements

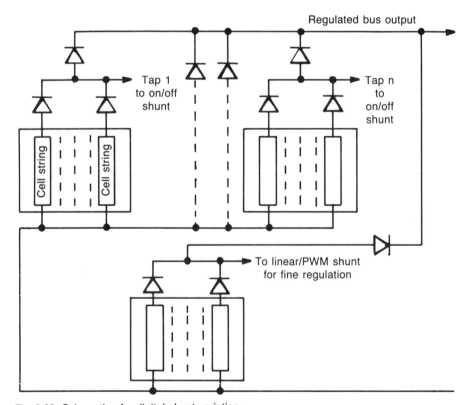

Fig. 3-38. Schematic of a digital shunt variation.

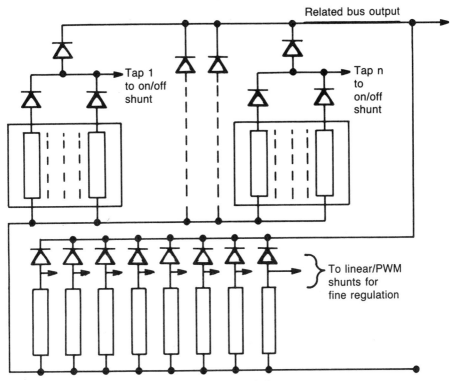

Fig. 3-39. Schematic of a digital shunt regulator variation.

at the bus voltage. Toward the end of the life, as the capability of the array declines, the regulator allows a better match of capability to requirements, such that at end-of-life a complete match is produced that allows each cell to operate at its peak power output. When the switches are used in shunt with the solar array sections, they can be connected to shunt the full solar array section or a portion of the section. Thus the name *full shunt* or *partial shunt*.

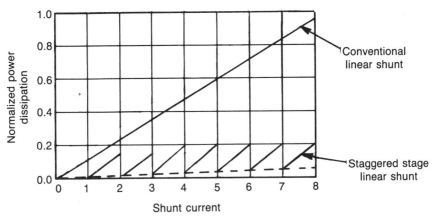

Fig. 3-40. Schematic of digital shunt regulator with staggered stage linear shunt.

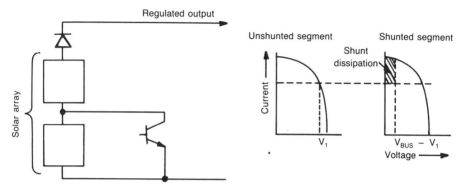

Fig. 3-41. Schematic of a partial shunt regulator.

In a partial shunt, not all of the excess solar array power appears as heat dissipation in the shunt. The unshunted and shunted sections of the solar array are mismatched and operate off the maximum power point, reflecting some of the excess power directly back to the solar array where it appears as heat and is reradiated to space. Since this occurs only when excess power is being dissipated, the small reduction in solar array efficiency due to increased temperature has no impact upon subsystem performance. In the case of a partial shunt, the shunt transistors can be located on the back of the solar array, which enable easy heat dissipation and it radiates directly into space. This requires the use of more signal slip rings than power slip rings. However, shunt transistors have to withstand large temperature excursions or cycles. Merits and demerits of partial- and full-shunt regulators are compared and presented in Table 3-11.

Table 3-11. Comparison of Full Versus Partial Shunt Regulators.

Electrical Ratings	Full Shunt	Partial Shunt
Voltage	Bus voltage	Less than Bus Voltage
Current	Solar Array section short circuit current	Solar Array section short circuit current
Thermal Dissipation	More	Less
Solar Array Harness	Less	More
Isolation Diodes	Less	More
Slip rings*	Less	More

*This does not apply if there is no relative motion or movement of solar array with respect to the spacecraft.

Comparison. The linear shunt regulators weigh more as they dissipate more. Hence they are not usually employed. The performance of total PWM Vs DSR with linear shunt for fine regulation is similar and is shown in Table 3-12. DSR has better regulation, which is obvious due to the large control loop bandwidth of the linear shunt and a negligible ripple voltage. The main difference is that DSR has a lower risk of possible EMI problem, at the expense of greater thermal dissipation.

3.5.4 Maximum Power Transfer Techniques

As mentioned earlier, most of the spacecraft use solar radiation as an energy source and solar cells as energy converters to convert the solar radiation into electrical energy. The I-V and P-V characteristics of the solar array are temperature dependent as shown in Fig. 3-42. When the spacecraft enters an eclipse, the solar array output vanishes completely. Further, when the spacecraft leaves the eclipse, the solar array output is extremely high for a long time, as the array temperature is very low. Referring to Fig. 3-42, it is evident that the array voltage at maximum power output varies. For the optimal utilization of the available power, it is necessary that the maximum power point is continuously tracked and the available power is used for all the loads as well as charging of the battery. Some of the principles usually employed to transfer the maximum power from the solar array to the loads and the battery are described below:

Principle-1. This principle is based on the characteristic of any solar array that the ratio of voltage at the maximum power point (MPP) and that at open circuit of the solar array is approximately constant (between 0.7 to 0.75). Hence, an auxiliary (small) array which is maintained in the same environment as the main array and not loaded is employed to monitor the open circuit voltage. This is used to control the operating point on the main solar array such that the maximum power is continuously transferred to the loads and the battery.

Principle-2. Usually, temperature is the main factor affecting the MPP of a solar array. Hence, temperature of the solar array is measured continuously

Table 3-12. Comparison of Shunt Regulators.

Parameter	PWM	DSR with Linear Shunt
Regulation	Good	Better
Weight (relative)	1	1
Thermal Dissipation (relative)	1	0.85 to 2.2
EMI	High (Continuous)	Low (Intermittent)
Ripple (relative)	10	1
Transient Response	1	1

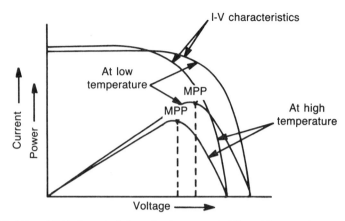

Fig. 3-42. I-V and P-V characteristics of solar array at different temperatures.

to control its operating point, such that maximum power is transferred to the loads and the battery.

Principle-3. Solar array is connected to the battery through an interface which provides maximum power transfer. As the battery voltage remains constant during a switching period of this interface, the current ripple is representative of the power ripple. As the scanning is continued, this ripple reaches a peak. The peak is sensed by a peak detector and the scanning is reversed. Thus, the operating point is maintained near the MPP.

Principle-4. A product of output current and voltage is generated continuously as the solar array is loaded. This is fed to a peak detector. When the peak is sensed, the loading is reversed. This enables operation close to MPP.

Principle-5. Referring to Fig. 3-43 it is seen that the ac and dc impedances of a solar array are equal at the MPP. This property is made use of in sensing the MPP. A ripple is injected into the solar array bus, and d_v/d_i and V/I are continuously measured and are compared to control the operating point such that the maximum power is transferred from the solar array.

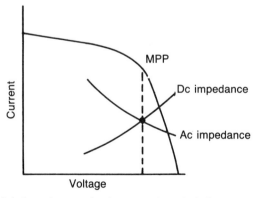

Fig. 3-43. Ac and dc impedances of solar array characteristic.

Principle-6. As any solar array is loaded gradually, the rate of change of power output with respect to output current changes its slope as the MPP is crossed. A low frequency perturbing signal is injected into the solar array, which causes small changes in the array power. The d_P/d_I is continuously measured and is used to control the operating point function for maximum power transfer.

3.5.5 PPT and DET Based Power Systems

As described earlier, the PCCS can be classified into dissipative and non-dissipative systems. Two of the commonly employed and versatile power system configurations are (a) peak power tracking (PPT) system and (b) direct energy transfer (DET) system. PPT-based power system falls under non-dissipative system whereas the DET-based power system still falls under dissipative systems although much improvement and more advances have been made in the past few years.

PPT-Based Power System. The power systems shown in Fig. 3-27 utilize the PPT approach. The main advantages of this approach are fewer thermal problems and improved efficiency of the overall power system. The maximum power available from the solar array is processed by the peak power tracker and its output is used to charge the batteries as well as to supply power to the batteries. When the batteries are fully charged up, the peak power tracker electronically moves the operation of the solar array off the maximum power point, toward the open circuit voltage. As the storage batteries are placed across the bus, the bus voltage varies as the batteries are charged or discharged. Thus this system is called the *unregulated bus power system*. In the power system shown in Fig. 3-27(B), an additional regulator is added to regulate the bus; resulting in a regulated bus power system. Regulated bus and unregulated bus power systems each have advantages and disadvantages as described in the previous section.

The MMS Modular Power Subsystem (MPS) uses the PPT approach and is currently flying on three satellites, i.e., SMM, LANDSAT-4, and LANDSAT-5. In addition, it is planned for use on the GRO, UARS, TOPEX, Explorer Platform, and AXAF. Also other spacecraft used this approach to the design of a power system.

DET Based Power System. The power systems shown in Fig. 3-26 utilize the DET approach. In this approach, the power from the solar array is directly transferred to the load without the use of any series connected regulator or converter. Hence the name *direct energy transfer*. However, to limit or maintain the bus at a predetermined voltage level, a shunt limiter or regulator is employed. Storage batteries are charged using charge regulators. If the bus is of the regulated type then a battery discharge (boost) regulator is also employed to regulate the bus during the orbital night. Thus, Fig. 3-26(A) shows the block schematic of a DET Regulated Bus Power Subsystem. The shunt regulator maintains the bus voltage at a fixed value while the charge regulator charges the batteries at a constant current during the orbital day. During the orbital night, the discharge regulator boosts the voltage of the storage batteries to the bus voltage level and maintains it. Thus regulated bus power is available continuously.

Figure 3-26(B) shows the block schematic of a DET unregulated Bus Power Subsystem. The shunt regulator limits the bus voltage at a fixed value while the charge regulator charges the batteries during the orbital day. The storage batteries are connected across the bus and thus the bus voltage varies as the batteries discharge. During the orbital night, the storage batteries directly supply power to the loads. Thus unregulated bus power is available continuously.

Miscellaneous Configurations. Some spacecraft power systems utilize a hybrid approach providing two power busses, one regulated and the other unregulated. Many variations and derivations of previously described types or approaches have been used.

Comparison of PPT- and DET-based Power Systems. At low load power levels the PPT-based power system dissipates less power. As the load power increases, the power dissipation of the DET-based power system increases but at a lower rate, thereby equal to the PPT based power system. As the power dissipation in a DET-based power system increases at a gradually lowering rate as the load power increases, the power dissipation in DET-based power system is lower at higher power levels.

In the PPT Approach, the PCCS (unregulated bus) consists of a) Peak Power Tracker and b) Power Distribution Unit, whereas in the DET Approach, the PCCS (unregulated bus) consists of a) Shunt Regulator, b) Charge Regulator and c) Power Distribution Unit. A comparison of the relative performance of DET- and PPT-based power systems is given in Table 3-13.

The power dissipation of the PPT-based power system equals the power dissipation of the DET-based power system at about 800 watts spacecraft load. The weight and volume of the shunt and charge regulators in a DET system weigh and occupy more volume than a single PPT-based system dissipating a power equal to the sum of the shunt and charge regulators. Hence, the crossover point might perhaps be at about 1200 watts. Now considering the cost aspects of fabrication, testing, etc., for peak power tracking converter versus charge and shunt regulators and the smaller solar array in the case of PPT-based power system, the crossover point may be at about 1400 watts spacecraft load range. This is true only if the orbit is in the neighborhood of 500 km and the orbit contains eclipses. At higher load power levels, the crossover point moves slightly. However, the qualitative results are the same in that the PPT-based power system is optimum for spacecraft in LEO orbit with power requirement levels of 500 watts. As the spacecraft power increases or the orbital altitude increases, gradually the DET-based power system performance exceeds the performance of the PPT-based power system. Thus, it is highly advantageous to choose the PPT approach to the design of an Electrical Power System for Low Earth Orbit spacecraft with power requirements of lower than 1000 watts. For spacecraft in higher orbit and/or with higher power requirements, the DET-based power system is the optimum.

DET-based power system performance can be further improved by

**Table 3-13. Comparison of PPT Approach
Versus DET Approach for Electrical Power System.**

Parameter	PPT Approach	DET Approach
Thermal & Mass Constraints	Heat dissipation is limited to a low and nearly constant value.	in periods of excess power, heat is usually dissipated in the shunt regulator which therefore requires heavy heat sinks.
Radiation Damage	More power is available at the BOL to operate additional equipment or increase its operating time. Solar array degrades gracefully in this approach.	Additional power at the BOL is not properly utilized. Solar array degrades very rapidly in this approach beyond design life.
Solar Array Design	Solar array voltage can be selected independently and the design is simple.	The choice of solar array voltage depends on electronics, battery, thermal design, etc. array design freeze is delayed.
Solar Array Loading	Improved array loading Efficiency	Low Efficiency especially at the beginning of sunlight period when battery voltage is low and array voltage is high.
Solar Array - Battery Lock-up	Does not exist.	Can exist in the unregulated main bus system. Requires special provision to get out of lock-up whenever it occurs.

designing the charge regulator such that the charge regulator itself maintains the bus at a fixed voltage (shunt regulator is OFF now) and charges the batteries at a variable current utilizing all the power available from the solar array minus the load. When the batteries are completely charged, the shunt regulator is then turned on and the charge regulator goes into trickle charge mode. This approach increases the efficiency of the overall power system.

3.5.6 Power Distribution

Whether the spacecraft power bus is regulated or unregulated, the spacecraft subsystems require different positive and negative voltages with varying regulation requirements. Therefore, the bus voltage is further regulated, levelled up, levelled down, and/or inverted using regulators and dc-dc converters. If this process of further regulation, etc., is carried out at each load end separately, then such an approach is termed as a *Decentralized Regulation Approach* (DRA). In this case obviously, there is no need to opt for the regulated main bus approach, as battery size increases as the inefficiencies of the boost and other converters have to be accounted for. Hence, a spacecraft using DRA for power distribution, usually employs the unregulated main bus approach. On the other hand, if this process of further regulation is carried out in the main power system for all the loads, then such an approach is termed as a *Centralized Regulation Approach* (CRA).

The decentralized regulation system has the advantage of individually tailor-designing each power subsystem for each subsystem or load needs without any compromise. In a centralized regulation system, regulated voltages are determined as a compromise for all subsystems and invariably it was found that subsystems had additional regulators and dc-dc converters for achieving further regulation and less ripple. Table 3-14 presents a comparison of CRA Versus DRA to Power Distribution.

For large spacecraft, a lengthy harness is needed to interconnect the subsystems. The longer the harness, the larger the voltage drop in the harness and the poorer the regulation if a regulated bus concept is utilized. The length of harness depends upon the relative locations of subsystems or payloads with respect to the power system. In addition, to have an easy interface between the power system and all other subsystems, proper isolation is very essential, and to have proper isolation, each user has to use either a dc-dc converter or a dc-ac inverter at their input. Thus, considering the advantages and merits, CRA is preferable for spacecraft with average power requirements of about 100 watts. However, at higher power levels, it is highly advantageous to follow DRA to the power distribution.

Table 3-14. Comparison of CRA Versus DRA to Power Distribution.

Parameter	Centralized Regulation Approach	Decentralized Regulation Approach
Regulation requirements	Compromise.	No compromise.
Isolation	Needs independent filters at the input of loads.	Filters are with the individual converters.
Protection	Needs current limiters on each line.	Protection is designed into each converter.
Preferable	At low power levels and for smaller spacecraft.	At high power levels and for bigger spacecraft.

3.5.7 Power Control Circuits

The Power Control Unit controls the power flow in the spacecraft and distributes the power to subsystems. Normally, only the telecommand system is hardwired to the regulated bus or storage battery to realize high uplink reliability for emergency corrective measures. Other subsystems are connected via the relay controlled switches. Usually, each power switch, either of solid-state type or electromechanical type contains an over-current cutout circuit on all the lines. Power consumption is less for the electromechanical type of switches.

The solar array output is used to provide power to loads and to charge the storage battery during the orbital day; the storage battery supplies power to the loads during the orbital night or when the load demand exceeds the solar array capability. Voltage detectors are used to detect the high and low levels of the battery and the solar array. Whenever the solar array output goes low (spacecraft enters eclipse or load demand exceeds the solar array capability) the corresponding voltage detector output signal is used to route the power to the subsystems from the battery. Whenever solar array output is high (spacecraft in sunlit period) the corresponding voltage detector output signal is used to charge the battery as well as power the subsystems from the solar array. When the battery reaches its end of charge, the corresponding voltage detector output signal is used to stop the charging of the battery and start the trickle charge. When the battery reaches its end of discharge while it is being discharged (in normal circumstances, battery does not reach its end of discharge voltage) the corresponding voltage detector output signal is used to avoid further discharge of the battery and isolate all the loads from the battery except Telecommand. The control circuit also usually houses a separate power supply to feed power to all the control and housekeeping circuits of the system.

3.5.8 Power System Auxiliary Circuits

Other circuits in the power system, like current sensors, voltage sensors, relay drivers, etc., are commonly known as power system auxiliary circuits. The current sensors are usually employed for current sensing and current limiting. The currents are measured in two ways, viz., (i) by sending the current through a resistor of known value and amplifying the voltage drop across the resistor, and (ii) by measuring the induced magnetic field because of the current.

The auxiliary circuits include the circuits to interface with the telemetry system enabling to transmit down the necessary information to monitor the status and health of the power system. It also includes circuits to interface with the telecommand system to receive commands from ground and to turn relays on or off. The relay and relay drivers are peak-power consuming circuits because of their transient operation and are most often responsive to noise. Constancy of threshold levels against temperature changes and high noise immunity are important criteria for the design of relay drivers.

3.5.9 Ac Power Generation and Distribution

The dc power generation and distribution discussed earlier have some major drawbacks for high (kilowatt range) power requirements. They are, (a) requirements of complex interfaces such as, voltage conversion, monitoring, protection, redundancy, etc., between the power conditioning system and the loads; (b) increase of harness mass and losses because of the low distribution voltage of dc system; and (c) low overall system efficiency.

The use of an ac power distribution system minimizes some of the above drawbacks. A square-wave ac power generation and distribution system is a better solution for meeting the requirements of high efficiency, good voltage regulation and low filter weight compared to sinusoidal ac power generation and distribution.

3.5.10 Recent Trends

As spacecraft become larger and more complex in their uses, and PCCS designs tend to become more sophisticated, the replacement of conventional printed circuit board technology by ceramic thick-film circuit technology offers the following advantages: high reliability, high voltage and power capability, low volume and weight, space and nuclear radiation stability, vacuum cleanliness, temperature stability, high mechanical strength, and low costs.

The increasing power demand in spacecraft during recent years has resulted in the development of power systems that guarantee optimum and reliable utilization of the on-board electrical power. All the units, of the power system are designed and optimized individually, according to specific requirements. In the long run, however, such a procedure can become too expensive. Modular approach to the design and development of the power system appears to hold some promise to overcome this constraint. More specifically, modularization of the power systems offers a number of advantages such as, high reliability by easy incorporation of redundancy, low development effort and low cost, optimization of individual system units, minimum project time, low weight, wide spectrum of power levels by combining a number of power modules, and increase in power handling capability of currently available space-qualified electronic components. A modular power system was developed utilizing a nondissipative unregulated main (power) bus approach for NASA Multi-Mission Spacecraft.

3.6 POWER SYSTEM DESIGN

3.6.1 Power System Design Using Peak Power Tracker Approach

A power system is conceptualized and designed to meet the power requirements of a spacecraft under consideration and the power requirements are presented in Table 3-15. A Peak-Power Tracker-based Power Conditioning and Control system, as shown in Fig. 3-27(B), is baselined and following are the design factors considered in sizing the major components of the power system:

Mission Life = 3 years
Orbit   = 550 km altitude and
                45 degrees inclination

Table 3-15. Spacecraft Power Requirements.

Subsystems	Power (Watts)	
	Normal	Peak**
PAYLOAD		
Payload-1	35.0	55.0
Payload-2	55.0	65.0
Payload-3	100.0	100.0
Payload-4	15.0	25.0
Payload-5	80.0	80.0
Payload-6	100.0	240.0
Subtotal	385.0	565.0
Support Systems		
TT&C	60.0	80.0
EPS	40.0	50.0
ACS	90.0	90.0
Mechanisms	2.0	5.0
Thermal Control	25.0	30.0
Miscellaneous	5.0	5.0
Harness	16.0	24.0
Subtotal	238.0	284.0
Margin (10%)	63.0	85.0
Spacecraft Total	686.0	934.0

**Peak can occur anytime during orbital day or night for 8 minutes per orbit.

Efficiency of power conversion, E(PC)   =   95%
Average operating point is away from
   peak power point by, E(OFF)   =   95%
Battery end of charge voltage, V(BC)   =   1.45 V
Battery end of discharge voltage, V(BD)   =   1.15 V
Battery recharge factor, BRF   =   1.10
Maximum eclipse duration, T(e)   =   35.6 minutes
Minimum day time or sunlit period, T(d)   =   60 minutes.

Alternately if T(e) and T(d) are not given or available, then T(e) and T(d) can be read from Table 3-16 or Figs. 3-44 to 3-47 or one can calculate using the formulas given in Chapter 1.

Solar Arrays are oriented perpendicular to the sun and if they are not then the final solar array area computed is equal to the projected area of the solar array rather than actual area of the solar array.

Table 3-16. Orbital Altitude Versus Ratio of Maximum Eclipse/Minimum Sunlight Period.

Altitude (km)	Maximum Eclipse/ Orbital Period	Minimum Sunlight/ Orbital Period	Maximum Eclipse/ Minimum Sunlight	Orbital Period (Hrs)
200	0.4213	0.5787	0.7280	1.4790
300	0.4042	0.5958	0.6784	1.5128
400	0.3901	0.6099	0.6396	1.5470
500	0.3779	0.6221	0.6074	1.5813
600	0.3670	0.6330	0.5798	1.6159
700	0.3572	0.6428	0.5557	1.6508
800	0.3483	0.6517	0.5344	1.6859
900	0.3400	0.6600	0.5152	1.7212
1000	0.3324	0.6676	0.4979	1.7568
1500	0.3003	0.6997	0.4292	1.9584
2000	0.2754	0.7246	0.3800	2.1259
3000	0.2338	0.7662	0.3052	2.5176
4000	0.2107	0.7893	0.2669	2.9309

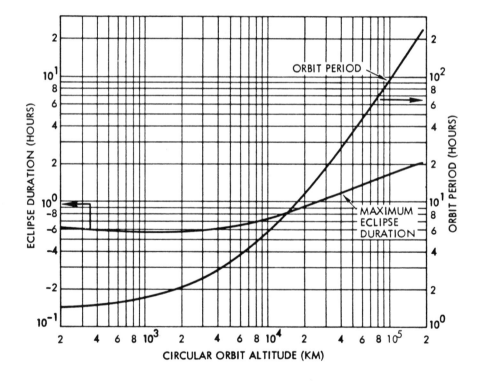

Fig. 3-44. Orbit period and eclipse duration of circular earth orbits. (Reproduced from "Solar Cell Array Handbook", JPL SP 43-38, with permission from Jet Propulsion Laboratory, California).

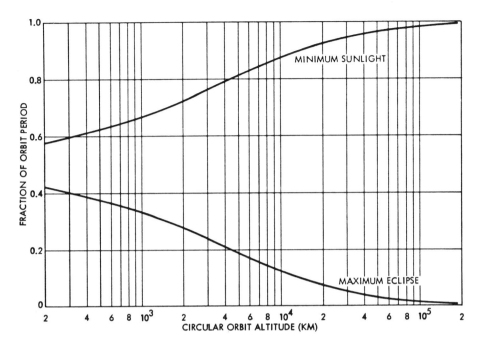

Fig. 3-45. Fractional sun time of circular earth orbits (Reproduced from "Solar Cell Array Handbook", JPL SP 43-38, with permission from Jet Propulsion Laboratory, California).

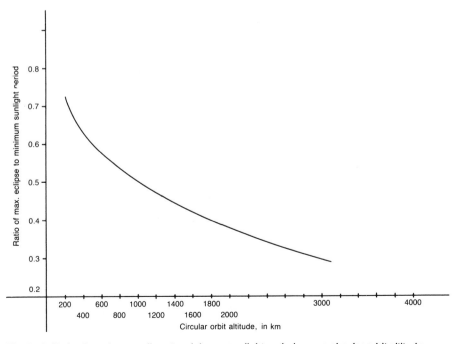

Fig. 3-46. Ratio of maximum eclipse to minimum sunlight period versus circular orbit altitude.

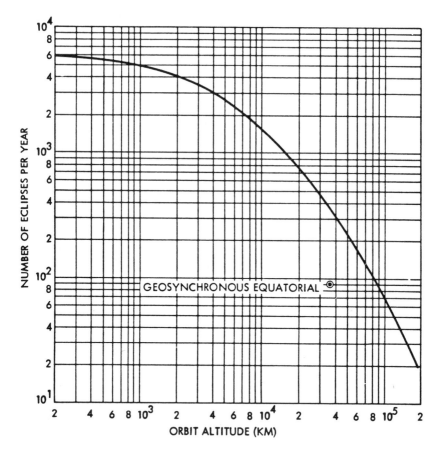

Fig. 3-47. Maximum number of annual satellite eclipses in circular earth orbits (Reproduced from "Solar Cell Array Handbook", JPL SP 43-38, with permission from Jet Propulsion Laboratory, California).

Power Requirement from Solar Array. Following is the relationship between the power requirement from solar array and the load power.

$$P(SA) = \frac{P \text{ (Day Ave)}}{E(\text{OFF}) \times E(\text{PC})}$$

$$+ \quad \frac{[\, P(\text{eclipse ave}) \times T(e) + P(p) \times T(p)\,] \times V(BC) \times BRF}{E(\text{OFF}) \times E(\text{PC}) \times T(d) \times V(BD)}$$

Where $P(p)$ is the average power requirement during the peak. $T(p)$ is the duration of the peak in minutes.

As the peak power can occur during orbital day or night, for solar array sizing, it is assumed that it occurs during orbital night. Once the solar array is sized this way, then it meets the power requirements even if the peak occurs during day. For the spacecraft power requirements given in Table 3-15, the power requirement from the Solar Array is computed to be about 1436 watts.

Peak Power Tracker Ratings. The solar array produces more power at the beginning of life as well as when the spacecraft comes out of eclipse as the array is cool. However, if the load demand is constant throughout the mission, the peak power tracker is rated for an average power as given below:

$$P(BOL) \times E(PR) \times E(OFF)$$

The peak power rating of the peak power tracker usually is at least 150% of average value calculated above. However, the duration of operation at this peak power depends upon the thermal control system. But the peak power tracker has to be electrically rated for an input voltage equal to the solar array open circuit voltage at minimum temperature and for an input current equal to the short circuit current at the maximum temperature.

Energy Storage or Battery Capacity Requirements. The maximum eclipse duration, Tem, is calculated using the following formula:

$$Tem = (T/180) \{ \cos^{-1} \sin \cos^{-1} [R/(R+H)] \} \text{ in hours.}$$

and

$$T = 2.7722 \; 10E\text{-}6 \; (R+H)^{1/2} \text{ in hours.}$$

where   R = Mean radius of the Earth in km
         H = Orbital altitude in km

Using the above formula, maximum eclipse duration can be computed at various altitudes and Fig. 3-44 presents such a plot of orbital altitude versus maximum eclipse duration. The maximum eclipse duration, *Tem*, is about 0.6 hours for 550 km orbit. Hence, the required energy, *E*, is given by

$$E = P \times Tem \text{ watt-hours}$$

The battery capacity requirements are calculated considering the use of Ni-Cd cells. The number of eclipses experienced by a spacecraft in an orbital altitude of 550 km are 5900. The allowed depth of discharge varies as a function of charge-discharge cycles over the mission life and the allowed depth of discharge is obtained from Fig. 3-48. The battery watt-hour (*WH*) capacity, *WH*, is given by

$$WH = E/[\text{Depth of Discharge}]$$

Average battery voltage depends upon the life of the mission and the number of charge-discharge cycles. It decreases as the number of charge-discharge cycles increases. Sometimes, the battery voltage can be brought to normal levels by reconditioning. For a nickel cadmium battery with 22 cells

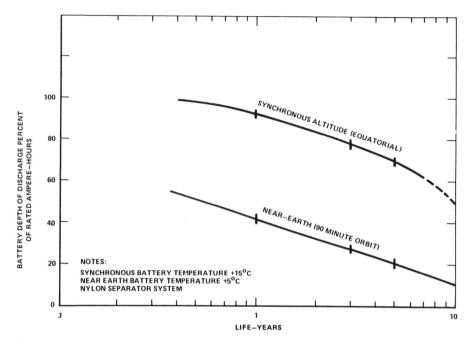

Fig. 3-48. Maximum design utilization of nickel-cadmium batteries for spacecraft applications (courtesy of NASA-GSFC).

in series and for about 5900 charge-discharge cycles, a battery average voltage of 27.5 volts is assumed. Thus, the battery ampere-hour capacity, *AH*, is given by

$$AH = WH/27.5$$

Thus watt-hour and ampere-hour capacities are computed for spacecraft eclipse power requirement. This analysis is more conservative and assumes a worst case beta angle; however beta angle changes in an orbit and inclination can be different. For a particular case, the energy storage requirement or battery rating can be calculated using effective shadow rather than maximum shadow, the formula for calculating the effective shadow is given below:

Minimum Shadow $T(ss) = Tem \{1 - [(R+H)/R] \sin^2 (i + 23.5)\}$

Effective Shadow Duration, $Tee = [(Tsm^2 + Tss^2) 2]^{1/2}$

It is assumed that the sun is contained in the orbit plane some time during the mission and where $R > (R+H) \sin (i + 23.5)$ and this analysis is not valid for sun synchronous orbits.

Now considering the power requirements given in Table 3-15, the power required during eclipse is 686 watts for 27.6 minutes plus 934 watts for 8 minutes and the energy requirement during eclipse is computed to be about 440 WH. As the allowed D.O.D. for Ni-Cd cells is 26% referring to Fig. 3-48, the watt-hour battery capacity of the storage battery shall be about 1693 WH. Assuming an average voltage of about 27.5 volts (battery of 22 cells), the ampere-hour capacity of the battery is 61.6 AH. Referring to Fig. 3-49 of nickel-cadmium cells for satellite applications, to meet this requirement two 35 AH Ni-Cd batteries each with 22 cells in series have been selected.

Solar Array Sizing. There are various types of solar cells as presented in Table 3-17. Depending upon the radiation levels that the cells are going to experience and their degradation rates, a trade has to be carried out to select the best type of cells for the application. Also degradation depends upon the solar cell cover glass thickness as shown in Fig. 3-50. Figures 3-51 to 3-53 show the radiation fluence impinging on the solar cells. Some cells may be efficient at lower radiation levels than the others and vice-versa as shown in Fig. 3-54. This figure indicates that at radiation levels lower than 10E14 1-MeV fluence, k6 ¾ cells produce more power than k4 ¾ cells. However, at radiation levels higher than 10E14 1-Mev fluence, k4 ¾ cells produce more power. Thus selection of solar cells for each spacecraft mission shall be carried out judiciously. Detailed data about radiation fluence is available in the JPL *Solar Cell Radiation Handbook* and 1-MeV equivalent electron fluence is calculated as shown below. However, if total shielding is equal to the value in Figs. 3-51 to 3-53, then the radiation fluence can be directly read from these figures.

Radiation Fluence (550 km altitude, 45 degrees inclination)

	Equivalent Thickness of Fused Silica (Mils)	1-MeV Fluence for one year
Front Side	6	5.91 E12
Back side	12	3.97 E12
One year total		9.88 E12
Three year total		2.96 E13

From JPL handbook on solar cell radiation data, the degradation due to this radiation is obtained. At these radiation levels, k6 ¾ type cells are selected as these are more efficient. If k6 ¾ or k4 ¾ cells are selected then the radiation degradation can also be read from Table 3-18. The details of solar cells selected are given below:

Solar Cells	k6 ¾ TYPE
	10 ohm-Cm
	8 mil thick
	Back Surface Field
	BSR Type
	DAR
Coverglass	6 mil
Backside shielding	12 mil equivalent

GENERAL ELECTRIC NICKEL-CADMIUM SATELLITE CELLS

Standard Cells

Size (Ah) Capacity	Catalog Number 4280—	A Height	B Overall Height	C Base Width	D Base Lg.	Neg. Plate Treat	Weight (Grams) Calc. Max.
1	01AB01	1.560	1.695	.402	1.295	—	50
2	02AB03	1.790	2.200	.650	2.000	Ag	115
3	03AB07	2.030	2.483	.813	2.117	—	155
4	04AB36	2.330	2.790	.830	2.137	Ag	190
5	05AB01	2.710	3.159	.831	2.132	Ag	230
6	06AB49	3.180	3.640	.846	2.155	Ag	2080
7	07AB09	3.500	3.960	.831	2.132	Ag	310
8	08AB05	2.800	3.406	.891	2.988	Ag	380
10	10AB08	3.330	3.926	.891	3.000	Ag	455
12	12AB24	4.030	4.630	.903	2.988	Ag	585
15	15AB19	4.720	5.320	.891	2.983	Ag	670
17	24AB01	4.534	5.134	.886	2.988	Tfe	635
20	24AB06	6.350	6.950	.891	3.000	Tfe	950
20	24AB03	6.530	7.136	.903	3.000	Ag	935
24	24AB02	4.720	5.320	.891	3.763	Ag	845
30	30AB10	7.130	7.726	.903	3.000	Tfe	1115
50	50AB12	5.650	6.250	1.297	5.007		2000
100	100AB52	7.290	8.208	1.522	7.177		4020

Nickel cadmium cells
for satellite applications

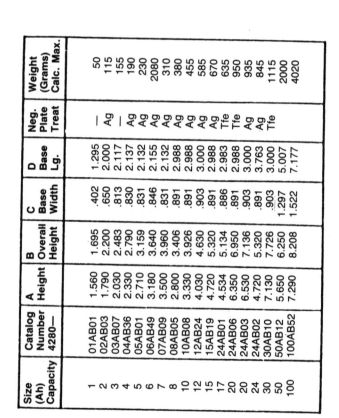

Low Profile Cells

Size (Ah) Capacity	Catalog Number 4280—	A Height	B Overall Height	C Base Width	D Base Lg.	Neg. Plate Treat	Weight (Grams) Calc. Max.
5	0AB101	2.330	2.790	.819	2.128	Ag	205
5	05EB11	2.740	3.200	.828	2.137	Ag	275
6	06AB58	2.789	3.187	.890	2.131	Tfe	265
12	12AB31	3.121	3.651	.892	2.972	Tfe	425
15	15AB22	4.469	4.892	.886	2.983	Tfe	635
17	17AB02	3.937	4.370	.906	3.413	Tfe	610
21	21EB01	4.272	4.695	.971	3.663	—	760
22	22AB01	4.547	4.970	.953	3.663	Tfe	820
22	22AB02	4.547	4.970	.962	3.672	Ag	840
24	24AB24	3.937	4.370	1.124	3.730	Tfe	875
24	24AB25	4.162	4.585	.896	4.347	Tfe	895
25	25AB01	5.241	5.664	.962	3.672	Tfe	985
30	30AB13	4.820	5.268	1.036	4.021	Tfe	1120
34	34AB01	5.440	5.862	1.304	3.404	Ag	1310
35	35AB01	4.745	5.203	1.025	3.988	Tfe	1050
40	40AB03	6.240	6.669	1.304	3.404	Ag	1485
50	50AB20	5.679	6.447	1.328	4.941	Tfe	2025

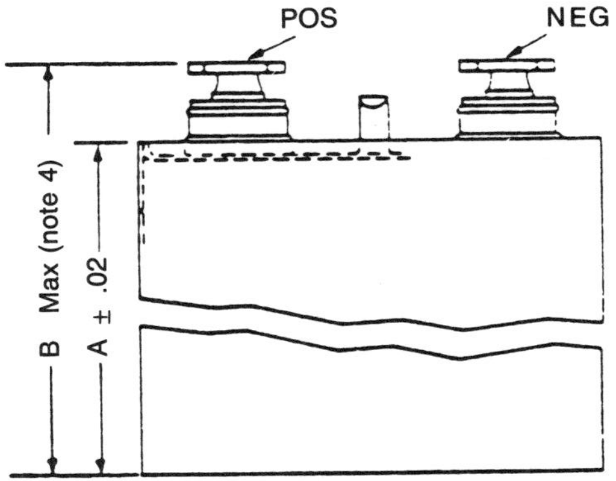

1. Cell thickness dimension to be measured with cell restrained on broad faces, torqued 5.5 to 6.3 in.-lbs. on (4) #10-32 bolts.
2. Those dimensions indicated by this note represent the condition prior to the application of solder.
3. Capacity rating is minimum rating at C/2 at 25° C.
4. Negative plate treatment for any design is optional.

Fig. 3-49. Nickel cadmium cells physical dimensions (Reproduced with Permission from General Electric Company Nickel-Cadmium Battery Catalog).

Table 3-17. Solar Cell (Bare) Electrical Performance Parameters Mechanized Cell Line.

Cell Size	2 Cm × 4 Cm (Nominal)
Contacts	Solderless or Solder Coated
Temperature	28 Degrees Centigrade, AMO, 135.3 mW/Cm²

Cell Type		Thickness (mils)	Isc (mA)	Voc (mV)	I_{MP} (mA)	V_{MP} (mV)	P_{MP} (mW)
k4 ¾	- 2 Ohm-Cm	10	310	590	292	485	141.6
		8	306	590	288	485	139.7
	- 10 Ohm-Cm	10	313	544	294	454	133.5
		8	310	544	291	454	132.1
k6 ¾	- 2-10 Ohm-Cm	10	346	595	322	485	156.2
		8	343	595	319	485	154.7
k7	- 10 Ohm-Cm	10	368	595	340	485	164.9
		8	365	595	337	485	163.4

(Courtesy of Spectrolab, Inc. "Products/Capabilities Overview" Catalog)

The k6 ¾ cell produces an output power of 191.4 watts per square meter at 28 degrees centigrade. However, this power reflects the power output at cell level and there are various loss and degradation factors to be considered

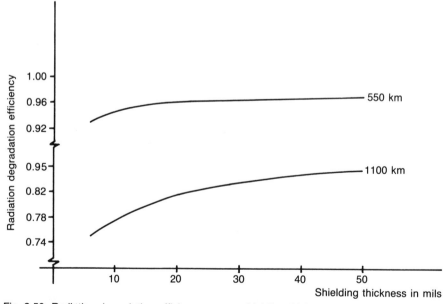

Fig. 3-50. Radiation degradation efficiency versus shielding thickness.

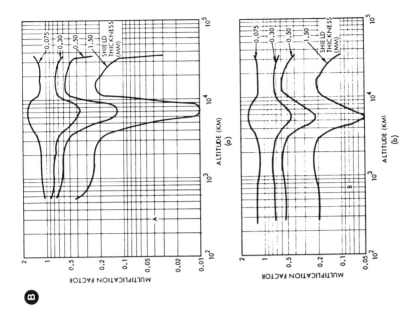

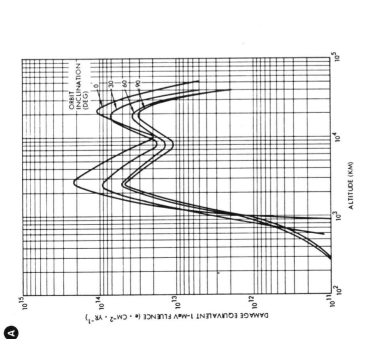

Fig. 3-51. (A) Damage equivalent 1-MeV fluence in circular earth orbit due to trapped electrons for I_{SC} and P_{mp} of silicon cells protected by 0.15-mm thick fused silica covers and infinitely thick back shields (Reproduced from "Solar Cell Array Handbook", JPL SP 43-38, with permission from Jet Propulsion Laboratory, California). (B) Multiplication factors for damage equivalent 1-MeV fluence shown in Figure A for four different thicknesses for (a) zero degree inclined orbits and (b) 90-degree inclined orbits (reproduced from "Solar Cell Array Handbook", JPL SP 43-38, with permission from Jet Propulsion Laboratory, California).

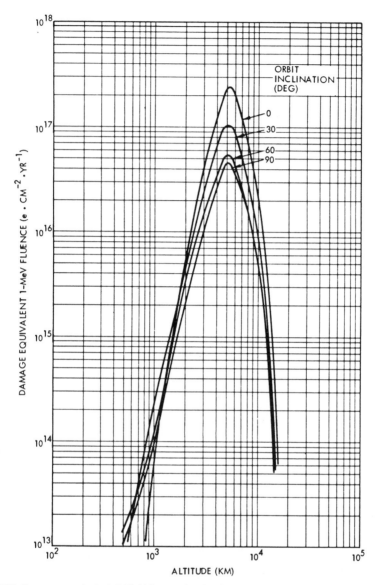

Fig. 3-52. Damage equivalent 1-MeV fluence in circular earth orbit due to trapped protons for I_{sc} and P_{mp} of silicon cells protected by 0.15-mm thick fused silica covers and infinitely thick back shields (Reproduced from "Solar Cell Array Handbook", JPL SP 43-38, with permission from Jet Propulsion Laboratory, California).

to obtain the power output of an array. All the factors including radiation effect are summarized below:

Initial losses

Cell mismatch	0.9900
Measurement errors	0.9900

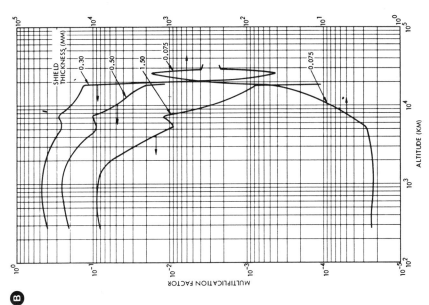

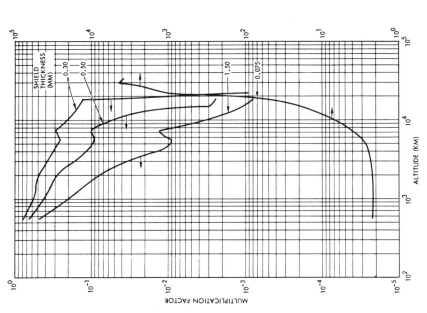

Fig. 3-53. Multiplication factors for damage equivalent 1-MeV fluence shown in Fig. 3-52 for four different thicknesses for (A) zero degree inclined orbits and (B) 90-degree inclined orbits (Reproduced from "Solar Cell Array Handbook", JPL SP 43-38, with permission from Jet Propulsion Laboratory, California).

Table 3-18. Radiation Degradation Versus Orbital Altitude.

Orbital Altitude (km)	Proton + Electron Radiation (3 years) (1 MeV Eq / sq cm)	Radiation Degradation	
		k4¾ cells	k6¾ cells
555	2.937 + 13	0.973	0.905
1111	4.239 + 14	0.850	0.715
1852	4.215 + 15	0.650	0.516
2778	3.075 + 16	0.465	0.350
4167	1.353 + 17	0.340	0.228
5093	1.992 + 17	0.305	0.191
6482	1.935 + 17	0.305	0.191
7408	1.602 + 17	0.325	0.210
8334	1.251 + 17	0.345	0.230
9260	9.027 + 16	0.375	0.258
10186	6.330 + 16	0.410	0.292
11112	3.828 + 16	0.454	0.326
12964	1.086 + 16	0.570	0.442
14816	2.561 + 15	0.700	0.558
16668	6.709 + 14	0.810	0.670
18520	4.710 + 14	0.840	0.700
20372	4.287 + 14	0.850	0.710
31484	9.270 + 13	0.940	0.830

Note:- 45 degrees inclination orbit is assumed in the above radiation fluence computations.

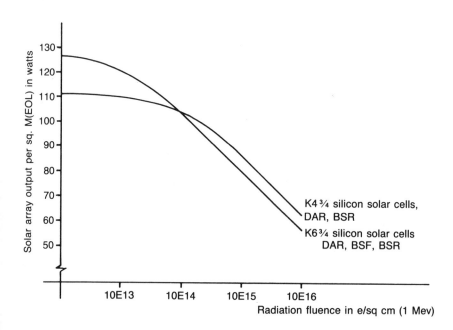

Fig. 3-54. Solar array output versus radiation fluence.

Wiring & diode losses	0.9400
Temperature correction	
(70 degrees centigrade)	0.8236
Subtotal	0.7587

End of third year

UV Degradation	0.9800
Micrometeorite Damage	0.9900
Random failures	0.9900
Sun Intensity variation	0.9650
Radiation effects	0.9050
Subtotal	0.8388
Total	0.6363

Then a cell output of 191.4 watts per square meter at 28 degrees centigrade will degrade to 121.8 watts per square meter at maximum temperature, at the end of 3 year mission. Thus, to obtain 1436 watts, the solar cell area shall be about 11.8 square meters. As the cells cannot be mounted completely on the solar panels, the typical packaging efficiency achievable are in the neighborhood of 85%. Thus the solar array area is about 13.8 square meters.

Weights. The weights of nickel cadmium storage cells are given in Fig. 3-49. Typically the battery weight is about 115% to 135% of the weight of the cells and this variation is due to the thermal mass requirement and it varies as a function of load, orbit altitude, charge and discharge rate, etc.

Solar Array, typically rigid type, weighs about 4.5 to 5.5 kg per square meter.

3.6.2 Power System Design Using Direct Energy Transfer Approach

In the above design, a PPT-based power system is selected. However, if a DET-based power system as shown in Fig. 3-26(A), is selected then the following relationship is used to calculate the solar array power output.

Charge Regulator output:

$$P(ch) = P(eclipse) \left[\frac{T(e) \times V(BC) \times BRF}{T(d) \times V(BD)} \right]$$

Shunt Regulator output:

$$P(sh) = P(day) + \frac{P(ch)}{E(ch)} - P(day\ min)$$

Solar Array Output:

$$P(SA) = P(sh) + P(day\ min) + P(sh\ diss\ min)$$

where   P   = Power System Output
         P(day)   = Power required by load during day
         P(eclipse)   = Power required by load during eclipse
         T(e)   = Eclipse duration in minutes
         T(d)   = Duration of orbital day in minutes
         P(ch)   = Charge Regulator Output
         P(sh)   = Shunt Regulator output
         P(day min)   = Minimum load power during day
         E(ch)   = Charge Regulator efficiency
         P(sh diss)   = Shunt Regulator dissipation

3.6.3 BOL Maximum/EOL Minimum Power Ratios

Total EOL overall loss/degradation factor to find out EOL solar array power from solar cell output power at BOL and at 28 degrees centigrade is = 0.6363.
PPT Approach. BOL Maximum Power loss factor:

> Diode Loss   = 0.94
> Temperature (0° C)   = 1.1176
> Overall loss factor   = 1.0505

Ratio between BOL max/EOL min power = 1.0505/0.6363 = 1.651
DET Approach. Loss factors, for DET regulated bus operation:

> Temperature (70 degrees °C)   = 0.8236
> Correction due to 0 degrees °C   = 1.129
> Diode loss factor   = 0.94
> Overall loss factor   = 0.874

Ratio between BOL max/EOL min Power = 0.874/0.6363 = 1.3736.

3.6.4 Launch Power

When spacecraft are launched using rockets, usually some housekeeping power is required for essential spacecraft operation. If this power is taken from the on-board battery then the on-board battery has to be sized considering the launch as well as on-orbit power requirements. This, sometimes might increase the size of the battery unnecessarily. Hence, an expendable battery is also carried which will be thrown out once the spacecraft is in orbit. Thus, consider a case where the on-orbit power requirements are, say, 100 watts. Maximum eclipse period (for geosynchronous communication spacecraft) is about 72 minutes. Assuming a worst case situation that eclipse can occur immediately after the spacecraft is in orbit. Hence the energy required for spacecraft operation is 120 watt-hours. Let us assume that the launch energy requirement is about 200 watt-hours. Thus, the total energy requirement is about 320 watt-hours.

☐ **Case-1.** If the on-board battery is sized for this one cycle energy requirement and assuming that for only once a 75% discharge is allowed, then the battery size will be 427 watt-hours or a battery with a capacity of about 15 AH with an average battery voltage of 28 volts is needed. This battery weighs about 38 lbs.

☐ **Case-2.** Alternately, if the on-board battery is only sized for on-board needs of 120 watt-hours, then with a depth of discharge of 60%, a 200 watt-hour battery or about 7 AH battery with an average battery voltage of 28 volts is needed. As 75% of this energy can be used only once, i.e., 150 watt-hours, the remaining 277 watt-hours is obtained by employing an expendable short life high energy density storage battery like a silver-zinc battery. Accordingly, the batteries chosen are as follows:

$$Ni\text{-}Cd \text{ long life Battery } = 200 \text{ WH}$$
$$Ag\text{-}Zn \text{ short life Battery } = 280 \text{ WH}$$

The Ni-Cd battery weighs about 19 lbs and the Ag-Zn battery weighs about 8 lbs. Thus, there is a net saving of 11 lbs of on-board weight by using an expendable short life battery for launch operations.

3.6.5 Temperature Factor/Correction

Temperature Factor/correction, $F_{T(op)}$, for silicon solar cells is given by

$$F_{T(op)} = \frac{V_{MPO} + \Delta V}{V_{MPO}}$$

and

$$\Delta V = BV [\ T(op) - T(data) \ ]$$

Example:

$$T(data) = 28 \text{ degrees C}$$
$$T(op) = 70 \text{ degrees C}$$
$$BV = -0.0021 \text{ for silicon solar cells}$$
$$V_{MPO} = 0.5 \text{ for k6}\tfrac{3}{4} \text{ BSR, DAR, BSF}$$

Therefore

$$\Delta V = -0.0021(70\text{-}28) = -0.0882$$
$$F_{T(op)} = (0.5 - 0.0882)/0.5 = 0.8236$$

The following table gives the temperature factors for temperature starting from -60 to 70 degrees centigrade.

$T(\text{op})$	$F_{T(\text{op})}$
-60	1.3696
-40	1.2856
-20	1.2016
0	1.1176
10	1.0756
20	1.0336
28	1.0000
40	0.9496
50	0.9076
57	0.8782
70	0.8236

3.6.6 Deviation from Cosine Law or Cosine Value Correction

When the solar panels are not perpendicular to the sun, they produce less power and the power production follows cosine values, i.e., the product of power generated by the solar panels when they are perpendicular to the sun and the cosine of the angle between sun incidence and normal to the solar panels. However, there is a deviation from cosine law to be accounted for as the solar panels do not produce following the cosine value between about 45 degrees and 90 degrees. Thus Fig. 3-55 and Table 3-19 present the normalized solar cell current and cosine value versus the angle in degrees for two types of solar cells. The degree of deviation of output power generated from the cosine value depends upon the type of the solar cell, its thickness, manufacturing process, cover glass, etc.

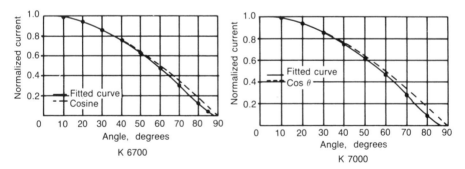

Fig. 3-55. Normalized solar cell current and cosine value versus incidence angle (Reproduced with Permission from Spectrolab, Inc. "Products/Capabilities Overview" catalog).

3.6.7 Nickel-Cadmium Cells—Efficiency

In the power system design, the battery overall efficiency is accounted as the product of battery ampere-hour recharge factor and the ratio of the average voltage to discharge voltage. It also, can be accounted for in terms of charge, discharge, transition and overcharge efficiencies as below. Any method

Table 3-19. Normalized Solar Cell Current and Cosine Value Versus Incident Angle.

Angle (degrees)	Normalized Current		Cosine Value	Deviation of k7000 type (%)
	k6700	k7000		
0	1.0000	1.0000	1.0000	0.0
10	0.9848	0.9848	0.9848	0.0
20	0.9397	0.9397	0.9397	0.0
30	0.8660	0.8660	0.8660	0.0
40	0.7600	0.7566	0.7660	−1.3
50	0.6300	0.6230	0.6428	−3.1
60	0.4690	0.4670	0.5000	−6.6
70	0.3000	0.2900	0.3420	−15.2
80	0.1100	0.1000	0.1736	−42.4
87	0.0000	0.0000	0.0523	−99.8
90	0.0000	0.0000	0.0000	0.0

can be followed if the charging is carried out at a constant rate. However, if the charging is carried out with a variable charge rate and it varies from one instant to another instant in a single orbit (like the case of a spacecraft where the solar array generates power in the form of a sine wave) then it shall be accounted in terms of charge, discharge, transition, and overcharge efficiencies individually.

Discharge Efficiency (Exothermic). Approximately 15% of the total energy released during the discharge is heat, 5% is pressure-volume work and side reactions, and 80% is useful electricity. In exothermic mode, it releases heat in addition to electrical power.

Charge Efficiency (Endothermic). Charge efficiency varies as a function of recharged battery capacity, charge rate and operating temperature and is typically higher than 97%.

Transition from Endothermic Charge to Exothermic Overcharge. During the transition, the charge efficiency changes from 97% to 0%.

Over Charge (Exothermic)/Trickle Charge. In overcharge phase, the heat generated is equal to the electrical power input. Thus the efficiency depends upon the achievable trickle charge rate, which in turn depends upon the charge rate, temperature of the battery, charge regulator operation, etc. If the spacecraft power requirement is constant over its mission duration, then the batteries will be recharged much before the eclipse starts. Thus they get into taper charge (overcharge) if V/T limit charge approach is used and during this period all the overcharge will convert into heat by the batteries. Hence, thermal control is taken into consideration to maintain the batteries within their desired operating temperature limits.

Overall Battery Efficiency. Thus, overall battery efficiency is typically in the range of 65 to 80%.

Chapter 4
Attitude Control System

A NY SATELLITE IN ORBIT REQUIRES STABILIZATION TO INCREASE ITS USEFUL-
ness and effectiveness. For example, when the satellite is not stabilized,
it has to use omni-antennas so that receivers on ground can receive the
information sent by satellites irrespective of the orientation. This necessitates
a high power transmitter as only a small portion of the total power is radiated
in the direction of the earth. On the other hand if there are means to stabilize
the satellite to point the antennas towards the earth, then directional antennas
can be employed to increase the effective transmitter power. This reduces the
actual transmitter power required compared to a satellite without stabilization.

In addition, there are many other aspects which necessitate the need for
satellite stabilization, some of which are described below. Solar arrays generate
maximum power if they are perpendicular to the sun, and any deviation results
in solar arrays generating less power. In addition, some satellites carry some
scientific payloads which have to observe some point in celestial space—a
particular star, the sun, a comet, etc. Moreover, when an apogee kick-motor
is also incorporated in the system, stabilization becomes a prime necessity.
Attitude control is also needed during the transfer orbit for optimum solar array
illumination, for tracking purposes, and for reorientation to properly align the
satellite center-line before firing the apogee motor. Also, a satellite is subjected
to a variety of disturbing torques generated by atmospheric drag, solar wind
and radiation pressure, magnetic field, gravitational field, micrometeorite
impact and even movement of components within the satellite itself. Spacecraft designers are finding ways to convert some of these disturbing torques

into useful control torques. Attitude control becomes important due to the above requirements.

4.1 ATTITUDE CONTROL

The concept and mechanism of attitude control is broadly the same whether the spacecraft is in low earth orbit (LEO) or geosynchronous orbit (GEO) and irrespective of its specific purpose. However, depending upon the satellite orbit and whether it is launched by a rocket or by shuttle, the sequence of attitude control through various phases of the spacecraft before its final attitude is achieved may vary. If a satellite orbit is LEO and if the satellite is launched directly into this orbit, then the satellite will possess a low spin rate which is imparted by the launcher/rocket. As the satellite is already in its orbit (final altitude and inclination), its attitude (orientation) is changed to achieve its final attitude. Depending upon the nature of the mission, the final attitude can be spin stabilized (at a higher spin rate) with the spin axis oriented differently from launch phase or it can be three-axis stabilized, in which case the appendages (i.e., antennas of various types, solar arrays) will deploy after achieving zero or appropriate spin rate of the spacecraft. The desired attitude is achieved by using momentum wheels, magnetic torquers, booms, thrusters, etc.

On the other hand, if the satellite orbit is GEO, then the satellite can have many phases before it attains or reaches its final orbit. A satellite whose final orbit is GEO, can be launched into a low altitude elliptic orbit, circularized later using propulsion rockets or thrusters. Then the satellite is put into a highly elliptic orbit with its apogee altitude equal to its final satellite orbit altitude (in this case, it is 35786 km). Then by firing the apogee kick motor at apogee, the satellite orbit is circularized. As all these operations are related to orbit attainment, the system used to carry out these functions is called the *Orbit Control System* (OCS). Once the satellite is in orbit, the satellite attitude is modified into its final desired attitude (for example, in the case of a three-axis stabilized communication satellite, the solar arrays are oriented and kept perpendicular to the sun and the directional antennas are oriented towards the earth) by the attitude control system (ACS). When both the functions of orbit and attitude control are combined into a single system then it is called an *Attitude and Orbit Control System* (AOCS). Figure 4-1 defines the pitch, yaw,

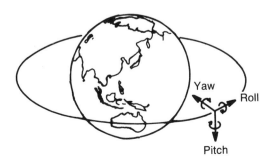

Yaw

Roll

Pitch

Fig. 4-1. Definitions of pitch, yaw, and roll coordinates.

roll coordinates of a satellite in orbit. As shown in the figure, the pitch axis is normal to the orbit plane, yaw axis points to the center of the earth, and roll axis is along the velocity vector of the spacecraft.

As mentioned above, the final attitude can be spin stabilized or three-axis stabilized. The satellite can have gimbaled antennas or gimbaled instruments, gimbaled solar arrays, etc. Depending on the spacecraft orbit and the mission, the AOCS system configuration changes and it uses different control techniques. The attitude control techniques can be classified into two types, i.e., active and passive. Active systems use torquing devices that require power, such as reaction jets or wheels. Passive systems derive the torque from the environment, such as the gravity gradient or the Earth's magnetic field.

Some satellites must meet simultaneously two opposite functional requirements. Pointing an antenna towards the Earth and orienting a solar cell array towards the sun. In such cases, it becomes necessary to adopt a three-axis stabilization system. Most often, in this case solar cell arrays are mounted on gimbals.

4.2 TYPES OF STABILIZATION

Table 4-1 lists the types of stabilization and Table 4-2 presents the merits and demerits of stabilization methods. The three principal stabilization techniques are, a) gravity gradient stabilization, b) three-axis stabilization, and c) spin stabilization with a large despun antenna platform.

Gravity gradient stabilization (Fig. 4-2) is more practicable for very large satellites. This requires very long booms and heavy tip masses for accurate control and thus the satellite weight increases enormously. In this type of stabilization, the satellite does not encounter damping in a gravity field and hence damping is added to the system. This method is based on the tendency of the body in orbit to align itself with its long axis (i.e., the axis of minimum

Table 4-1. Various Types of Stabilization.

Types	Comments
Three Axis Stabilization	It is an active control method, requires attitude actuators, more fine pointing control is possible
Spin Stabilization	Simple, used for some scientific satellites
Gravity Gradient Stabilization	Stable with respect to main central body, requires long booms
Solar Radiation Stabilization	Used in high altitude or interplanetary orbits, it is a passive control method
Magnetic Stabilization	Can be used close to the Earth, Coarse control, slow

Table 4-2. Merits and Demerits of Various Stabilization Methods.

Stabilization Method	Merits	Demerits
Three-Axis	Fine control, fast with gas jets or reaction wheels, flexible	Uses consumables, too fast for some experiments, expensive, limited life
Spin	Simple, any orientation is possible	Stable and rigid structure, needs proper balancing, Antennas, sensors, solar panels cannot point inertial targets, limited life
Gravity Gradient	Stable orientation with respect to central body, might decay or drift	Limited orientations, requires large booms, control limited to about one degree, might wobble

moment of inertia) pointing towards the earth because of the variation in gravitational attraction with distance. This system enables an earth-oriented satellite to remain within desired orientation. Gravity gradient stabilization has several other problems due to heat and perturbation caused by the oblateness of the earth and other planetary motions. Thus stringent orientation accuracies may not be achievable using gravity gradient stabilization alone.

The three-axis stabilization ensures precise control and high reliability. However, weight and power requirements make it practicable only when tradeoffs recommend so.

In the spin stabilized system, the satellite is spun around a rotationally symmetrical axis. The momentum imparted by the spin tends to keep the spin axis fixed in inertial space. Simultaneously, the antenna system is mechanically despun against the rotation of the outer body so that it remains fixed towards a point on earth. Compared to the three-axis system, it requires a lesser number of control elements (as the stationkeeping can be done with only two jets, one radial and the other axial).

Antennas located on the spin axis will cause a doughnut shaped beam to remain in a relatively fixed position with respect to the earth. However, spin

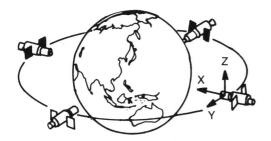

Fig. 4-2. Gravity gradient stabilized spacecraft in orbit.

stabilization does not permit the use of narrow beam antennas that can provide significant economies in transmitting power. For example at medium altitudes (10,000 km), an earth-pointing satellite requires only one sixth the transmitting power required by a spin-stabilized satellite and at synchronous altitude (35,786 km), this ratio can be one twelfth.

4.3 TYPES OF MISSIONS

4.3.1 Type-1

Communications satellites are considered under this type, whose primary purpose is to handle communications with earth stations with the satellite's transmitting and receiving antennas pointed towards the coverage zones on the surface of the earth. The satellite may be a spin-stabilized cylindrically shaped spacecraft (with body mounted solar panels) having its spin axis normal to the orbit plane as it requires the least amount of energy for controlling the spacecraft or a three-axis stabilized spacecraft with deployable solar panels for generating power.

Two-Axis Control (Spinning). In two-axis control, the spacecraft spin axis is oriented normal to the orbit plane and antennas are despun mechanically or electrically to earth-point them. The solar array can be either body mounted or mounted on fixed paddles. The fixed paddle solar array configuration usually interferes with the antennas and may cause spacecraft stability problems. Therefore, the body mounted array is preferred.

When the spacecraft spin axis is oriented in the orbit plane, antennas are earth pointed by precessing the spacecraft attitude once per orbit by employing flywheels and/or gas jets. The solar array, just like in the previous case, can either be body mounted or mounted on paddles. The spherical configuration is usually preferred by power system designers due to constant projection efficiency of solar array throughout the year whereas the projection efficiency changes in cylindrical spacecraft over a range and it requires a larger number of solar cells.

Two-Axis Control (Non-Spinning). Non-spinning two axis control is utilized when one axis of a spacecraft is to be earth pointed. Active attitude control of such a non-spinning earth-oriented spacecraft is achieved by employing gas jets and flywheels. Motion about the third axis, which is parallel to the local vertical, is not controlled. The solar array can be body mounted or mounted on fixed paddles. As the third axis is not controlled, either solar array configuration undergoes larger thermal excursions compared to the spin stabilized spacecraft. Thus, the solar array performance is less efficient. To overcome this inefficiency, solar array requires control of the third axis.

Three-Axis Control. When the third axis of the previously described two axis non-spinning stabilized spacecraft is controlled, it becomes three-axis control. Either a zero angular rate is maintained about the third axis by employing inertial sensors, or a varying attitude is maintained to point the solar array towards the sun. Thus, three axis control enables the use of body mounted, fixed paddles or fully oriented solar arrays. Fully oriented paddles are usually used advantageously as the projection efficiency of the solar array approaches one in this case. As this control utilizes active devices it is known as a three-axis active control system.

On the other hand, earth pointing of a spacecraft can be maintained by properly utilizing gravity gradient torques as a result of a properly designed spacecraft configuration and active control is required only for the third axis. The solar array can be body mounted or fixed paddle mounted. The temperature extremes of the solar array in this case may be less compared to the non-spinning two-axis controlled spacecraft.

4.3.2 Type-2

Low earth orbit spacecraft are considered here. Polar orbits provide the greatest earth coverage for mapping or navigational applications. Sun synchronous orbit has a nodal regression rate equal in magnitude and sense to the mean rate of revolution of the earth about the sun (about 0.985 degrees per day) and the spacecraft maintains its initial orientation relative to the sun. Such an orbit is retrograde and lies between inclination angles of 95.7 degrees and 180 degrees at altitudes up to 5900 km. Twilight sun synchronous orbits have continuous sun throughout the orbit and storage batteries may not be required.

Spin Stabilized Control. In the spin stabilized spacecraft, the spin axis is normal to the orbit plane. Synchronizing the antennas, cameras, or experimental sensors with earth view determines the spacecraft configuration. The spin axis of the spacecraft lying within the orbit plane is similar to the control described above. In this case, if the spacecraft is not precessed, it will maintain a fixed orientation in inertial space. Thus, in a sun synchronous orbit, solar array can always be perpendicular to the earth-sun line without an active control.

Two-Axis Control (Non-Spinning). This is similar to the non-spinning two-axis control for Type-1 missions. An additional improvement in solar array performance is obtained for polar orbits because the ± 23 degrees change in earth-sun line from the equatorial plane has no effect.

Three-Axis Control. This is similar to the three-axis control presented for Type-1 missions.

Thus, the satellite control system maintains the satellite in the desired attitude and orbit position as the satellite is subjected to a variety of disturbing torques. Thus almost every satellite has some kind of attitude control. Figure 4-3 shows a typical attitude control system. The attitude of the satellite is measured using various sensors on-board the satellite and attitude actuators are used to correct the attitude such that the satellite is brought back or reoriented into the desired attitude and maintained. Various attitude measuring sensors are described in the next section. Then follows the description of various attitude actuators and some examples of attitude control systems.

4.4 ATTITUDE MEASURING SENSORS

As mentioned above, the attitude measuring sensors vary depending upon the mission and accuracy requirements of a satellite. Earth/horizon sensors,

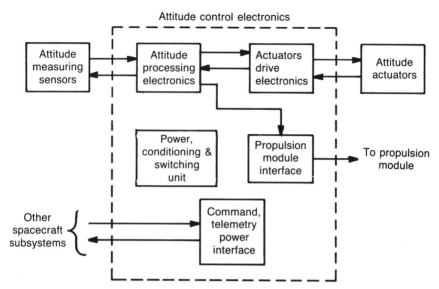

Fig. 4-3. Block schematic of Attitude Control System.

sun sensors, star sensors, magnetometers, and gyroscopes, are some of the sensors.

4.4.1 Earth Sensors

Earth sensors are used to scan across the earth, measuring rotation angles to define the spacecraft's attitude relative to the Earth from the spacecraft's altitude. For a spacecraft in low earth orbit, the earth is the second brightest celestial object and covers up to 40% of the sky. The earth presents an extended target to a sensor compared to the Sun and stars due to their relative distances from the spacecraft.

Earth emanates infrared radiation and the IR intensity in the 15 micrometer spectral band and is relatively constant. This portion of the earth's IR spectrum as viewed from outer space is dominated by carbon dioxide (CO_2) absorption bands, and a major portion of the emission at 15 micrometers originates from altitudes above the earth's troposphere. Hence, this radiation is measured/detected using a sensor or bolometer or other type. An earth sensor usually consists of a scanning mechanism, an optical system, a detector and signal processing electronics. The optical system of an earth sensor consists of a narrow-band filter to limit the observed spectral band and a lens to focus the target image on the detector.

Though photo diodes, thermopiles, pyroelectric detectors, bolometer thermistors can be used as detectors, the bolometer type is devoid of all the demerits. A bolometer is a sensitive resistance thermistor consisting of fused conglomerates or sinters of manganese, cobalt, and nickel oxide bonded to a heat sink. Incident radiation heats and alters the resistance. Typical temperature coefficient is 3.5% per degree kelvin. Constant current passed through the thermistor, generates a voltage that changes as the thermistor gets heated up and the variation in voltage is sensed. This type of detector can sense temperature changes of 0.001 degrees K due to radiation despite ambient temperature changes four orders of magnitude greater.

The sensor can be mounted on a nonmovable surface or on a scanning device such that the sensor scans an angle. Some types of sensors in the scanner class are, a) sensors with field of views (FOV) that scan by means of spacecraft spin motion; b) sensors that scan through mirrors or germanium prism lenses that rotate in conjunction with reaction and angular momentum control wheels; and c) sensors that scan by constant speed motors. Some of the typical sensors manufactured by Ithaco Inc. are described here.

4.4.1.1 Horizon Crossing Indicator

The *Horizon Crossing Indicator* (HCI) is intended for use on a spinner spacecraft for attitude determination. It senses the thermal discontinuity between earth and space as the line of sight crosses the horizon. Figure 4-4 shows the optical diagram of the HCI. Features include a digital signal processor for accurate horizon location independent of radiance levels, crossing angles, and field of view effects. Figure 4-5 shows HCI scan pattern and its outputs. Table 4-3 presents Ithaco's HCIs specifications.

When steerability is added to this sensor, it is known as a *Steerable Horizon Crossing Indicator* (SHCI). The SHCI's line of sight is adjustable in 3 increments by command over the full 360 degrees. Figure 4-6 shows SHCI mounting configuration on a spinner spacecraft. A photo of SHCI and its optical diagram are shown in Figs. 4-7 and 4-8 respectively. A 45 degree fold mirror is mounted to a ball-bearing suspended housing driven by a stepper motor. A position sensor in conjunction with an up-down counter provides the line of sight position information for all incremental position settings. The SHCI's possible applications are shown in Fig. 4-9 and its signal outputs (line of sight position and the horizon location) are used for attitude determination of the spacecraft. Table 4-4 presents Ithaco's Steerable Horizon Crossing Indicator Specifications.

4.4.1.2 Conical Earth Sensor

The *Conical Earth Sensor* (CES) can be used for attitude determination in virtually any orbit from 100 km to super synchronous, whether circular or highly

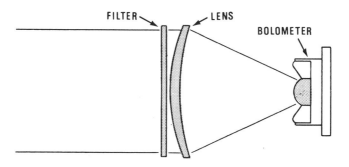

Fig. 4-4. Optical diagram of a Horizon Crossing Indicator (courtesy of Ithaco Inc.).

Fig. 4-5. Horizon crossing indicator scan pattern and its outputs (courtesy of Ithaco Inc.).

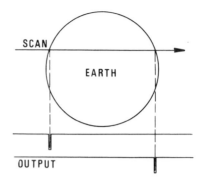

Table 4-3. Horizon Crossing Indicator Specifications (Courtesy of Ithaco Inc.).

Specifications

APPLICATION		For spinning space vehicle launch phase, transfer orbit, on-orbit
SPACECRAFT SPIN RATE		1 – 200 rpm
OPTICAL PASSBAND		14 – 16μ
ACCURACY		0.1° (3σ)
SIZE	Sensor –	9.9 x 5.9 cm (maximum diameter)
	Electronics –	12.8 x 10.3 x 4.3 cm
WEIGHT		0.65 kg
POWER		< 0.7W

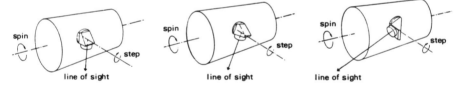

Fig. 4-6. Steerable horizon crossing indicator spacecraft mounting configuration (courtesy of Ithaco Inc.).

elliptical. Its attitude coverage is unique: 90 degree maneuvers are permitted in low orbits. The CES is used on several current programs including LANDSAT-D, P 80-1, and SPACELAB.

The Conical Earth Sensor consists of a sensor head and an electronics box. The sensor head is illustrated in Fig. 4-10. A hollow shaft motor contains within its shaft an optical barrel that includes a germanium lens and a germanium immersed bolometer detector. A wedge-shaped prism rotates in front of the lens to deflect the image of the detector by 45 degrees. A coated window not only establishes the 14-16 micron passband, but also hermetically seals the unit.

Operation. The Conical Earth Sensor concept is illustrated in Fig. 4-11. The field of view of the sensor is deflected by 45 degrees and rotated continuously, which results in the scan pattern shown in Fig. 4-12. When the scan path intersects the earth, the infrared detector senses the difference in

Fig. 4-7. Photo of a steerable horizon crossing sensor and its electronics (courtesy of Ithaco Inc.).

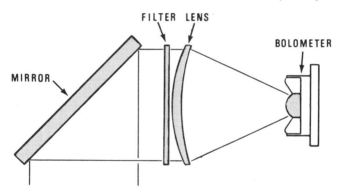

FILTER LENS

BOLOMETER

MIRROR

Fig. 4-8. Steerable horizon crossing indicator optical diagram (courtesy of Ithaco Inc.).

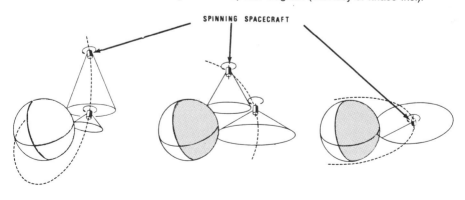

SPINNING SPACECRAFT

LAUNCH PHASE,
TRANSFER ORBIT

ON-ORBIT

Fig. 4-9. Possible applications of steerable horizon crossing indicator (courtesy of Ithaco Inc.).

Table 4-4. Steerable Horizon Crossing Indicator Specifications (Courtesy of Ithaco Inc.).
Specifications

APPLICATION	For spinning space vehicle launch phase, transfer orbit, on−orbit (elliptical or circular, any inclination)
SPACECRAFT SPIN RATE	1 to 200 rpm
INSTANTANEOUS FIELD OF VIEW	$1° \times 1°$
FIELD OF VIEW COVERAGE	$0° − 360°$
INCREMENT (120 steps) @	$3°$
STEP RATE	0 to 200−steps/sec
OPTICAL PASSBAND	$14 − 16\mu$
ACCURACY	$0.1°$ (3σ)
ALTITUDE RANGE	150 − 60000 km
SIZE Sensor − Electronics−	13.34 × 9.14 cm (max dia) 15.24 × 15.24 × 5.23 cm
WEIGHT	1.6 kg
POWER	2.5W static +7.5W when stepped

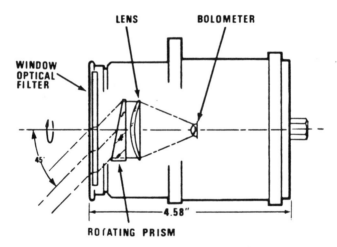

Fig. 4-10. Conical horizon sensor internal details (courtesy of Ithaco Inc.).

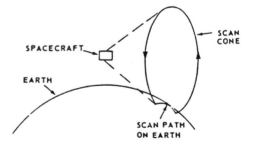

Fig. 4-11. Conical earth sensor in-orbit
usage concept (courtesy of Ithaco Inc.).

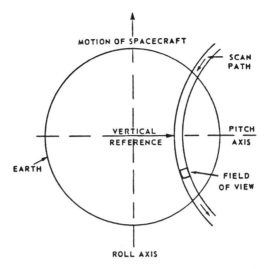

Fig. 4-12. Conical earth sensor scan pattern (courtesy of Ithaco Inc.).

temperature between the earth and space. The resulting pulse (Fig. 4-13) can be processed to determine both pitch and roll attitude. A vertical reference pulse on the sensor defines where the center of the Earth (E) pulse will be at pitch null. The width of the pulse is a function of the attitude error in roll.

In low altitude orbits (100 km to 1500 km), the sensor spin axis will normally be parallel to either the pitch or roll axis of the spacecraft. At higher orbits, the sensor is tilted down in order that the scan path intersect the earth. At geosynchronous altitude, a 40 degree tilt angle provides, as shown in Fig. 4-14, good geometrical gain (Gain is equal to the change in pulse width per attitude angle). Because the width of the earth pulse is also a function of orbit altitude, the null width must be predicted if a single sensor is to be used. This reference is built into the unit and can be adjusted by command. If the orbit is elliptical, two back-to-back sensors may be required, depending on the accuracy requirement (Fig. 4-15). This two sensor configuration accommodates highly

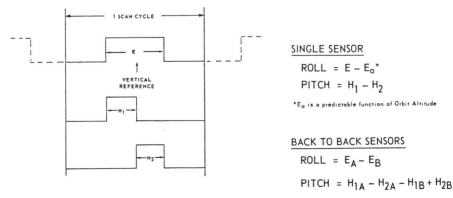

Fig. 4-13 Conical earth sensor electrical output pulses (courtesy of Ithaco Inc.).

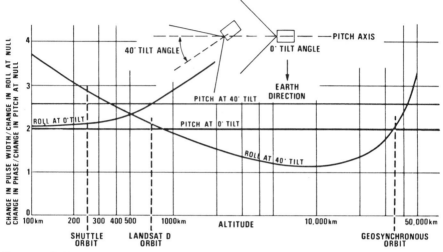

Fig. 4-14. Roll geometric gain versus spacecraft altitude (courtesy of Ithaco Inc.).

elliptical orbits without error because the altitude related term cancels out. Table 4-5 presents the Ithaco's Conical Earth Sensor Specifications.

4.4.1.3 Boresight Limb Sensor

The *Boresight Limb Sensor* (BLS) maximizes science return by providing real time, dedicated aiming knowledge to limb looking experiments. It is an infrared sensor that measures the aiming position of a spacecraft payload with respect to the earth's limb. It measures both the displacement angle and the

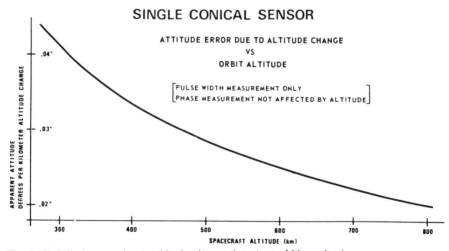

Fig. 4-15. Attitude error due to altitude change (courtesy of Ithaco Inc.).

Table 4-5. Conical Earth Sensor Specifications (Courtesy of Ithaco Inc.).

Specification Summary

APPLICATION	Two axis attitude determination for spacecraft		
ALTITUDE RANGE	100 km to super—synchronous		
ACCURACY	Geosynchronous	< .05°	
	Low Orbits*	< .10°	
	SENSOR	ELECTRONICS	TOTAL
SIZE	4.0'' x 3.0'' dia.	6'' x 7'' x 3''	
WEIGHT	2.1 lbs (0.95 kg)	3.4 lbs. (1.55 kg)	5.5 lbs (2.5 kg)
POWER	4 Watts	4 Watts	8 Watts

*Does not include the predictable effects of earth's oblateness and seasonal radiance changes in low orbit applications. Predictable seasonal effects will be about 0.1° and oblateness can produce an error of up to 0.3° for a geocentrically pointed spacecraft, depending on the orbit altitude. These effects can be removed either by spacecraft or ground processing.

rotational angle as illustrated in Fig. 4-16. In general, the BLS is mounted on the payload, boresighted at the horizon at the point viewed by the experiment and it provides real time position information at a rate of four updates per second.

It has a rotating scan and a germanium optical system, as illustrated in Fig. 4-16. The scan angle is determined by a rotating prism and it can be anywhere below 45 degrees. The 11 degree angle shown provides excellent performance for displacements of up to ±5 degrees. For larger displacements, larger scan angles can be used. An optical pickup provides the reference for the rotational motion measurement.

The BLS operates in the 14 to 16 micron passband, which is free of diurnal effects and is by far the most stable from the standpoint of radiance effects. Table 4-6 presents the specifications of this sensor.

High Accuracy Attitude Reference. The BLS can also be used to very accurately reference a remote payload back to a Central Attitude Determination System (CADS). This is done by using two BLS's one at the payload, the other at the CADS, and measuring the differences between the two BLS outputs. The pair of BLS's would be aimed at the same point on the horizon so that the effects of horizon anomalies are minimized or eliminated. Figure 4-17 illustrates the use of a pair of BLS's in this manner.

4.4.2 Sun Sensors

The Sun subtends an angular radius of 0.267 degree at 1 AU (the distance between the earth and the Sun) and is nearly orbit independent as the orbital altitudes of our interest are very much smaller than 1 AU and for most applications the Sun can be treated as a point source. A simple Sun sensor can be used to detect Sun reference as the Sun is relatively very bright. A sun sensor can be used for a variety of spacecraft applications. The basic element that is most often used as the sensing element is the silicon solar cell. Whenever light falls on the sensing element, it converts the solar radiation into an electrical signal, whose magnitude depends upon the area of the sensor, incident light (magnitude), angle of incidence, etc. However, depending upon the application, the output of this signal can be given to the user directly or

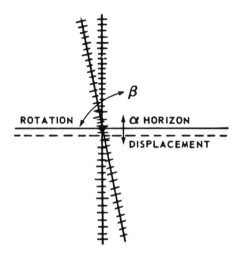

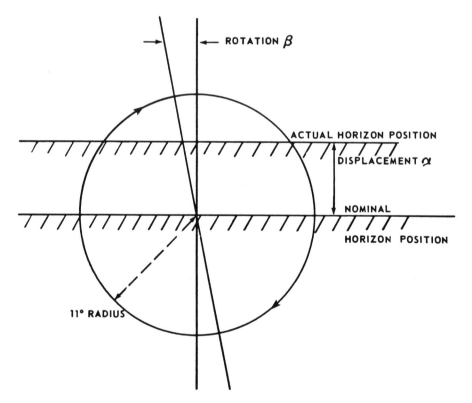

Fig. 4-16. Boresight limb sensor rotating scan illustration (courtesy of Ithaco Inc.).

Table 4-6. Boresight Limb Sensor Specifications (Courtesy of Ithaco Inc.).

Specification Summary

APPLICATION	Two—axis attitude determination for limb pointing payloads on the Space Shuttle, free flyers, and sounding rockets.		
ATTITUDE RANGE	100 km and higher		
ABSOLUTE ACCURACY	0.1° (250 km alt.) .07° (500 km alt.) .05° (1000 km alt.)		
RELATIVE ACCURACY	0.01° (between 2 sensors, see page 4)		
	SENSOR	ELECTRONICS	TOTAL
SIZE	4.0'' x 3.0'' dia.	6'' x 7'' x 1.7''	
WEIGHT	2.1 lbs	2.6 lbs	4.7 lbs
POWER	4.4 Watts	3.2 Watts	7.6 Watts

after processing. Thus, the Sun sensor assembly can have a constant level and an analog or digital output. Depending upon the sensor assembly and the implementation, they can be of fine or coarse type and can output analog or digital signals (including various digital coded).

Sun Presence Sensor. This sensor has a constant output whenever the Sun is present in the field and it is commonly known as a Sun presence sensor.

Analog Sun Sensor. This sensor has an analog output that varies as a function of the incidence angle (assuming that the sun intensity is constant as the sensor is in the orbit) and is commonly known as an analog sun sensor. The relationship between the sensor output, I, and the angle of incidence is given by

$$I(\theta) = Io \times \cos \theta$$

where $I(\theta)$ is the output of the sensor for an incidence angle of θ, $I\theta$ is output of the sensor when the Sun is perpendicular to the sensor.

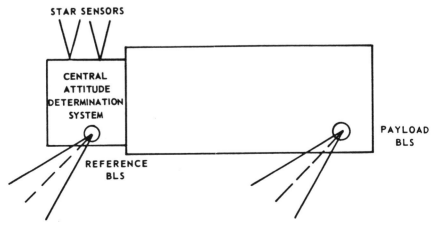

Fig. 4-17. Use of two boresight limb sensors (courtesy of Ithaco Inc.).

Depending upon the number of sensors within an assembly, the sensor can be a single-axis or two-axis system and the field of view can be varied depending upon the need. Resolution depends upon the signal processing electronics and the sensing element.

Digital Sun Sensor. This produces a digitally coded output that can be used directly by the attitude determining electronics. This sensor uses a number of solar cells arranged in a digital code form. In addition, some more cells are used for easy signal processing and for obtaining the sign of the incident angle. Thus, the slits are divided into a series of rows with a solar cell underneath each row and are related to a) an automatic threshold adjust (ATA), b) a sign bit, c) digitally coded bits and d) fine bits.

Because the solar cell voltage is proportional to the incident cosine angle, θ, a fixed threshold is inadequate for determining the voltage at which a bit is turned ON. This is compensated for by use of the ATA slit, which is half the width of other slits. Consequently, the ATA solar cell output is half that from any slit. A bit is turned ON if its cell voltage is greater than the ATA cell voltage and consequently ON denotes that a reticle slit is illuminated more than half.

The sign bit or more significant bit determines which side of the sensor the Sun is on. The coded bits provide a discrete measure of the linear displacement of Sun image relative to the sensor center line. Though several codes are used, Gray code is famous because of its merits. The fine bits are used by an interpolation circuit to provide increased resolution.

The sun sensors have been developed with fields of view ranging from several square arc-minutes to 128 by 128 degrees and resolutions of less than an arc-second to several degrees.

4.4.3 Star Sensors

Star sensors sense stars. As the stars are relatively farther than Earth and the Sun, their intensity is relatively several orders of magnitude less. These sensors usually contain a sensing element or detector, tracking platform on which the sensor is mounted or the mirror to track, a sun shade, an optical system, field of view limiter and an electronic signal processing unit, which processes the output of sensor to identify the star in association with other orbital information. Thus, these sensors provide outputs to specify the location and the visual magnitude of the star.

Star sensors can be equipped with a shutter which can be commanded to open or close. This also can be controlled to automatically shutdown when a bright object/star, like, the sun, earth, moon, etc., is moving closely.

Star Scanner. It uses the spacecraft rotation to provide the searching and sensing function and the sensor is mounted on the spacecraft at an appropriate location.

The CS-207 solid-state star scanner manufactured by Ball Aerospace Systems Division, employs a pair of linear silicon diode detector elements on a common substrate. One of the detector elements is aligned parallel to the plane of the spin axis, the other is inclined at 20 degrees. Output signals from

**Table 4-7. Solid-State Star Scanner Specifications
(Courtesy of Ball Aerospace Systems Division).**

Specifications

ELECTRO-OPTICS

Detector: Two element silicon diode array

Lens: 50 mm, f/1.4

PERFORMANCE

Field-of-view: 12° elevation

Star sensitivity: +1.6 to -1.4 silicon
magnitude (8 commandable thresholds)

Probability of Detection: 0.99
(+1.6 silicon magnitude)

Spacecraft Spin Rate: .12 to 8 rpm

Sensor Optical Axis: 14° from vehicle
spin axis

Sun Shade Protection: ±70°

Power Consumption: Less than 2.0 Watts
from 28 VDC

Accuracy

Position: ±10 arc minutes (1σ)

Star Intensity: ±0.25 magnitude

MECHANICAL CONFIGURATION
(including sunshade)

Size: 15" × 14" × 7.875"
(381 mm × 356 mm × 200 mm)

Weight: 6 lbs. (2.7 Kg)

ENVIRONMENTAL

(operating levels)

Temperature: -20C to 45C

Pressure: Surface to 10^{-10} torr

Acceleration: Thrust Axis: 30 g's, 3 minutes duration
Lateral Axis: ±5 g's, 3 minutes duration

			Sine		
			Lateral Axis:	5-18 Hz	0.9" D.A. (23 mm)
Vibration (acceptance levels)				18-50 Hz	2.5 g
Random				50-100 Hz	3.75 g
Lateral Axis:	20-40	+9 dB/oct		100-200 Hz	3.0 g
(11.5 g rms	40-100	0.4 g²/Hz		200-400 Hz	3.75 g
overall)	100-200	-9 dB/oct		400-2000Hz	7.5 g
	200-2000	0.05 g²/Hz			
			Thrust Axis:	5-13	0.9" D.A. (23 mm)
Thrust Axis:	20-40	+9 dB/oct		13-35 Hz	8.0 g
(12.2 g rms	40-100	0.7 g²/Hz		35-100 Hz	22.5 g
overall)	100-260	-9 dB/oct		100-130 Hz	9.0 g
	260-2000	0.04 g²/Hz		130-2000 Hz	7.5 g

both detectors are processed by a single electronics channel. Table 4-7 presents the technical specifications. Good field uniformity over 12 degrees field of view is achieved by stopping down a 50 mm, f/0.95 lens to F/1.4. The signal processing electronics of the CS-207 consume less than 2.0 watts in generating a star position pulse and a corresponding star intensity signal from detected star crossings. Eight commandable thresholds permit selection of the stars to be detected within the range of − 1.4 to + 1.6 silicon magnitude.

Gimbaled Star Tracker. It searches out and acquires stars using mechanical action of the gimbal on which the sensor is mounted.

Fixed Head Star Tracker. It has electronic searching and tracking capabilities over a limited field of view.

Star Tracker tracks stars within designed field of view (FOV) and over a visual spectral magnitude range. For example, the star tracker employed on Landsat had a field of view of 8 by 8 degrees and tracked stars over 2 to 6 visual magnitude. In the acquisition mode, first the star tracker is held in one position, then the spacecraft or the gimbal on which the star sensor is mounted, is rotated such that the star tracker acquires a star of sufficient (magnitude) brightness. Then the star tracker enters the tracking mode and tracks the star until it leaves the FOV. Once it loses a star, it again goes into acquisition mode. Once a star is acquired, it is evaluated to match with star table stored in an on-board computer or on ground. This process is continued until desired star is located.

4.4.3.1 NASA Standard Fixed Head Star Tracker

NASA selected the Ball Aerospace Systems Division Standard Star Tracker (SST) as a standard component because of its versatility, sensitivity and flight-proven design. The tracker incorporates all the landmark features, plus the convenience of a self-contained power converter, digital position outputs, and several performance options which increase its utility. Its combination of large field of view and high sensitivity enable it to detect and track stars in any portion of the sky, thereby placing no constraints on spacecraft orientation. This tracker is equally useful for closed loop attitude control or star field mapping for precise attitude determination.

Performance. Table 4-8 presents SST technical specifications. The tracker is an all electronic, strap down device that automatically searches its 8-degree field of view for a target star, then switches modes to track the star throughout the field. Stars of 6th magnitude or brighter can be tracked at vehicle rates of up to 0.3 degrees per second. The two-axis, 12 bit digital and/or analog output signals represent star position anywhere in the field to better than 3 arc minutes and 30 arc seconds at null. By applying the calibration data supplied with the SST, the accuracies can be improved to 10 arc seconds (1 sigma). When the Internal Error Correction Option is installed, the output signals are accurate to 1 arc minute over the field of view.

Operation. When power is applied to the tracker, it automatically searches for a star brighter than the command threshold (+3, +4, +5 and +6 magnitude levels). When a star exceeds the threshold, the track mode is engaged. Tracking of the star continues until it leaves the field of view or a "break track" command is received. When either of these events occurs, the search mode is resumed. A repeated "track mode/break track" sequence will result in an accurate star map.

4.4.4 Magnetometer

This measures magnetic field along with its input axes and can be designed to measure the magnetic field to milligauss accuracy. Magnetometers can be used for obtaining both the direction and magnitude of the magnetic field. However, magnetometers are not accurate inertial attitude sensors because the magnetic field is not completely known and the models used to predict the magnetic field direction and magnitude at the spacecraft's position are subject to relatively substantial errors. Furthermore, because the earth's magnetic field

Table 4-8. NASA Standard Fixed Head Star Tracker
Specifications (Courtesy of Ball Aerospace Systems Division).

Specifications

ELECTRO-OPTICS

Photo-Sensor
Type: ITT F4012RP image dissector

Cathode Spectral Type: S-20

Lens
Type: refractive, four element, low scatter
custom design
Characteristics: 70mm, f/1.2

Bright Object Sensor (option)
Protection Angles: 20° to 60° for solar input;
9° for −12 magnitude source

Shutter (option)
Location: mounted on opposite side of SST
mounting structure
Type: two overlapping blades driven by
brushless DC torque motor

PERFORMANCE

Field of View: 8° x 8°

Star Sensitivity: +6 magnitude or brighter

Search Mode
Scan Type: raster scan
Maximum Star Acquisition Time: 10 sec.

Track Mode
Scan Type: unidirectional cross-scan
Total Scan Period (two axes): 100 msec.
Output Data Rate: 10 updates/sec (each axis)

Power
Input Voltage: 21 to 35 Vdc
Consumption: 18 Watts

Accuracy
Noise Equivalent Angle: 16 arc seconds
(+6 magnitude star)
Total Accuracy: 10 arc seconds (1 σ) cali-
brated, anywhere in field of view
Null Accuracy: 30 arc seconds uncalibrated

Output Signals
Analog: star magnitude, temperature
Digital: two 16 bit serial words, each giving
star position (12 bits) plus status

MECHANICAL

Size: 6.6" x 7.1" x 12.2"
(168mm x 180mm x 310mm)

Weight: 17 lbs (7.71 kg)

ENVIRONMENTAL (Operating Levels)

Temperature: −10 C to +50 C
(−30 C to +60 C survival)

Shock: 5g sine sweep, 200-2000 Hz,
4 octaves/min.

Vacuum: surface to 10^{-10} torr

Acceleration: 20 g, 1 minute, all axes

Vibration (all axes)

Type	Frequency(Hz)	Level
Sinusoidal	5-28	0.33" D.A.
(4 oct./min)	28-100	±14 g
	100-200	±10 g
Random	20-130	+6 dB/oct.
(1 min/axis)	130-1000	0.08 g²/Hz
	1000-2000	−3 dB/oct.

strength decreases with distance from the Earth as $1/r^3$, residual spacecraft magnetic biases eventually dominate the total magnetic field measurement, generally limiting the use of magnetometers to spacecraft below 1000 km. Earth's magnetic field in LEO is about 0.5 gauss.

A magnetometer consists of a magnetic sensor and an electronics unit that processes the sensor output into a usable signal. Magnetic sensors are broadly divided into two types. One type uses fundamental atomic properties such as Zeeman splitting or nuclear magnetic resonance. A second type uses magnetic inductance and is based on Faraday's law (an electromagnetive force (EMF), E, is induced in a conducting coil placed in a time-varying magnetic flux). Among the inductance-type magnetometers there are two types. One is known as a search-coil magnetometer, which uses a solenoid coil of N turns wound over a ferromagnetic core with a known magnetic permeability and core cross-sectional area. The EMF induced in the coil when placed in a magnetic field produces a voltage proportional to the magnetic field. The second type is known as a fluxgate magnetometer.

4.4.5 Gyroscopes

A gyroscope is an instrument that uses a rapidly spinning mass to sense the inertial orientation of its spin axis. Rate gyros and rate integrating gyros are attitude sensors used to measure changes in the spacecraft orientation. Rate gyros measure spacecraft angular rates and rate integrating gyros measure spacecraft angular displacements directly.

4.4.5.1 Solid-State Rate Transducers

Complete angular dynamics information can be obtained using the solid-state rate sensor shown in Fig. 4-18 manufactured by Humphrey, Inc. When the sensor is combined with the appropriate operational amplifier, it becomes an angular position, rate, or acceleration transducer.

Figure 4-19 shows a crystal pump moving gas through a nozzle a distance L to impinge on the sense wires. When the unit is rotated at a rate, W_i, the gas jet lag distance is D given by $L \times T_g \times W_i$, where T_g is the transit time of the gas. Since L/T_g is the velocity of gas, V_g, then D is equal to $[L^2 \times W_i]/V_g$. This indicates that the sensitivity increases as the square of the jet length. T_g is the factor to know in the simulations presented, and T_g is proportional to L/E(pump). Figures 4-20 and 4-21 show the pins and wiring connections. The protective diodes are not shown on the wires.

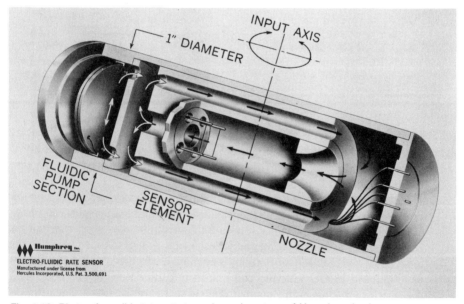

Fig. 4-18. Photo of a solid-state rate transducer (courtesy of Humphrey Inc.).

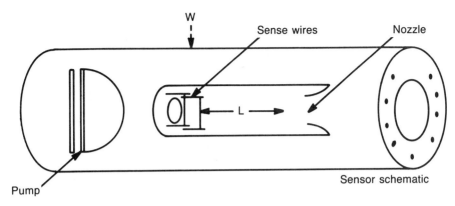

Fig. 4-19. Solid-state rate sensor schematic (courtesy of Humphrey Inc.).

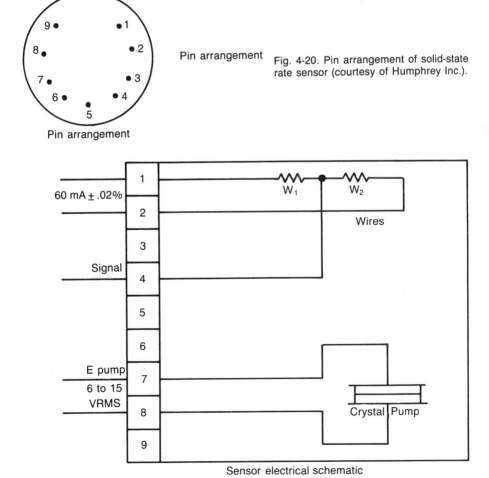

Fig. 4-20. Pin arrangement of solid-state rate sensor (courtesy of Humphrey Inc.).

Sensor electrical schematic

Fig.4-21. Solid-state rate sensor electrical schematic (courtesy of Humphrey Inc.).

Power Supply Requirements.

Bridge Current:	60 mA, ±.02% regulation, through W1, and W2. This will determine nominal bridge voltage.
Bridge Voltage:	1 Vdc nominal (as determined when proper current flows).
Crystal Pump:	±6 Vdc to ±15 Vdc ±.02% regulation to pump oscillator (Fig. 4-22) (2″ units use ±6 Vdc, 4″ units use ±15 Vdc)
Amplifier:	±6 Vdc to ±15 Vdc ±.02% regulation

The output voltage from the basic sensor is $e_o = [K \times W_i]/(1 + T_g \times S)$. Here K is the scale factor of volts/radian/second, T_g is the gas time constant and is determined by the length and the velocity of the gas jet. W_i is the input rate and S the operator. The addition of an operational amplifier with an output voltage E_o, a feedback impedance transfer function Z_f, and an input impedance transfer function Z_i, makes the output voltage $E_o = -[e_o \times Z_f]/Z_i = -[Z_f \times K \times W_i]/Z_i(1 + T_g \times S)$. The basic sensing is a lag of the gas onto the lagging wire proportional to the rate of turn. The foregoing allows the exact duplication of a second order rate gyro with inertia, springs, and damper. Or, the procedure may be modified to duplicate an integrating rate gyro, or an angular accelerometer. Using one Quad 741 chip, one may have a preamplifier and rate, position, and angular acceleration outputs. The following pages will show that the damping and natural frequency of the rate gyro synthesized is largely set by resistor and capacitor values resulting in extremely good stability.

Applications. Humphrey, Inc's RT01 Series Rate Sensors with a gas pump and preamplifier become Rate Transducers. The RT01 series are available as is, or as RT03 single-axis Rate Transducers, and as RT02 three-axis Rate Transducers.

□ **The Bridge.** The sensor wires W1 and W2 are compared with two fixed resistors R1 and R2 as shown below in Fig. 4-23 (numbers refer to sensor pins). Remove the diodes for final calibration. Also e_o null is in the microvolt range requiring a precision meter. Calibration of the bridge must be done while gas is flowing on the wires, and should not be nulled finally until the pump voltage is set.

□ **The Gas Pump.** The pump crystal runs at about 4 kHz. The crystal may be run by an audio oscillator on either sine wave or square wave. Care must be taken to adjust the frequency to the crystal for maximum current flow. To avoid this problem a simple feedback oscillator as shown in Fig. 4-22 may be used. It is necessary to set the pump voltage for maximum bridge null stability and sensitivity.

□ **Preamplifier.** It is necessary to amplify the very low signal from the gas passing over the wires with an input rate. Amplifier gains from 400 to 1000 are not uncommon. The nature of the bridge signal and reference voltage requires an instrumentation amplifier. There are many forms of instrumentation amplifiers and one simple form is shown in Fig. 4-24 using a Dual 741 or 747.

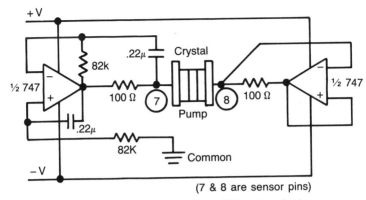

(7 & 8 are sensor pins)

Fig. 4-22. Simple pump oscillator (courtesy of Humphrey Inc.).

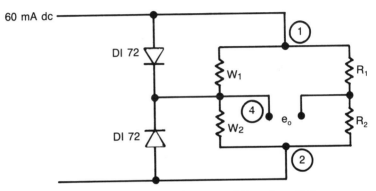

Fig. 4-23. Sensor bridge connection (courtesy of Humphrey Inc.).

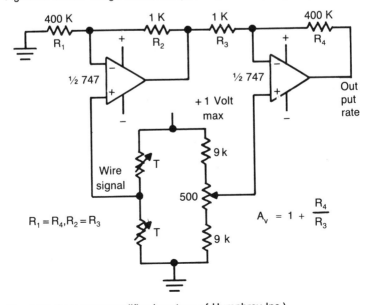

Fig. 4-24. Sensor preamplifier (courtesy of Humphrey Inc.).

The amplifier shown has a gain of 400 and the resistors must be matched to better than 0.1% for good common mode. The requirement for a good instrumentation amplifier prevents the simultaneous use for operational purposes. The avoidance of dissimilar metals and thermo-couple junctions is very important in selecting components and their location.

☐ **Simulation.** A single-axis rate transducer is a sensor with a pump, bridge, and preamplifier combination (see Fig. 4-25). If e_i, is the rate, then e_o is given by $-[R_2 \times e_i]/[R_1 \times (1 + R_2 \times C_1 \times S)]$, where $R_2 \times C_1$ is the time constant of the gas, and R_2/R_1 is the amplification of input rate to output volts. Typical gas velocity is 90 inches/second. Other parameters vary with the particular sensor.

☐ **Rate Gyros.** By adding an operational amplifier to the amplifier shown in Fig. 4-24, which is a part of the RT03 transducer, the second-order transfer function can be acquired. A rate gyro is a second-order transfer function of the form:

$$\frac{E_o}{W_i} = -\frac{K}{(1 + \dfrac{2\zeta}{W_n}S + \dfrac{S^2}{W^2_n})}.$$

Using the RT03 and network in Fig. 4-26, the above function can be simulated:

$$\text{Gain} = \frac{E_o}{e_i} = \frac{R_3 K}{R_1 + R_2}$$

The natural frequency W_n will be

$$W_n = \sqrt{\frac{R_1 + R_2}{T_x R_1 [R_2 C_1 + C_2 (R_2 + R_3)]}}$$

The damping ratio will be:

$$\zeta = \frac{W_n}{2}\left[T_x + (\frac{R_1}{R_1 + R_2})[R_2 C_1 + C_2 (R_2 + R_3)]\right]$$

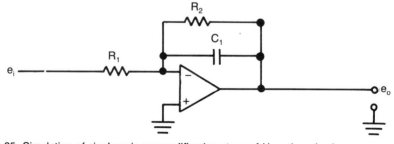

Fig. 4-25. Simulation of single axis preamplifier (courtesy of Humphrey Inc.).

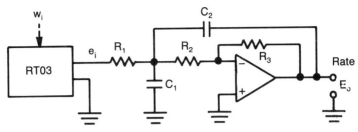

Fig. 4-26. Rate gyro simulation (courtesy of Humphrey Inc.).

The transfer function is shown in Fig. 4-27. This network and transducer will respond to steady-state rates as well as oscillatory input rates. The figure also shows a rate gyro simulation that responds to oscillatory rates, but does not pass steady-state rates. The natural frequency, W_n, of the network shown in the figure is:

$$W_n = \sqrt{\frac{1}{T_e R_1 C_1}}.$$

The damping ratio:

$$\zeta = \frac{W_n}{2}(T_e + R_1 C_1).$$

The transfer function for Fig. 4-27 reduces to:

$$\frac{E_o}{W_i} = -\frac{C_1}{C_2}\frac{K}{[1 + (T_e + R_1 C_1)S + T_e R_1 C_1 S^2]}$$

□ **Integrating Rate Gyro.** The rate signal is integrated to get angular displacement from an RT03 and referring to Fig. 4-28:

$$\frac{E_o}{W_i} = \frac{K}{R_1 C_1 S}\frac{(1 + R_2 C_1 S)}{(1 + T_e S)}.$$

$$\text{Let } R_2 C_1 = T_e, \text{then } E_o = \frac{K}{R_1 C_1}\left(\frac{W_i}{S}\right).$$

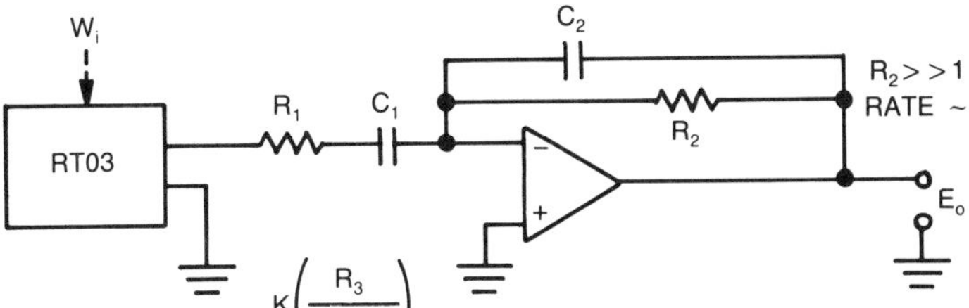

$$\frac{E_o}{W_i} = \frac{K\left(\dfrac{R_3}{R_1 + R_2}\right)}{1 + \left[T_2 + \dfrac{R_1}{R_1 + R_2}[R_2 C_1 + C_2(R_2 + R_3)]\right]S + \dfrac{T_2 R_1}{R_1 + R_2}[R_2 C_1 + C_2(R_2 + R_3)]S^2}$$

Fig. 4-27. Oscillatory rate transducer (courtesy of Humphrey Inc.).

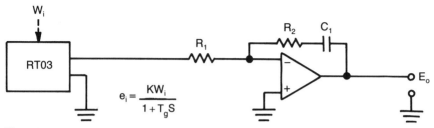

Fig. 4-28. Angular displacement measurement (courtesy of Humphrey Inc.).

☐ **Angular Accelerometer.** The rate signal is differentiated to compute angular acceleration. A typical network is shown in Fig. 4-29 and its transfer function is given by:

$$E_o = \frac{KW_i}{(1 + T_g S)} C_2 S 2 R_1 \left(1 + \frac{R_1 C_1 S}{2}\right).$$

$$\text{Let } \frac{R_1 C_1}{2} = T_r, \text{ then } E_o = K C_2 2 R_1 (SW_i).$$

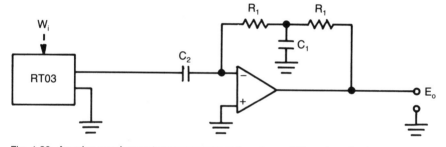

Fig. 4-29. Angular accelerometer measurement (courtesy of Humphrey Inc.).

4.5 ATTITUDE ACTUATORS

Attitude actuators are used to correct the attitude of a spacecraft such that it attains and stays in the desired attitude. Several attitude actuators are available that can be used depending upon the mission and payload requirements. These can be passive or active and within active types some use expendables such as hydrazine gas. Momentum or reaction wheels or gyros, magnetic torquers, reaction control engines or gas jets, and nutation dampers are some of the actuators.

4.5.1 Momentum and Reaction Wheels

Momentum and reaction wheels are devices for the storage of angular momentum, which are used on spacecraft for several purposes, namely to add stability against disturbance torques, to provide a variable momentum to allow operation at one revolution per orbit for Earth-oriented missions, to absorb cyclic torques, and to transfer momentum to the satellite body for the execution of slewing maneuvers. These devices depend on the momentum of a spinning wheel, $h = I \times W$, where I is the momentum of inertia about the rotation axis and W is the angular velocity. In general, a momentum wheel consists of a housing containing a flywheel, bearing assembly and electric drive motor together with electronics for driving the wheel and controlling and measuring the wheel angular rate. The motor produces a net acceleration torque about the flywheel axis. At its nominal speed, the wheel has certain angular momentum. The wheel speed controller operates in one of two modes: constant wheel speed or torque control. In the constant wheel speed control mode, the wheel angular rate is automatically maintained at a desired value. In the torque control mode, wheel speed is modulated through closed loop control in response to an error signal in order to maintain pointing towards a particular direction within the specified limits. The loop controller provides continuous control of wheel speed over a range. Momentum wheels can be broadly divided, depending upon the application, into the following categories.

Flywheel or Inertia Wheel. It is defined as a rotating wheel or disk used to store or transfer momentum.

Momentum Wheel. It is a flywheel designed to operate at a biased nonzero momentum. It provides a variable-momentum storage capability about its rotation axis, and is usually fixed in the spacecraft.

A momentum bias design is common for dual-spin Earth orbiting spacecraft. The momentum wheel is mounted along the pitch axis of the spacecraft and the pitch axis is controlled such that it is normal to the orbit plane. This allows the instruments (with certain field of views) to scan over the Earth.

Reaction Wheel. It is a flywheel designed to operate at zero bias. It contains a two-phase ac servomotor and can store angular momentum. These are designed to exhibit a relatively constant torque speed curve. This also contains a tachometer to measure its speed.

Momentum Unloading. As explained earlier, momentum wheels are used to get rigidity about a particular axis of the spacecraft. Thus, the momentum wheel axis is aligned to this axis and the wheel is run at a particular rate. But whenever magnetic torquer bars are operated to react or counteract earth's magnetic field effect, the momentum wheel absorbs momentum to maintain the spacecraft in the specified direction by increasing its speed. This process continues and the momentum wheel reaches its designed highest speed. Now the momentum wheel speed has to be brought down to its minimum value or the momentum wheel will be damaged or will fly away. Thrusters are operated to unload or reduce the speed of the momentum wheel in a few steps.

4.5.1.1 Application of Momentum Wheels

Thus in principle, momentum wheels are gyroscopic actuators designed for compensating periodic disturbing torques that act on three-axis stabilized spacecraft, particularly on geostationary communications satellites. A satellite is affected by various disturbing torques that act to change the given satellite attitude. For accurate pointing of the satellite antennas, disturbing torques have to be compensated. This is accomplished either by control thrusters consuming considerable amount of pressurized gas or by momentum wheels that are capable of compensating periodic disturbing torques during the whole operating time of the satellite.

One single momentum wheel can provide three-axis stabilization of a satellite with the two axes in the orbit plane being held in their position by the gyroscopic effect of the momentum wheel (passive stabilization). An active attitude control about the third axis, which is orthogonal to the orbit plane, is obtained by increasing or descreasing the momentum wheel speed, that is, by acceleration or deceleration torques. This is normally performed in a range of $\pm 10\%$ about the nominal speed or momentum bias.

The momentum wheel remains inoperative during the launch phase, but once the satellite is injected into orbit, the momentum wheel is spun up. The generated reaction torque can possibly be used to contribute to despin the satellite or it is compensated by thrusters. In the course of the mission, the thrusters are operated only if the nonperiodic disturbing torques have summed up and exceed the control range of the momentum wheel. Thus, the use of a momentum wheel results in a considerable decrease in propellant (gas) consumption as compared to an attitude control by thrusters only. The reduction in switching operations provides both higher accuracy and reliability of the attitude control system.

Such momentum wheels manufactured by TELDIX are employed for the stabilization of the communications satellites: SYMPHONIE (DR 25), INTELSAT V (DR 35), APPLE (DR 22), ECS and MARECS (DR 16 AND DR 25), TELECOM-1 (DR 25), TV-SAT, TDF-1 and -2, TELE-X, SKYNET-4, ETS-V, DFS, INMARSAT as well as the Maritime Observation Satellite MOS-1 and the interplanetary probes MS-T5 and the Halley's Comet mission PLANET-A. In the aforementioned type designations, the letters DR stand for DRALLRAD, the numbers indicate the nominal angular momentum measured in Newtonmeter seconds (Nms). Figure 4-30 shows a photo of a DRALLRAD Momentum Wheel manufactured by TELDIX.

4.5.1.2 Application of Reaction Wheels

Reaction wheels serve as actuators to control the angular attitude of a satellite in orbit. Their nominal speed or momentum bias is zero. By accelerating a wheel of a sufficiently high moment of inertia in one or the other direction, a reaction torque acts upon the stator of the drive motor which is fixed to the satellite structure. Thus, the satellite will start to rotate accordingly in the opposite direction.

A set of three reaction wheels mounted in an orthogonal arrangement provides attitude control about all three axes of the satellite. With a skewed configuration of the fourth wheel, full redundancy can be obtained. One or two reaction wheels can also be combined with one momentum wheel to provide defined tilting of the momentum vector and, in consequence, of the satellite.

Reaction wheels are especially advantageous for satellites that require a variable though well defined attitude such as astronomical or earth observation satellites. Such reaction wheels manufactured by TELDIX have been employed

Fig. 4-30. DRALLRAD momentum wheel (courtesy of TELDIX).

on the Dutch/American Infra-Red Astronomical Satellite IRAS (RSR 14), ESTEC and DFVLR (RSR 12) and Nippon Telephone & Telegraph NTT (RSR 12) and the German X -Ray Satellite ROSAT. (RSR stands for Reaktions-Schwung -Rad, the numbers indicate the angular momentum storage capacity in Nms).

4.5.1.3 Design

The main subassemblies of momentum and reaction wheels are the rotor, the bearing unit, the drive motor with its electronics and the housing with baseplate.

Rotor. If a particular angular momentum storage capacity is required and the maximum speed is selected to keep the motor power consumption within reasonable limits, the rotor mass and shape have to be optimized such as to obtain a high ratio of moment of inertia versus mass. A good solution consists of an annular mass designed as a disc wheel or as a spoked wheel.

The small disc wheel as shown in Fig. 4-31 is a special design for the lower angular momentum range up to about 20 Nms. The spoked wheel as shown in Fig. 4-32 is used with an angular momentum range from 14 to about 200 Nms. It is normally equipped with damping rings being clamped to the spokes. They considerably reduce the amplitude and thus the load on the ball bearings when the flywheel is vibrated, particularly at resonance frequencies.

Bearing Unit. In order to obtain a long life of the rotating device in orbit the design of the bearings is of the utmost importance. Teldix has developed a special bearing unit with preloaded angular contact ball bearings for the

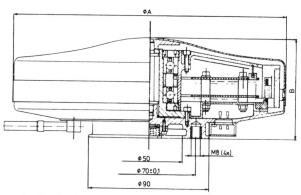

Fig. 4-31. Example of a flywheel with disc-type rotor (courtesy of TELDIX).

momentum and reaction wheels DRALLRAD, which is capable of providing a lifetime of 10 years or more at extremely low friction.

Precision ball bearings show a theoretical life of several decades if lubricated adequately. In applying the cleansing and lubricating procedure developed by Teldix a few milligrams of highly refined mineral oil will suffice to operate the momentum wheels for many months. After that time it will be desirable to start supplying additional lubricant. For this purpose, that bearing unit contains the integrated lubrication system (known as TELDILUB). With this system, the centrifugal force causes additional lubricant to creep from the reservoir slowly towards the ball races. It will arrive there only after several months of operation, i.e., when it may actually be needed to maintain the lubrication film between the moving parts. Henceforth, a well defined amount of lubricant (a few milligrams per year) will flow continuously to the bearings preventing any wear and without noticeably increasing the friction. The oil remains within the bearing unit because its migration or outgassing is prevented by labyrinth seals with an oil repelling surface. Thus, the actual lifetime approaches the theoretical value.

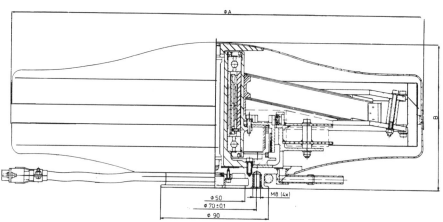

Fig. 4-32. Example of a flywheel with spoked rotor (courtesy of TELDIX).

The ball bearings are exactly the same for flywheels from 25 to 35 cm in diameter, while the length of the bearing unit is matched to the size of the wheel. Due to the comparatively large diameter of these ball bearings, the balls reach even at a rather low rotational speed of the wheel a sufficiently high rolling speed to produce a lubrication film. Thus, the metallic contact between balls and races is limited to a very short time at the beginning of run up. These bearings withstand the vibrations at launch so that no caging mechanism is required. The bearing units can be equipped with a thermistor for temperature monitoring.

Drive Motor With Electronics. The permanent magnets on the rotor provide a strong magnetic field. Opposite to them on the stator there are three-phase windings embedded in fiber reinforced resin (ironless version) or wound on sheet metal cores. Commutation is accomplished electronically, including speed reversal in case of reaction wheels. The associated electronic circuits are included in the wheel housing. Output signals for speed and speed direction can be provided.

These brushless dc motors have been successfully employed in a great number of applications. The design assures a minimum torque ripple at all speeds. Especially for reaction wheels, the ironless version is recommended to avoid any magnetic cogging. An asynchronous ac motor is also employed and it could be advantageous for reaction wheels depending on the attitude control concept.

Housing. A lightweight evacuated housing protects the wheel from adverse environmental influences.

Wheel Drive Electronics. The wheel drive electronics as shown in Figs. 4-33 and 4-34 constitutes the interface between the spacecraft's attitude control electronics and the momentum/reaction wheels. It receives the commands from the control electronics and determines the torques, speeds, and speed directions of the wheels with an accuracy of up to 0.1%. The electronics can be equipped with 1, 2, or 3 channels as needed for spacecraft equipped with up to four momentum or reaction wheels or with a combination of both.

The wheels may be operated either with torque control or with speed control. If speed control is required a digital speed control loop is included in the wheel drive electronics. The electronics is equipped with isolated dc/dc converters to produce the required voltages, if the source voltage is not regulated. Table 4-9 presents the technical data of wheel drive electronics.

Sizes. In order to provide near optimum solutions for all kinds of satellites Teldix DRALLRAD wheels are available in five different diameter classes (diameters measured in centimeters) and Table 4-10 presents the typical technical data. The sectional parts of the outline drawings show the disc type rotor for the small wheels (20 and 26 cm diameter) and the spoked rotor for the large wheels (35, 50, and 60 cm diameter). The mounting interface is identical for all wheel classes. The mounting surface is orthogonal to the angular momentum vector within one minute of arc. Digital or analog tachometer signals as well as ball bearing temperature signals are available.

Fig. 4-33. Wheel drive electronics (courtesy of TELDIX).

4.5.1.4 Magnetic Bearing Wheels

Reaction wheels with ball bearings have been developed using the space proven bearing design. Although the stiction torque of these precision bearings is extremely low there may be applications where the torque step at zero speed

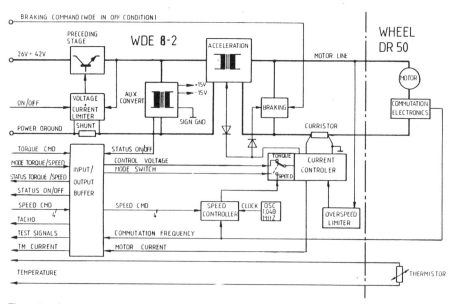

Fig. 4-34. Schematic of typical wheel drive electronics (courtesy of TELDIX).

Table 4-9. Wheel Drive Electronics Specifications (Courtesy of TELDIX).

WDE typ Application Associated wheels		WDE 1-0 (dual channel) INTELSAT V DR 35	WDE 5-0 (dual channel) TV-SAT/TDF-1 TELE-X, DR 50	WDE 8-2 (single channel) DFS DR 50	WDE 9-0 (dual channel) ROSAT RSR 25
Weight	kg	2.25	3.1	1.9	3.9
Dimensions					
– length	mm	210	205	205	204
– width	mm	160	145	133	184
– height	mm	130	160	118	170
Supply voltages	VDC	+ 5 ± 15 + 50	+ 50 ± 2% main bus ***	26–42 main bus ***	25–33
Signal interface		digital/analog	digital/analog	digital/analog	digital
Operational modes			speed command or reaction torque command		
Power consuption at constant speed of the wheels					
– WDE	W	5	5	6	5
– wheels	W	10*	8*	9	12
– total	W	15	13	15	17**
Peak power consumption at max. torque and max. speed of the wheels					
– WDE	W	19	20	30	20
– wheel	W	60*	70*	70	130**
– total	W	79	90	100	150
Environmental conditions			suitable for satellites compatible with launchers such as ARIANE or Space Shuttle		

*) One wheel active, second wheel cold redundant
**) Two wheels active simultaneously
***) Isolated DC/DC converters

Table 4-10. DRALLRAD Momentum Wheel Technical Data (Courtesy of TELDIX).

Wheel diameter	cm	20	26	35	50	60*)
Angular momentum range	Nms	1.8...6.5	5.0...20	14...80	50...300	200...1000
Max. reaction torque	Nm	0.2	0.2	0.2	0.3	0.3–0.6
Speed***)	min⁻¹	6000	6000	6000	6000	6000
Loss torque at max. speed**)	Nm	≤ 0.012	≤ 0.013	≤ 0.015	≤ 0.022	≤ 0.07
Power consumption:						
– steady state (depending on speed)	W	2...7	2...8	2...10	3...15	10...50
– max. power rating	W	≤ 60	≤ 80	≤ 100	≤ 150	≤ 500
Dimensions:						
– diameter A	mm	203	260	350	500	600
– height B	mm	75	85	120	150	180
Weight	kg	2.7...3.4	3.5...6.0	5.0...8.0	7.5...12	20...37
Environmental conditions:						
– operating temperature – vibration (sinusoidal) – vibration (random) – linear acceleration			suitable for satellites compatible with launchers such as ARIANE or Space Shuttle			

*) under development
**) with ironless motors
***) Max. speed of reaction wheels, nominal speed of momentum wheels (control range ± 10%)

must be much smaller or even zero. For such applications the *Magnetic Bearing Reaction wheel* offers significant advantages. Such a wheel developed by TELDIX is described in detail.

Design. Some of the Design goals of the Magnetic Bearing wheel are pancake profile, low volume, low weight, high tilting stiffness, minimum stiction and ripple torques, no caging for launch, low cost, and high reliability.

Bearings. In principle, suspending a rotating mass by means of magnetic forces requires a five-degree of freedom suspension and control system. Magnetic forces can be generated either using permanent magnets or by electronically controlled electromagnets.

The wheel uses a passive suspension in axial direction, which, at the same time, stabilizes the wheel against tilting (i.e., one translatory and two rotatory degrees of freedom). In radial directions, however, active control loops are applied for the two remaining translatory degrees of freedom. The design is shown in Figs. 4-35, 4-36, and 4-37. Passive suspension is achieved by four samarium-cobalt magnets mounted between pole shoes. The flux generated by the axially magnetized permanent magnets extends across the pole shoes, the air gap, and the U-shaped feedback ring, which is part of the rotor. Secondary forces being normal to the radial attraction forces adjust the feedback ring in axial direction. Thus torques that oppose tilting movements of the rotor about its diametral axes are also obtained.

The same pole shoes are used as cores for eight solenoids, the flux of which is superimposed on that of the permanent magnets. Thus, a quasi-linear relationship between the current in the solenoids and the radial forces is obtained that facilitates the active control of the rotor's radial position. A set of four pickups senses the rotor position and controls the solenoid current via servo amplifiers. Since it is possible to generate the required force by use of only two coils per axis, sufficient redundancy is attained.

Pole shoes and the feedback ring are made of soft iron. This results in a fairly homogeneous field in the air gap that keeps friction torques (due to eddy currents) to a minimum. This simple pancake shaped design offers a

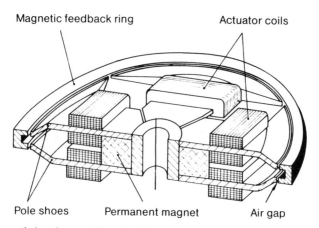

Magnetic feedback ring   Actuator coils

Pole shoes   Permanent magnet   Air gap

Fig. 4-35. Magnetic bearing reaction wheel isometric sectional view (courtesy of TELDIX).

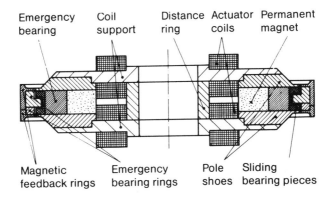

Emergency   Coil   Distance Actuator Permanent
bearing   support   ring   coils   magnet

Magnetic   Emergency   Pole   Sliding
feedback rings bearing rings   shoes bearing pieces

Magnetic Bearing Assembly for Reaction wheel Type MRR
(Magnet and coil arrangement forms part of the stator,
feedback ring is fixed to the rotor)

Fig. 4-36. Magnetic bearing assembly for reaction wheel type MRR (courtesy of TELDIX).

remarkably high tilting stiffness and is especially suited for one-side mounting in the satellite.

Emergency Bearing. In order to avoid damage of stator and rotor in case of power failure, an emergency bearing is provided that assures a smooth run down of the rotor. It consists of a T-shaped sliding ring mounted inside the feedback ring protruding into the groove of its counterpart between the pole shoes of the stator. After a power interruption at maximum speed, the reaction wheel is safely decelerated and can be levitated and started again with the power returning. In addition, the emergency bearing is designed such as to act as a mechanical stop for the rotor during launch and to damp its vibrations. Thus, a separate caging mechanism is not required.

Rotor. The angular momentum storage capacity of the reaction wheel type MRR-2 is 2 Nms and the maximum speed is $\pm$ 3200 revolutions per minute. With these figures, a high ratio of moment of inertia versus mass can only

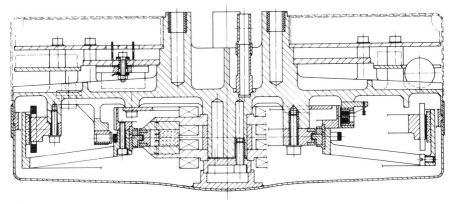

Fig. 4-37. Magnetic bearing reaction wheel (sectional view) (courtesy of TELDIX).

be achieved if a high fraction of the rotor mass effectively contributes to the moment of inertia. For this reason, the rotor consists of an annular rim that is linked to a hub by means of spokes. The hub accommodates the feedback ring of the bearing as well as the reference surface and segments for the position sensors and commutation sensors, respectively.

The permanent magnets for the motor are mounted on the inside of the rim, thus contributing to the moment of inertia, too. The reaction wheel can be enclosed in a hermetically sealed housing, which is evacuated in order to avoid air drag during testing on the ground.

Drive Motor. The drive motor is an ironless and brushless dc motor which has been successfully employed in several similar applications. The $SmCo_5$ magnets on the rotor provide a strong magnetic field. Opposite to them on the stator there are the three-phase windings embedded in fiber reinforced resin. Commutation is accomplished electronically, including speed reversal. The ironless design of the motor assures minimum torque ripple at all speeds and avoids magnetic stiction.

Built In Electronics. The electronic circuits for motor commutation and reversal as well as for the magnetic bearing are installed in the lower compartment of the wheel housing. The electronics receives a digital speed direction command input. Reaction torque is controlled by the motor current.

Wheel Drive Electronics. The wheel drive electronics constitutes the interface between the satellite's attitude control system and reaction wheels. It receives all the necessary digital commands, interprets them and supplies the appropriate signals as well as the motor current to the wheel. Table 4-11 presents the TELDIX Magnetic Bearing Reaction Wheel Technical Data.

4.5.2 Magnetic Torquers

The magnetic torquer is a coil wound on a ferromagnetic rod and encapsulated with a protective covering. A controlled current passing through the coil develops the desired magnetic dipole moment.

These torquers are used to generate magnetic dipole moments for attitude and angular momentum control. They are also used to compensate for residual spacecraft biases and to counteract attitude drift due to environmental disturbance torques. These torquers are also used to react or counteract the earth's magnetic field.

TORQRODS. Ithaco Inc's TORQROD generates a magnetic moment with better than 1.5% linearity in the specified dipole moment range. The maximum dipole moment at increased nonlinearity and power is about 130 to 150% of the specified magnetic dipole moment. Ithaco has manufactured TORQRODS as small as 7 Am^2 to as large as 1,250 Am^2 and have designs for up to 4,000 Am^2. Table 4-12 presents details about the Ithaco's Torqrods.

Magnetic Moment Compensator. The primary application of magnetic moment sensor is to reduce disturbance torques in order to (a) conserve control gas, (b) improve the yaw accuracy of momentum biased spacecraft, and (c) reduce control power. The Ithaco's Magnetic Moment

Table 4-11. Magnetic Bearing Reaction Wheel Technical Data (Courtesy of TELDIX).

TECHNICAL DATA
Magnetic Bearing Reaction Wheel, Type MRR 2

Angular momentum (max.)	Nms	± 2
Speed (max.)	min^{-1}	± 3200
Reaction torque (max.)	Nm	± 0.1
Torque ripple	%	10
Drag torque at max. speed	Nm	$< 10^{-3}$
Stiction torque	Nm	$< 10^{-4}$
Bearings radial		electromagnetic (active)
axial		permanent-magnetic (passive)
tilting		permanent-magnetic (passive)
emergency		sliding
Bearing stiffness radial	N/mm	110
axial	N/mm	75
tilting	Nm/rad	> 75
Tilt damping		0.01
Slew rate	deg/s	$\leqq 5$
Caging for launch		none
Power consumption steady state	W	3
at max. torque	W	< 50
Dimensions of housing diameter	mm	< 220
height	mm	< 90
Weight incl. electronics	kg	3
Reliability (10 years) without redundancies		> 0.96
Environmental conditions operating temperature		suitable for satellites
vibration (sinusoidal)		compatible with launchers
vibration (random)		of the ARIANE class.
linear acceleration		

Compensator utilizes chargeable permanent magnets that are activated by ground command. Any moment between zero and 14,000 pole-cm in each of three axes can be achieved, with a resolution of about 300 pole-cm. Once the desired magnetic moment is achieved, the magnetic moment compensator will retain that moment indefinitely with power off. A Hall sensor monitors the moment in each magnet for telemetry. Table 4-13 presents specifications of various magnetic moment compensators.

4.5.3 Nutation Damper

The Nutation Damper is a passive, viscous damper consisting of a straight tube partially filled with mercury and is usually mounted through the spacecraft platform or around the deckplate. This is designed to dissipate nutations about the spin axis caused by torques imparted by the launcher last stage

Table 4-12. Details of Ithaco's TORQRODS (Courtesy of Ithaco Inc.).

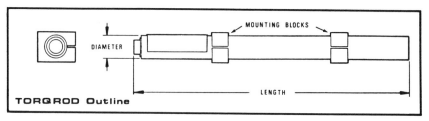

TORQROD Outline

DIPOLE MOMENT [Am²]		SIZE [cm]		WEIGHT [kg]	POWER * [W]	
Linearity				Includes mounting	Linearity	
1%	20%	length	diameter	blocks †	1%	20%
10	15	40	1.8	0.4 †	0.6	1.0
15	20	45	1.8	0.5 †	0.6	1.5
20	30	49	1.9	0.6 †	0.7	1.7
30	50	56	2.1	0.9 †	0.7	1.8
60	85	64	2.6	1.7 †	0.8	2.0
100	150	72	3.6	2.8	1.1	2.7
150	250	84	3.8	3.2	1.3	3.5
250	350	104	4.3	6.2	1.8	4.4
350	500	115	4.7	8.3	2.1	5.0
500	700	130	5.0	11.1	2.3	5.5
1,250	1,750	200	5.3	18.5	3.3	7.6
2,900	4,000	250	7.6	49.9	6.0	16.0

1 Ampere meter² = 1000 p−cm   *When a single winding is used, power doubles.

separation, thrust engine firing, apogee motor firing, and station acquisition by high thrust engine firing.

Fokker Nutation Damper. As part of its basic construction, a nutation damper always contains a constrained movable mass, which moves in a viscous medium. In fluid dampers the fluid is the viscous medium as well as the movable mass. The energy generated by the nutational motion is in this way transformed into heat and the motion of the rotating body becomes stable.

The configuration usually comprises, as shown in Fig. 4-38, two partially filled reservoirs connected by two tubes, one for liquid, the other for gas above the liquid. Use of advanced electron-beam welding techniques for the joints and severe leak testing in vacuum guarantee hermetical sealing. Figure 4-39

Table 4-13. Magnetic Moment Compensator Specifications (Courtesy of Ithaco Inc.).

Specifications

APPLICATION		Spacecraft residual dipole moment compensation
MAXIMUM DIPOLE MOMENT		14,000 p−cm (other sizes also available)
INCREMENTAL CHANGE		> 300 p−cm
SIZE	Magnets −	312.17 mm × 25.40 mm (diameter)
	Electronics −	178.56 mm × 153.16 mm × 44.45 mm
WEIGHT	MMC	1.1 kg
POWER	MMC	< 1.0 W

Fig. 4-38. Nutation damper schematic
(courtesy of Fokker).

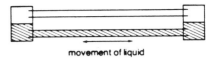

movement of liquid

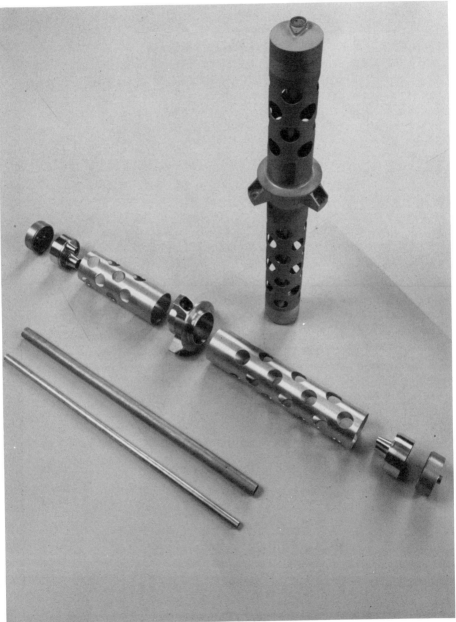

Fig. 4-39. Photograph of a nutation damper manufactured by Fokker (courtesy of Fokker).

shows the photograph of the nutation damper manufactured by Fokker. The damper is positioned either parallel (meridional mounting) to, as shown in Fig. 4-40, or perpendicular to the satellite spin axis (equatorial mounting).

Applications. For reasons of dynamic balance or redundancy two or more dampers are sometimes used. Damper dimensions and weight depend on spin rate, mass properties of the satellite, position of damper in the satellite and damping performance requirements. The British Scientific Satellite weighing about 130 kg, launched in 1974 and with a spin rate of 8 rpm, required a nutation damper with a damping time constant of 25 minutes. The damping time constant is the time in which the nutation angle is reduced by the factor e (= 2.7). Two dampers manufactured by Fokker with a length of 53 cm and a mass of 220 grams each were mounted parallel to the spin axis to meet above requirement. The ISEE-B scientific satellite weighing about 170 kg, launched in 1977 and with a spin rate of 20 rpm, required nutation damper with a damping time constant of 60 minutes. One equatorial damper manufactured by Fokker with a length of 21 cm and a mass of 225 grams was employed to meet the above requirement.

4.5.4 Thrusters

Thrusters are small rockets or jets whose ON and OFF times can be controlled continuously and these are part of the Propulsion system which is dealt with in Chapter 6.

4.5.5 Scanwheel

The concept of a scanwheel evolved from recognition of the fact that many earth-pointing three-axis attitude control systems employ both horizon sensors and reaction wheels. Combining the horizon scanner with the momentum wheel has clear advantages, including fewer bearings, lower weight, lower power, and lower cost.

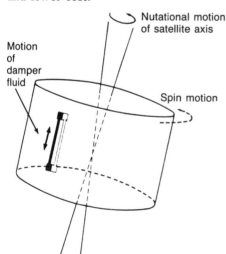

Nutational motion of satellite axis

Motion of damper fluid

Spin motion

Fig. 4-40. Meridional mounting of nutation damper (courtesy of Fokker).

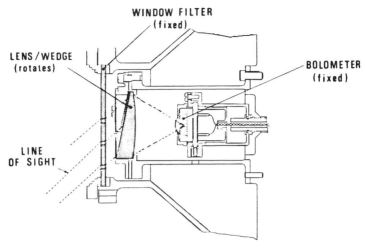

Fig. 4-41. SCANWHEEL optical diagram (courtesy of TELDIX).

The scanwheel is a variable momentum wheel, powered by a two-phase induction motor, with an integral IR system. The rotor is outside the stator, to maximize momentum. The angular position of the rotor is detected by a set of magnetic pickups. This information is utilized to compute attitude and to electronically blank the upper part of the scan that might intersect with spacecraft projections.

The optical system consists of a bolometer, a rotating prism, lens, and a germanium window that is coated with an interference filter to establish the optical passband. The scanwheel optical diagram is shown in Fig. 4-41 and the scanwheel on-orbit geometry is shown in Fig. 4-42. Table 4-14 presents the Ithaco's Scanwheel Specifications.

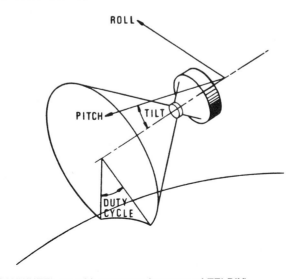

Fig. 4-42. SCANWHEEL on-orbit geometry (courtesy of TELDIX).

Table 4-14. Scanwheel Specifications (Courtesy of Ithaco Inc.).

SCANWHEEL Specifications

	TYPE A	TYPE B	TYPE C
Angular Momentum @2000 rpm	1.25 N-m-s	5.69 N-m-s	14.22 N-m-s
Operating Speed Range	300 – 2400 rpm	300 – 2400 rpm	300 – 2300 rpm
Motor	2 phase AC	2 phase AC	2 phase AC
Power @ 900 rpm @25°C	2.5W	3.5W	5.0 W
Stall Power (max)	20W	30W	30W
Torque Capability (typical)	25 mN–m	25 mN–m	25 mN–m
Scanner Type	Conical	Conical	Conical
Optics	14 – 16 μ	14 – 16 μ	14 – 16 μ
Altitude Range *	150 – 500000km	150 – 50000 km	150 – 50000 km
Weight (typical)	3.75 kg	6.8 kg	11.12 kg
Size	See outline drawings		

* should be tilted for altitudes > 1500 km

For high accuracy or elliptical orbits, two scanwheels are recommended. The Delta Pac, SAS-3, ELMS, HCMM, MS-T4, ASTRO B, and EXOS-C attitude control systems employed a single scanwheel each, whereas Nimbus, LANDSAT, P72-2, SAGE, SEASAT-A and ERBS employed two scanwheels each.

4.6 ATTITUDE CONTROL ELECTRONICS

The Attitude Control Electronics is the heart of the Attitude Control system. It provides the interface with attitude measuring sensors to obtain attitude information, interfaces with actuators to command them to operate in a controlled manner to change the spacecraft attitude to the desired (final) attitude and to maintain in that desired attitude. A typical functional block diagram of ACS electronics is shown in Fig. 4-3.

4.7 ATTITUDE CONTROL SYSTEM EXAMPLES

Following are some of the examples of the attitude control systems developed using the attitude measuring sensors and attitude actuators in different combinations. Ithaco, Inc. manufactures various components for the attitude control system. It has also developed various attitude control systems and some of them, namely Magnetic Acquisition/Despin System, Three-Axis Attitude Control System, and Spinner Magnetic Attitude Control System are described here. Also described is an attitude Control System for a communication satellite in geosynchronous orbit.

4.7.1 Magnetic Acquisition/Despin System

The *Magnetic Acquisition/Despin System* is an automatic despin and coarse acquisition system for three-axis stabilized spacecraft. It can be used either as a primary system, or as a backup acquisition system for a spacecraft that is inadvertently tumbling about undefinable axes at arbitrary rates. This system has been successfully used for initial despin and acquisition on both the HCMM (Heat Capacity Mapping Mission) and SAGE (Stratospheric Aerosol Gas Experiment) spacecraft as an integral part of a three-axis control system that uses no expendables.

Applications. Following are some of the applications: a) initial despin and acquisition for rotating or tumbling spacecraft; b) autonomous backup for inertial platform/star mapper control systems in the event that there is a temporary loss of control such as can be caused by a gas leak or by a computer "glitch"; c) when used in conjunction with a fixed-speed momentum wheel, the system aligns the wheel spin axis within a few degrees of the orbit normal, and reduces the pitch rate to a terminal value of twice per orbit (locked to the earth's magnetic field) and d) nutation damping for momentum biased or spinner spacecraft.

Hardware. The hardware consists of a three-axis magnetometer, three electromagnets (TORQRODS) and a control electronics assembly.

Operation. The Magnetic Acquisition/Despin System automatically reduces the rate of change of the earth's magnetic field with respect to the spacecraft in all three axes by reducing the spacecraft spin rate. This is accomplished by utilizing a three-axis magnetometer to measure the rate of change of the earth's magnetic field and then applying the proper torque to the spacecraft through the interaction of three orthogonal electromagnets with the earth's magnetic field.

Performance. This Magnetic Acquisition/Despin System is employed for initial despin and acquisition of a spacecraft in 600 km orbit with an inclination of 97.6 degrees and with moments of inertia of 14 slug ft^2 in roll, 25 slug ft^2 in pitch, and 14 slug ft^2 in yaw. The pitch wheel has 1 ft-lb-sec momentum, and the Torqrods saturate at 10,000 pole-cm.

When the acquisition mode is turned on, the spacecraft was tumbling at up to 1.5 rpm about an axis that was totally unpredictable. There was a constant pitch bias momentum in the wheel, and the spacecraft is nutating wildly. After this nutation energy is removed in about one orbit, the spacecraft pitch axis is maneuvered by 90 degrees into a position normal to the orbit plane. At the same time, momentum is removed from the spacecraft in the pitch axis. Acquisition sequence is completed in about three orbits, with the pitch axis near the orbit normal and the pitch rate at about 0.1 degrees per second (locked to the earth's field).

The acquisition time is directly proportional to the amount of angular momentum that must be removed from the spacecraft, and inversely proportional to the size of the Torqrods. The time will also increase somewhat as the orbit inclination decreases. If this acquisition time is not acceptable for a particular mission, then an expendable (gas) propulsion system can be added to the attitude control system.

4.7.2 3-Axis Magnetic Attitude Control System

The system described is a complete three-axis attitude control system for low orbit (up to 2000 km) spacecraft, and has a pointing accuracy of better than 0.5 degrees in all axes. It requires no expendables (gas), and has an acquisition capability that is essentially independent of initial conditions (i.e., error angle or spin rate about any or all axes).

Applications. This control system was employed as the primary control system, on the HCMM (Heat Capacity Mission) and the SAGE (Stratospheric Aerosol Gas Experiment). Both are performing successfully in orbit. It also can be employed as backup control system for high accuracy inertial platform/star tracker systems in the event that there is a temporary loss of control, such as can be caused by a gas leak or a balky computer.

Hardware Required.

☐ **Scanwheels:** For horizon sensing, momentum bias, and pitch control. For circular orbits, a single Scanwheel is used whereas a second Scanwheel may be required for elliptical orbits depending on roll/yaw accuracy requirements.

☐ **Magnetometer.** For three-axis magnetic field sensing.

☐ **Three TORQRODS.** For applying torque to the spacecraft through interaction with the earth's magnetic field.

☐ **Control Electronics.** For all electronic functions associated with attitude control.

Operation.

☐ **Despin/Acquisition Mode.** In the despin/acquisition mode, the system automatically reduces the rate of change of the earth's magnetic field with respect to the spacecraft in all three axes by reducing the spacecraft spin rate. This is accomplished by utilizing a three-axis magnetometer to measure the rate of change of the earth's magnetic field, and then applying the proper torque to the spacecraft through the interaction of three orthogonal electromagnets with the earth's magnetic field. At the start of the acquisition mode, the scanwheel is turned on at a fixed speed. This provides momentum bias in the pitch axis, which in effect provides a roll/yaw reference through the dynamics of the spacecraft motion. The terminal condition in the acquisition mode is for the pitch axis to approach the orbit normal, and the pitch rate to reduce to about 0.1 degrees per second (locked to the earth's magnetic field). The horizon scanner is not used in the acquisition mode.

☐ **On-Orbit Mode.** When the terminal conditions of the acquisition mode are approached, the on-orbit mode is switched in by command. The horizon sensor, an integral part of the scanwheel, is actuated and begins sensing pitch and roll. The pitch loop is closed around the scanwheel and the earth lock in pitch occurs within seconds. The roll error is combined with the magnetic field data from the magnetometer and the pitch TORQROD is actuated in the proper polarity to null the roll error. Yaw is controlled through the dynamic stiffness of the system that results from the bias momentum in the scanwheel. The scanwheel is simultaneously held at a near constant speed by monitoring its tachometer and applying the appropriate magnetic torque in the roll/yaw plane. Nutation is prevented by sensing the magnetic field rate in the pitch axis and utilizing the pitch TORQROD to oppose this motion.

☐ **Orbit-Trim Mode.** This control system configuration readily facilitates ±90 degree pitch maneuvers. Thus a single nozzle is used to circularize the orbit of the spacecraft by pointing it either fore or aft depending on the circumstances. The initial circularization trim is done over a several orbit period in short pulses in order to minimize the effects of nozzle offset errors.

Performance. This system is employed for a spacecraft in 600 km orbit with an inclination of 97.6 degrees and with moments of inertia of 14 slug ft^2 in roll, 25 slug ft^2 in pitch, and 14 slug ft^2 in yaw. The pitch wheel has 1 ft-lb-sec momentum, and the Torqrods saturate at 10,000 pole-cm.

☐ **Acquisition Mode.** As the Magnetic Acquisition/Despin System is employed for this mode, the performance is the same as described in Section 4.7.1.

☐ **On-Orbit Mode.** Pointing within 0.3 degrees in all axes is obtained using this system and pitch motion transients due to the sun of up to 0.8 degrees in pitch (much lower in roll and yaw). Normal spacecraft rates did not exceed 0.01 degrees per second. Table 4-15 presents the Ithaco's three-Axis Magnetic Attitude Control System Hardware Summary.

The acquisition time is directly proportional to the amount of angular momentum that must be removed from the spacecraft, and inversely proportional to the size of the TORQRODS. The time will also increase somewhat as the orbit inclination decreases. If this acquisition time is not acceptable for a particular mission, then an expendable (gas) propulsion system can be added to the attitude control system.

**Table 4-15. 3-Axis Magnetic Attitude Control
System Hardware Summary (Courtesy of Ithaco Inc.).**

Hardware Summary

	WEIGHT	POWER
SCANWHEEL A	3.7 Kg	1 Watt (600 RPM)
SCANWHEEL B	6.8 Kg	2.6 Watts 1900 RPM (4 ft lb sec)
ELECTRONICS (Single SCANWHEEL)	3.4 Kg	5.3 Watts
ELECTRONICS (Dual SCANWHEEL)	4.3 Kg	8.0 Watts
TORQROD (10,000 PCM) *	.35 Kg	.7 Watts
MAGNETOMETER	.5 Kg	.7 Watts
TOTAL:		
Single Type B SCANWHEEL System	12.0 Kg	9.5 Watts
Dual Type B and Type A SCANWHEEL System	16.4 Kg	13.0 Watts

* Other sizes available

4.7.3 Spinner Magnetic Attitude Control System

The spinner Magnetic Attitude Control System performs automatically all of the functions required to maintain the orientation and spin rate of a low orbit spinning spacecraft whose spin axis is to be normal to the orbit plane.

Hardware.

☐ **Horizon Crossing Indicators (HCI).** Two HCIs are required, mounted in the same axial plane, each pointing approximately 45 degrees away from the orbit normal in opposite directions as shown in Fig. 4-43. The functions of the HCIs are, sensing for spin axis orientation, commutation of magnetometer for magnetic field measurements and sensing of earth position in the rotation cycle of the spacecraft.

☐ **Magnetometer.** A three-axis magnetometer is required to measure the instantaneous magnetic field and/or field-rate components.

☐ **TORQRODS.** Two electromagnet TORQRODS are required, one with 30,000 pole-cm for pitch axis and is used for precession and nutation control in both the acquisition and on-orbit modes and the second with 10,000 pole-cm for spin rate. This electromagnet is used to maintain the spin rate of the spacecraft.

☐ **Control Electronics.** The control electronics contain horizon-crossing indicator processor circuits, magnetometer electronics, magnetic control logic, power supply, and roll and spin attitude computer electronics.

Hardware Configuration. Figure 4-40 shows the hardware to be configured on the spacecraft. The primary constraints are:

a. The sensors should be at equal and opposite angles (approximately 45 degrees) with respect to the orbit plane. This angle should be chosen to give the best scale factor for a given orbit altitude.

b. The sensors should be at the same angles as viewed from the axis of rotation of the spacecraft so that they will, under null conditions, intersect the earth horizon at exactly the same time.

c. The relative angles of the HCI to magnetometer mounting in the orbit plane should be such that when the sensors cross the space/earth horizon, the magnetometer is oriented such that the vertical and horizontal components of the earth's magnetic field can be measured by sampling the magnetometer outputs with the space/earth pulse.

d. The three-axis magnetometer probe must be physically separated from the magnets as much as possible, and placed in a position to minimize interaction between the pitch magnetometer and the pitch magnet. Trimming out residual effects after mounting on the spacecraft is provided for by means of a trim plug on the electronics box.

e. Control torques depend on orbit inclination, being strongest for polar orbits. As orbit inclination decreases, the available torque decreases and stronger magnets may be required. The system is not suitable for a spacecraft in an equatorial orbit.

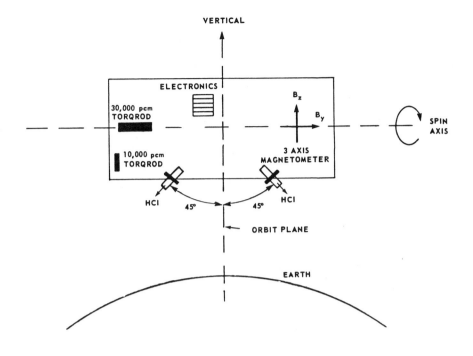

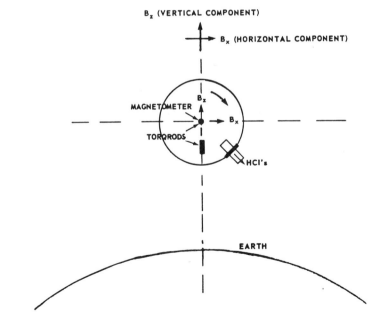

Fig. 4-43. Sensors mounting configuration at equal and opposite angles (courtesy of TELDIX).

Operation.

☐ **Acquisition Mode.** Automatically damps nutation and moves spin axis to a position that is within a few degrees of the orbit normal, regardless of the initial conditions.

☐ **On-Orbit Mode.** Automatically maintains the spin axis to within < 1% of the orbit normal. Automatically damps nutation to < 0.1 degree cone angle and automatically maintains spin rate to ± 1% of nominal value.

☐ **Backup Mode.** The spin axis (pitch) TORQROD is settable by command to any level from 0 to ± 100% of the rated moment in ten discrete steps. The spin axis orientation control loop is turned off in this mode. The spin rate control loop can be opened by command. Spin rate control can then be accomplished by single level plus and minus ground commands.

Though normally two horizon sensors are employed, a single sensor mode permits operation using a single sensor at reduced accuracy. This mode is selectable by ground command.

4.7.4 Attitude Control System for a Communications Satellite in Geosynchronous Orbit

In this case, the spacecraft attitude control system is designed to support two distinctly different mission phases. In the spinning phase, attitude control of the rigid body spacecraft is achieved by spinning the spacecraft at a predetermined spin rate. Gyroscopic stability is assured by proper mass distribution. During the in-orbit phase, when the satellite appendages are deployed, i.e., sun-tracking solar arrays, earth pointing antennas, etc., three-axis stabilization and control are accomplished by means of a set of actuators. The actuators commonly employed for this purpose are momentum wheels, magnetic torquers, thrusters, etc.

Spinning Phase. During the transfer orbit and station acquisition phases, the spacecraft is spun at a predetermined or calculated rpm about its maximum moment of inertia axis. Attitude is determined by employing appropriate attitude measuring sensors like a earth sensor and a digital sun sensor or some other combination of sensors and using orbital position (altitude, inclination) data. Axial and radial thrusters with appropriate thrust (kgf) rating are used to provide necessary attitude and orbit adjustments. Stabilization and control during the spinning phase can be easily achieved by proper mass distribution such that pitch axis/roll axis inertia ratio, (I_{zz}/I_{xx}), is greater than 1.3.

Attitude Acquisition Phase. During attitude acquisition phase, the spacecraft is transferred from spin-stabilized state to the on-station three-axis stabilized state. This is achieved by carrying out the following maneuvers:

a. Despin (together with passive or active nutation damping).
b. Deployment of solar arrays.
c. Acquisition of sun and sun autotracking.
d. Acquisition of the sun line into the spacecraft negative roll axis.

e. Spinup of the momentum wheel.

f. Acquisition of the earth by using earth sensor.

g. Thruster firings.

Digital sun sensor, three-axis gyro, and earth sensors are used during the acquisition phase.

Three-Axis Attitude Stabilized Phase. In the on-orbit operation phase for the duration of the mission, the spacecraft attitude is three-axis stabilized by the attitude control system with the yaw axis pointing toward the earth, the pitch axis normal to the orbit plane, and the roll axis along the orbital velocity vector. The primary means of attitude stabilization is achieved by the momentum wheel and thrusters combination. Pitch error information measured by the earth sensor is used to control spacecraft pitch attitude through momentum wheel torque control. Pitch thrusters are used to maintain the wheel momentum at the desired level.

Chapter 5

Telemetry, Tracking, Command, and Communication System

COMMANDS ARE NECESSARY TO MAINTAIN AND OPERATE THE SATELLITES. In order to transmit the proper commands, information is needed on the satellite position and condition. Telemetry, tracking, command and communication (TTC&C) provide the means of monitoring and controlling the satellite operations. These functions are usually integrated into a single subsystem known as the TTC&C System. In case of communication satellites, TTC&C is kept separated from the main communications payload. Telemetry is the system by which measurements made at a distance are transmitted to an observer. Tracking is observing and collecting data to plot the moving path of an object. And command is the means by which control is established and maintained.

A functional diagram of the TTC&C subsystem is shown in Fig. 5-1. A command originates at the satellite control center, is sent via land lines to an earth station and then transmitted to the satellite. The command receiver in the spacecraft, receives the signal, demodulates it, and processes it. After decoding a command, a verification signal is generally passed onto the telemetry and returned to ground. After verification, an execute signal sent from the control center is received, decoded, and distributed to the proper subsystem in the spacecraft. The telemetry subsystem gathers data from various other subsystems, processes them into a desired form, modulates this signal on an rf beacon, and transmits it to earth. Tracking is done by the earth station

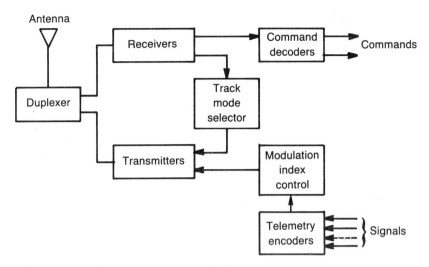

Fig. 5-1. Functional block schematic of TTC&C system.

antenna by a ranging signal sent up through the command link and returned by the telemetry transmitter.

Tracking function determines the satellite position and this information is used to calculate future orbital motion, pointing data for earth stations, and for firing thrusters to maintain satellite position. Angle tracking measures the azimuth and the elevation angles from the earth station to the satellite as a function of time, and requires only an rf beacon from the satellite. Ranging measures the actual distance (slant range) from the earth station to the satellite as a function of time.

5.1 TELEMETRY SYSTEM

Spacecraft are adequately instrumented with transducers or sensors to measure their environment and to report back to earth via the communication system. A sensor or a transducer is a sensing device that measures some physical parameter such as position, pressure, temperature, radiation, etc., and converts the measured quantity into an electrical signal. The measurement of local environment helps in determining the success of the design of individual subsystems and also can be used for ground calibration. Also this data can be used for failure diagnosis and provide information on which to base engineering changes. Thus, the telemetry system collects all the necessary data and transmits it to ground in a particular format.

5.1.1 Sensors or Transducers

The sensors can be broadly divided into (a) local environment sensors, which are used to make internal measurements onboard a spacecraft in order to evaluate the performance of the spacecraft, and (b) scientific measurement sensors, which are used to make the measurement of phenomena occurring

Table 5-1. Sensing Functions and Sensors or Transducers.

Function/Parameter	Sensor/Transducer	Comments
Position	Potentiometer	A constant current sent thru potentiometer, simple
Pressure	Bellows and potentiometer	A constant current sent thru potentiometer, simple
Temperature	Thermistor	Uses bridge amplifier, simple, low accuracy
	Platinum Wire	Uses bridge amplifier, simple, highly accurate
Speed	Tachometer	Direct, simple
Frequency	Counter	Simple
Structural stress	Strain gauge and bridge	
Power supply voltage	Resistor divider	a) Direct, simple, b) Voltage follower amplifier,simple, offers low output impedance
	Voltage controller oscillator	Relatively complex
	Magnetic oscillator	For isolated monitoring
Dc Current	Shunt	Analog amplifier, Simple
	Hall effect current amplifier	For isolated monitoring
Vibration	Piezoelectric crystal	
Acceleration	Accelerometer	
Angular displacement	Accelerometer	
Angular displacement	Free gyro	
Angular rate	Rate gyro	
Propellant flow rate	Flow meter	

either in the immediate vicinity of the spacecraft or in some other region of space. Some of the local environment measurement sensors are temperature, pressure, voltage, current, vibration, and acceleration sensors. Sensors or pickups can be either self-generating or may depend on the use of external electric circuits to supply a voltage. Among self-generating types are: (a) piezoelectric crystals, which are often used as vibration pickups because of their high frequency response, (b) electromagnetic generators used in tachometers and thermocouples, which produce a voltage when heat is applied to one junction. Non-self-generating types are such devices as variable resistances (potentiometers and strain gauges), variable inductances, and variable capacitors. There are also other special types such as photoelectric pickups. The capability of a pickup to produce an output that is a correct analog of the input is defined by a number of characteristics, i.e., a) linearity, b) sensitivity, c) accuracy, d) resolution, and e) frequency response. Some typical spacecraft functions and the appropriate transducers are presented in Table 5-1.

There are a fairly large number of scientific measurement sensors. As these sensors are housed and their signals are amplified before any further use, these sensors are also known as scientific measurement instruments. Some of them are: (a) high-energy particle and photon measuring instruments (Geiger-Mueller counter, and proportional counter), (b) magnetic-field measuring instrument, (c) electric-field measuring instrument, (d) plasma (low energy particles) measuring instrument, (e) ionospheric measuring instrument, and (f) micro-meteorite detecting instrument.

Having measured various parameters of interest, now some methods of processing these data are described. Since simultaneous transmission of several types of data is necessary, either time or frequency multiplexing or sometimes both are employed. Either analog or digital systems may be employed.

5.1.2 Multiplexing

Because of the large number of monitoring signals to be telemetered, it is necessary to multiplex or combine the signals for transmission. In telemetry, most of the signals vary slowly with time and require only a few cycles of bandwidth. It would therefore be very inefficient to provide a separate transmitter for each signal or sensor output. The number of signals or sensors will vary depending on the payload requirements or capability, but often as many as several hundred individual measurements are made. The outputs of the sensors are amplified and multiplexed before being transmitted down. There are two methods of multiplexing, (a) time-division multiplexing and (b) frequency-division multiplexing, which can be used separately or in combination. One shall make sure that frequencies of signals or information are inside the frequency response of the multiplexing method being used.

In time-division multiplexing, the information is transmitted as a series of pulses. The outputs of the sensors are fed to a commutator, which sequentially samples each sensor output several times per second. A long master synchronization pulse is provided at the start of each cycle to maintain synchronization with the ground decoding equipment. Electronic and solid state commutators can usually multiplex 60 input channels at a rate of 5 to 40 channel samples per second. This method is somewhat limited in frequency response and is used only with low-frequency information. The pulses can be sent directly to the carrier modulator or can be used with frequency-division multiplexing.

Frequency division multiplexing makes use of voltage-controlled subcarrier oscillators. Sensor or signal voltage of varying amplitude varies the output of oscillator frequency, such that the deviation in oscillator frequency represents the signal voltage. A certain maximum deviation of frequency is allowed from the center frequency and one subcarrier oscillator is used for each signal. The outputs of the subcarrier oscillators are fed to a single amplifier where they are combined into a complex waveshape containing all the frequencies produced by the oscillators.

5.1.3 Bandwidth

The total telemetry bandwidth requirements and telemetry multiplexing allocations are determined from telemetry instrumentation plus payload(s) or experiment(s) schedule. A summary of a schedule is considered where each set of data points are converted to an equivalent PCM bit rate and the complete list is summed up to give an estimate of the required telemetry bandwidth. The total number of data points serve to size the telemetry multiplexing.

5.1.4 Description

The telemetry system employs various modulation schemes, one of which is PCM/FM/PM. In this, the PCM Encoder Unit encodes analog and digital information from various subsystems in the spacecraft by means of time division multiplexing. It performs analog-to-digital conversion and organizes all the information into a serial bit stream that frequency modulates a voltage-controlled oscillator and finally phase modulates a transmitter. Further, the telemetry encoder accepts continuously variable input signals and converts them to binary form, typically into an 8-bit or a 16-bit binary code, depending upon the resolution needed. For example a 16-bit code with a full scale value of 5.1 volts results in a quantization step of 10 millivolts. The error in the input level, which gives a particular binary code, will not deviate by more than one bit from the theoretical level. It also accepts digital data in the form of a stream of binary pulses of various word lengths and incorporates them into the telemetry format. The telemetry encoder provides to all subsystems with a continuous train of shift clock pulses and a telemetry-enabling pulse gate that indicates when the digital data are to be received.

For better understanding consider a typical PCM telemetry system, the schematic of which is shown in Fig. 5-2. It consists of eight 128-channel analog multiplexers. The main multiplexer operates at 128 samples per second (sps) per channel. However, depending upon the data characteristic, how fast or how slow the data is varying, the subframe multiplexer can be operated at more

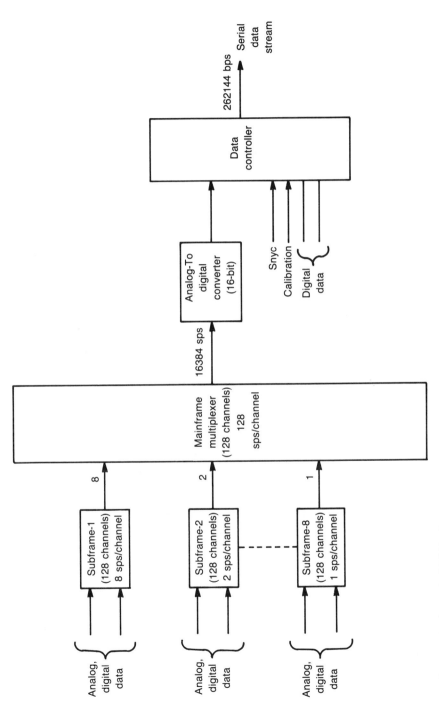

Fig. 5-2. Schematic of a typical PCM telemetry system.

or less than 128 samples per second per channel. Thus the main multiplexer puts out 16,384 analog samples per second, which are digitized to 16-bits for 0.5% resolution by employing an analog-to-digital converter. A data controller accepts this digitized analog-data train and interleaves direct digital data, and synchronization words, calibration and other words at the appropriate times into channels reserved in the main and subframes. Finally, the controller converts the parallel 16-bit words at 16,384 wps into a serial wavetrain at 262,144 bps.

Higher data rates are achieved by supercommutating in groups of 8 and 16 channels. Subframe-1 operates at 8 sps per channel and requires 8 channel supercommutation into the main frame. Subframe-2 requires four main-frame channels, and subframe-3 is directly multiplexed on a single channel. The serial data stream from PCM system is frequency modulated before being inputted into transmitter where it is phase modulated before being transmitted down.

For example, Intelsat I and II satellites used a PAM/FM/PM modulation scheme in telemetry system, whereas Intelsat VI used a PCM/PSK modulation scheme. Telemetry systems of Aryabhata and Bhaskara satellites employed PCM/FM/PM modulation schemes. Table 5-2 presents some of the typical telemetry monitoring data. All data and monitoring signals are routed to the telemetry system if the number of signals/wires are minimum. However, if the number of signals/wires are large, then it is advisable to have local multiplexers to multiplex large number of signals and run a set of wires to the telemetry system. This optimizes the overall telemetry system with respect to its performance, size, weight, etc.

5.2 TELECOMMAND SYSTEM

Mainly two types of commands are employed in spacecraft, namely, (a) value commands which are digital words of various lengths, and (b) pulse commands or also known as discrete ON/OFF relay commands. Pulse commands are those for which the command decoder will issue a single execute pulse of predetermined width typically 50 to 100 msec from the specified output channel immediately following receipt of the execute frame of the command message. Value commands contain typically, say, a 16-bit word in the command message, which is transferred by the command decoder to the specified register of a subsystem upon receipt of the execute frame. Value commands are identified as such by certain bits in the command word. The remaining bits specify the register to which the value bits are to be routed.

If commands have to be verified before they are executed, then the decoder provides outputs to the telemetry encoder to enable complete command information to be transmitted to ground for verification before the execute frame of the command message is transmitted to the spacecraft. Some of the typical commands in a spacecraft are presented in Table 5-3.

5.2.1 Description

The command system at the satellite control center/ground station employs various modulation schemes, one of which is a PCM/PSK/FM/PM. In this, the

Table 5-2. Some of the Typical Telemetry Monitoring Data.

Power Functions

Array/Load Power Disconnect Status
Battery On-Line Status
Battery Power Disconnect Status
Solar Array Bus Voltage
Unregulated Bus Voltage
Unregulated Bus Current
Battery Current
Power System Heater Power
Battery Voltage
Battery Temperature
Solar Array temperature
Other temperatures
Bus voltage
Bus current
Each subsystem current
Redundancy status

Telemetry Functions

Telemetry clock
Telemetry ready signal
Power from Power System
Redundancy Status
Temperature
Redundant System Status

Telecommand Functions

Power from Power System
Temperature
Redundancy Status
Status of Various Command Relays

ACS Functions

Power from Power System
Heater Power
Fine Error Sensor, Roll, (+)
Fine Error Sensor, Roll, (−)
Fine Error Sensor, Pitch, (+)
Fine Error Sensor, Pitch, (−)
Fine Error Sensor, Yaw, (+)
Fine Error Sensor, Yaw, (−)
Solar Array Index Signal
Solar Array Index Signal, Return
Direct Roll Control Output, (+)
Direct Roll Control Output, (−)
Direct Pitch Control Output, (+)
Direct Pitch Control Output, (−)
Direct Yaw Control Output, (+)
Direct Yaw Control Output, (−)
Gyro Heater Power,
Sun Sensor, Solar Array Pitch
Sun Sensor, Solar Array Yaw
Horizon Sensor
Pressure
Pulse Valve
Temperature
Attitude Integration error signal

Gyro Signal
Logic State Monitor
Command Storage Registers
Equipment Voltages
Miscellaneous
Gyro Temperature
Redundant System Status

Propulsion Functions

Power from Power System
Heater Power
Direct Roll Control
Direct Pitch Control
Direct Yaw Control
Temperature Sensor outputs
Propellant Tank Pressure
Propellant Latch Valve Status
Temperature
Pressure
Valve status
Thruster Status

Structure Functions

Heater Power
Temperatures

Antenna Functions

Temperature
Strain gauge
Receiver signal power
Receiver AGC
Transmitter Output Power
Miscellaneous

Tracking Functions

Fine Phase
Coarse Phase
Reference Phase
Analog Phase
Receiver Signal Power

Payload Functions

Power from Power System
Heater Power
Temperature
Redundancy Status

Miscellaneous Functions

Jettison Power
Heater Power
Pyrotechnic Power
Actuator Power
Separation Signals

Table 5-3. List of Typical Telecommands in a Spacecraft.

Power ON. Power OFF of various subsystems and components of subsystems

Power System

Solar Array Deployment
Speed control of Solar Array Drive
ON/OFF isolation commands for each battery
ON/OFF power isolation for each Solar Array wing
Charge Regulator ON/OFF control
Charge level selection
Charge to Trickle Charge change over control
Constant current charge selection
Voltage Limit battery charging
Solar Array Input, Enable
Solar Array Input, Disable
Power Output, Enable
Power Output, Disable
Battery Enable
Battery Disable
Command Input
Telemetry Output
Redundant functions/equipment commands

Telemetry Functions

Redundant functions/equipment commands

Telecommand Functions

Redundant functions/equipment commands

Communication Functions

Receiver Prime Enable
Receiver Redundant Enable
Heater Power ON/OFF
Command Input
Tracking mode ON/OFF
Tape Recorder ON/OFF
Redundant functions/equipment commands

Propulsion

ON/OFF commands for various thrusters for station keeping, attitude adjust, station acquisition, etc.
Redundant functions/equipment commands

Payload Functions

Power ON/OFF
Redundant functions/equipment commands

Miscellaneous Functions

AKM arm/fire control
Jettison Power ON/OFF
Redundant functions/equipment commands

ACS Functions

Gyro Heater Power ON/OFF
Redundant functions/equipment commands

basic command consists of a binary pulse train with a *pulse-code-modulation non-return to zero* (PCM/NRZ) format. The pulse code sequence is phase-shift keyed (PSK) onto a subcarrier oscillator, the phases of the ZERO and ONE states being 180° apart. A sinusoidal signal of the same frequency as the bit rate is also transmitted to assist in identifying the bit transition in the PCM train. Since the bit rate is exactly half the subcarrier frequency, the phase of the bit-rate sinusoid may be used to identify the reference or zero phase of the subcarrier. The phase of the sine wave will remain constant with respect to the bit train. Zero phase occurs when a positive-slope zero crossing of the sinusoid coincides with a bit transition. The composite PSK/PCM signal in turn frequency modulates (FM) a subcarrier oscillator and finally phase modulates (PM) a transmitter on ground.

The receiver on-board the spacecraft, receives this signal and demodulates it. The decoder accepts the phase-shift keyed (PSK) subcarrier from the command receiver, demodulates the subcarrier, and reduces the serial data stream of binary information into unique command signals. The main building blocks of the decoder are a demodulator, a clock pulse generator, several shift registers, a select matrix, buffers, control logic, and supporting power supplies, regulators, power switches, and EMI filters.

For example, Intelsat Standard-A Earth Station employs the following types of modulation: FDM-FM for telephone channels; SCPC-PCM-PSK for voice, low speed data and telegraph channels; SCPC-PSK for data channels and FM for television channels.

5.2.2 Command Format

Command formats differ from spacecraft to spacecraft and one of the typical formats is shown in Fig. 5-3, which consists of a synchronization code, command frame, execute frame, and command interval in its simplest form. Each command message is preceded by a synchronization code consisting of n ZEROs followed by m ONEs. This allows transients to settle in the bit-rate filter and uniquely indicates the beginning of a message. Until the beginning of the synchronization code, the decoder is in a quiescent state and draws minimum power. The decoder continuously watches for the sync code, once the sync code is present, it immediately processes the following bits. The com-

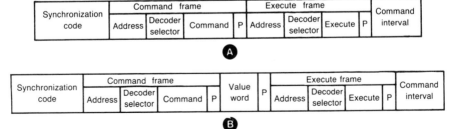

Fig. 5-3. Typical spacecraft command format, (A) for regular commands and (B) for value commands.

mand frame, which follows synchronization code, consists of spacecraft address, decoder selector bits, command word, and parity check.

The address is unique to each spacecraft and must be recognized by the decoder before the decoding process is allowed to continue. Decoder selector bits are used for selecting the particular decoder that has to provide the command output from its decoding matrix. The command word has certain number of binary bits and the decoder decodes by using a logic matrix and activates one of the output channels. Parity check is a single bit that follows the command word gives ONE parity of the previous bits. Decoding is carried out only after correct parity.

The value word follows immediately certain types of command frames (Fig. 5-3B) and is transferred into a register or stored in another subsystem. The last bit is a ONE parity digit for the preceding bits. The decoder stores this number for eventual transfer in parallel form to a register designated by the previous command frame.

The execute frame is the final frame of the command format, and consists of the spacecraft address followed by an execute command word, which is uniquely identified by the decoder. On recognition of the execute command word, the decode matrix is energized and the command channel already specified by the command word is activated for 50 msec to 100 msec duration. After transmission of the execute frame, there will normally be a minimum command interval and this returns the decoder to the quiescent state.

5.3 COMMUNICATION

5.3.1 Transmitters

The combined outputs of the sensors are used to modulate the high-power carrier frequency produced by the transmitter. Space communications generally use frequency modulation or pulse modulation. In FM/FM modulation scheme, the subcarrier oscillators are frequency modulated by the sensors and these, in turn, frequency modulate the carrier. The most common modulation scheme employed is PCM/PSK/PM in which PCM data is phase-shift keyed which in turn phase modulates the carrier. The major factors in transmitter design that influence the input power requirements are rf output power, operating frequency, data rate, and bandwidth. Output power is a function of antenna gain and directionality, maximum range, and the ground station receiver characteristics. Transmitter power requirements increase with bandwidth, which is a direct function of the data transmission rate, type of modulation, and stability requirements.

Figure 5-4 shows a schematic of a typical transmitter and it consists of a phase modulator and amplifier, limiter, mixer, filter, TWT amplifier stage, filter, master oscillator, frequency multipliers, TWT power supply, and an antenna. Encoded data is fed to a phase modulator, which also receives an intermediate frequency. For example, in the case of a C-band transmitter, the typical i-f is in the range of 45 to 90 MHz and transmitting frequency is around 4000 MHz. Output of the phase modulator is passed through a limiter into the mixer where the phase modulated data is translated up into the transmitting

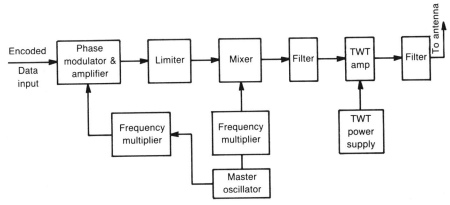

Fig. 5-4. Schematic of a typical transmitter.

frequency levels while simultaneously amplifying the signal strength. Further amplification, if needed, is achieved by the use of a travelling-wave-tube amplifier before being fed to an antenna via a filter for transmission to the ground station.

5.3.2 Receivers

Receivers are used in the spacecraft to detect and amplify the command signals transmitted from the ground station. The commands can be both real- and stored-time commands. The real-time commands are executed as they are decoded. Stored-time commands would be sent to a programmer and executed at the proper time later in the orbit. Stored-time commands may include turning the instruments/experiments on and off at various times throughout the orbit.

Figure 5-5 shows the schematic of a typical receiver. The rf signal is picked up by the receiving antenna and passed through an rf amplifier. The amplified signal is then sent to a mixer, which combines it with a signal from the local oscillator. Combining these two signals in the mixer produces a difference or intermediate frequency. The primary advantage in converting to this lower or intermediate frequency is the ability to obtain large gains and selectivity per stage of amplification than is possible at the higher frequencies. The

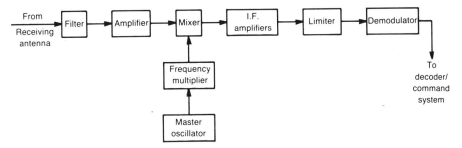

Fig. 5-5. Schematic of a typical receiver.

intermediate frequency (i-f) signal is passed through several amplifiers and is ready for demodulation. Any induced amplitude modulation is eliminated by passing the signal through a limiter. This limited signal is then fed to a discriminator, which produces the demodulated baseband signal.

The band-pass filter between the antenna and the rf amplifier provides additional selectivity and reduces adjacent channel interference. The isolator takes care of mismatch between the filter and the rf amplifier. Whenever the receiver input signal fades, then the automatic gain control circuit (AGC) maintains the output power constant over a range of fading.

The figure of merit of the receiving system is defined as the ratio of antenna gain (GA) to System noise temperature (T), where the system noise temperature is the sum of equivalent antenna noise temperature and equivalent receiver noise temperature.

Another type of receiver uses a phased-lock loop approach and employs a phase-lock carrier tracking loop, a separate wideband (modulation) phase detector, and a coherent AGC detector. In operation, the receiver is phase-locked by sweeping the unmodulated ground transmitter frequency across a portion of the uplink frequency assignment. When phase-lock is acquired, the ground transmitter frequency will be set to the nominal center frequency of the transponder and the transponder will not lose phase-lock when subjected to the normal doppler frequency offsets. When locked to the incoming signal, the receiver furnishes, (a) the demodulated command signal to the command decoder; (b) a reference frequency, coherent with the receiver carrier to drive the transmitter for tracking in transponder mode; and (c) the demodulated uplink (ranging and command) signal to the transmitter modulators.

Thus, phase-lock designs make the receiver tuned to the incoming frequency and thereby minimize bandwidth and resultant power requirements by automatically compensating for component drift and doppler shift. This type of design permits coherent retransmission of the received signal at the transmitter frequency for purposes of determining range rate information. The use of GaAs amplifiers in spacecraft receivers has yielded improved designs of low noise and reduced power requirements and they amplify the received wideband signals without converting to a lower intermediate frequency as is required by a conventional system. After amplification, the nominal uplink (typically 6 GHz) received signals are converted to the downlink (typically 4 GHz) transmit frequencies and further amplified in the two-stage TWT transmitter. Command signals are detected and decoded after amplification.

5.3.3 Modulation and Demodulation

The desirability of various modulation schemes for TTC&C is determined by the complexity that is acceptable, the link margins available, the data rates involved, the requirement of carrier component, etc. Much attention is devoted to the utilization of frequency or phase modulation. The general guideline is that the uplink modulation shall be simple for demodulation purposes. The on-board receiver should not be complex because of weight and reliability problems. The down-link modulation shall be such that it requires minimum power to transmit the required TM data rate and also provide a carrier component if required for tracking purpose.

Down Link. The carrier frequency for down-link (spacecraft to ground) is usually between 2 GHz and 4 GHz. At a 2 GHz frequency the lowest system noise temperature can be attained. At higher frequencies, rainfall and atmospheric attenuation increases. At lower frequencies cosmic noise becomes significant and below 1 GHz, Faraday rotation has to be overcome by employing polarization diversity or circular polarization on the ground system, which reduces the system sensitivity. The modulation scheme generally preferred is phase modulation as it does not suffer from the disadvantage of threshold effect of frequency modulation.

Up Link. The selection of the frequency for uplink (ground to spacecraft) is relatively easy as the desired signal-to-noise ratio can be easily achieved using high-power transmitters on the ground. Unless a coherent relationship is required for transponder operation, frequency modulation is generally used on the uplink because of the simplicity of the demodulation process compared to phase modulation.

A common antenna for uplink and downlink both on the ground and on-board the spacecraft may be used if the two frequencies are reasonably apart. In general the spacecraft transmitter frequency is lower than the receiver frequency of operation. Typical frequencies are listed below:

Band	Down frequency	Up frequency
C	4 GHz	6 GHz
VHF	136 MHz	148 MHz

5.4 SATELLITE TRACKING

Tracking is the process of observing the motions of one object relative to another object. There are various techniques, i.e., optical, radio, radar or infrared, etc., that can be employed for tracking the moving objects or objects in motion.

Optical techniques can be employed if the visibility is good. Even relatively cheap optical apparatus can give accurate results comparable to that obtained by radio tracking systems. The sophisticated optical techniques yield accuracies as good as those obtained by any other technique. Radio and radar trackers can work in any weather conditions. In radio tracking, the signal transmitted by the object to be tracked is received and processed. In radar tracking, on the other hand, a signal is transmitted, when this transmitted signal is received by the object being tracked, it reflects or sometimes retransmits back, which is received and processed. Usually reflected signals are weaker. However, retransmitted signals are stronger and hence they allow long range tracking. Infrared tracking depends upon the heat energy radiated by the objects being tracked. Infrared trackers unlike optical trackers, are not dependent on the observed surface being sunlit.

Among radio tracking, there are two well known methods, i.e., *doppler* and *interferometer*, which are described below:

5.4.1 Doppler Tracking

As a spacecraft passes an observer, its velocity relative to the observer changes continuously. The component of this velocity along the line joining the spacecraft and the observer will initially be directed towards the observer, and will be zero at the time of closest approach of the spacecraft and will subsequently be directed away from the observer. The nearer the spacecraft approaches, the greater is the maximum rate of change of velocity as illustrated in Fig. 5-6.

If the spacecraft is transmitting radio waves, the frequency detected at the observer will differ from that transmitted by an amount directly proportional to the relative velocity described above. This is the well known doppler-shift frequency, (f_d), which is commonly illustrated by the rising and falling pitch of a train whistle as the train at first approaches and then recedes. The doppler-shift frequency can be calculated from the equation.

$$f_d = \frac{\text{Relative velocity component}}{\text{Wavelength of transmitted signal}}$$

The frequency of the received signal will decrease throughout the pass, being higher than that of the transmitted signal as the spacecraft approaches, the same at the point of closest approach, and lower as the spacecraft leaves. From measurements of the varying received frequency, it is possible to derive the spacecraft velocity near the crossing point and the range at the closest approach. By making similar observations again after a known number of orbits and assuming a stable frequency from the spacecraft it is possible to calculate the characteristics of the orbit.

The doppler-shift frequency is typically of the order of only 0.003% of the transmission frequency. Thus if a spacecraft transmission frequency is 136 MHz, then the shift frequency is about 4 kHz. Such a small change in frequency is detected accurately by beating the incoming frequency with a highly stable

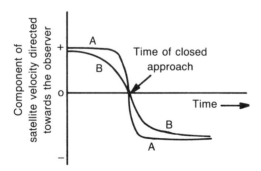

Fig. 5-6. Doppler tracking—component of satellite velocity directed towards observer versus approach time.

local oscillator with a frequency equal to the transmission frequency to result in a low-frequency signal that can be processed and analyzed. Use of a phase-locked loop for this application produces an excellent signal-to-noise ratio.

In the above discussion, only one observing station is used. By using three stations about 100 miles apart at an approximately equilateral triangle the position of a spacecraft can be determined to within about 0.2 mile.

5.4.2 Interferometer Tracking

In the interferometer system, to find the direction of the spacecraft in one plane, two identical aerials, A and B, are set up a few feet above ground at an accurately measured separation, typically 50 wavelengths of the spacecraft transmission frequency as illustrated in Fig. 5-7.

Energy from a spacecraft at a distance of several hundreds of miles away reaches A and B along parallel paths. If the spacecraft is in a plane perpendicular to and bisecting AB, then the signal reaches A and B simultaneously (in phase), the signals A and B are identical and when added give a maximum resultant signal. If the spacecraft is slightly offset from this plane, the path length to A and B will differ. If the path length is a half wavelength, then the signals received at A will be 180° out of phase than the signal received at B and will cancel each other when added. Thus by processing the signals received at A and B the spacecraft's orbit details can be determined.

With a system of fixed aerials such as explained above, the period during which observations can be made is confined to the short time while the spacecraft is within the narrow interferometer beamwidth. To increase the observation time the individual aerials can in principle be made steerable and it is in fact entirely feasible to use steerable aerials that can be locked together to point in the same direction and electronically locked to add their received signals in the correct sense.

5.4.3 Radar Tracking & Ranging

A radar tracker with a transmitter and a receiver can track "quiet" objects. Some of the earliest spacecraft tracking records were made on sputniks

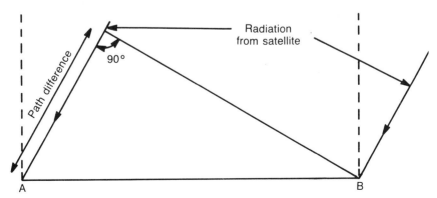

Fig. 5-7. Interferometer tracking set-up using two aerials "A" and "B."

and their carrier rockets using the 250 foot diameter telescopes at Jodrell Bank and a 45-foot diameter telescope at the Royal Radar establishment. Although one generally thinks of radar systems using a spacecraft as a passive reflector, most tracking systems, nowadays use a transponder on the spacecraft that is triggered when the radar transmits. This gives an increased range and more accurate range measurement. Thus, during the periods of tracking, the on-board receivers and transmitters are configured such that they work as transponder and facilitate spacecraft tracking.

High angular accuracy and long range are derived from narrow pencil beams using large diameter radar dishes. There is always a compromise between accuracy measurements and mechanical problems as the dish diameter is increased, the weight increases and the problem of controlling the dish accurately at a reasonable tracking rate becomes increasingly difficult.

One of the tracking systems features a digital ranging technique to determine ranges using full-code modulation. Once the range has been established, the full-code modulation is disabled. A number of different frequency tones are modulated on the uplink carrier. The spacecraft returns the signals on the downlink, and the ranging system determines the range by measuring the phase shift of the tones. The higher frequency tones provide system accuracy, and the lower frequencies provide range ambiguity resolution. The frequencies range from 10 Hz to 500 kHz.

5.5 ANTENNAS

Antennas are used for both transmitting and receiving radio frequency (rf) energy and are designed to match the proper impedance of the transmitter/receiver so that maximum rf energy is exchanged. Antennas with proper concentrators or reflectors are used to increase the strength of the transmitted signal by concentrating the rf energy into a narrow beam. The basic and simple antenna is the isotropic radiator which radiates rf energy uniformly in all directions and thus has a gain of unity. The simplest antennas used on spacecraft are rods of one-quarter the wavelength of the carrier frequency. Usually four of these are employed to obtain nearly an omni-directional radiation pattern. Attitude stabilized spacecraft or antennas with steering and pointing mechanisms can use directional antennas due to their higher gains.

Detection of the limited power radiated necessitates an antenna with a large effective aperture and high signal-to-noise ratio. Noise sources are (a) the receiver itself introduces noise, (b) thermal dissipation in waveguides, diplexers, etc., introduce noise, (c) thermal radiation from the sky, and (d) back radiation from the earth, since it is a thermal source at 300 degree K and is the major contributor of aerialnoise. It is thus important to reduce aerial pickup towards the earth while simultaneously achieving high gain in the required direction. The figure of merit of the aerial is defined as the ratio of aerial gain to system noise temperature.

5.5.1 Various Types

Paraboloidal Reflectors. The paraboloidal reflector offers a higher gain characteristic. A collimated beam can be obtained with a single feed at the focus of the paraboloid with a gain proportional to the area of the aperture. High performance can be guaranteed over a wide range of frequencies by changing the primary feed of the system. The simple arrangement is the front-feed paraboloid as shown in Fig. 5-8.

The gain and aperture efficiency are determined in the ideal case by the polar diagram of the primary feed, and the ratio of the focal length to the dish diameter, i.e., f/D ratio. Maximum gain is achieved with uniform illumination of the aperture. In practice, the gain is lowered because the edges of the aperture are less effectively used and also because some of the energy is not intercepted by the dish to some extent due to blockage of the feed and its supports and is radiated away at wide angles to the main beam.

In many front feed systems a focal plane dish has been used with f/D = 0.25 either in order to prevent interference with neighboring installations or pick-up from nearby sources of electrical radiation. Aperture efficiencies are sometimes sacrificed for low noise and freedom from interference problems. For example, at the receiving wavelength of 7 cm, the aperture was less than 30% effective, due partly to profile inaccuracies, and partly to the high taper of the dish illumination. Both gain and figure of merit increases with f/D ratio, although the variation is not great. An aerial with an f/D of about 0.36 and half aperture angle of 70 degrees, coupled with a much more accurate dish profile, has resulted in an aperture efficiency in excess of 60% at 7 cm wavelength.

Figure 5-9 shows the MARECS Satellite with its paraboloidal antenna. The main advantage of the front-fed paraboloid antenna is the simplicity of the arrangement and the resultant operational flexibility. Its efficiency rarely exceeds 60% unless gain is sacrificed. To achieve a system with a low-noise operation, low-noise receivers can be located at the focus. However, it is complex and inconvenient.

Cassegrain Aerials. To overcome the performance deficiencies of front-feed paraboloids and the mounting of low-noise receivers at their focus, cassegrain aerials are considered. In a cassegrain arrangement, a hyperboloid

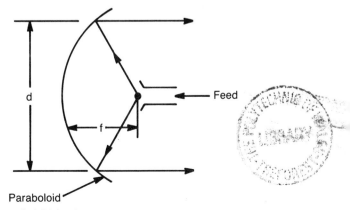

Fig. 5-8. Front feed paraboloidal antenna.

Fig. 5-9. MARECS satellite with paraboloidal antenna.

secondary reflector, placed symmetrically inside the focal distance gives a focus near the dish surface as shown in Fig. 5-10. The horn feed and low-noise receiver can be located near the apex of the dish and can be easily installed. Any spillover from the horn is directed towards the cold sky. The secondary reflector might require bulky supports. The scattering from these obstacles partially offsets expected improvements in noise performance. Thus the cassegrain system becomes worthwhile with large antennas only.

As the focus is near the dish surface there is greater freedom to mount primary feeds, particularly as space behind the dish can be easily utilized for waveguide and horn components. Thus, foghorn feeds are being used in some

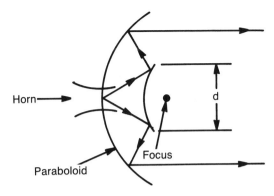

Fig. 5-10. Cassegrain antenna.

cassegrain arrangements, to provide a convenient mechanical support for low-noise receivers.

Horn Reflectors and Open Cassegrain Types. The performances of the front fed paraboloid, and of the cassegrain arrangement are impaired by the blocking of the aperture by feeds, secondary mirrors, and the supporting structure. However, by using only a portion of the paraboloid the feed can be kept clear of the reflector beam and thus aperture blocking can be reduced to zero as shown in Fig. 5-11. Although the structure is somewhat massive, the electrical performance is very good.

In an improved version, the horn is triply folded as shown in Fig. 5-12. Some deterioration in the performance is expected from the triple reflecting plates. In another configuration, a cassegrain reflector is used in place of the horn to produce a focus near the paraboloid surface. Since the sub reflector and its supports do not block the aperture in this configuration, it is known as "open" cassegrain. Aperture efficiencies of better than 65% are achievable with these antennas.

An Example. The flexrib reflector used, as shown in Fig. 5-13, on the ATS-F and -G is an example of the paraboloidal reflector antenna. During launch phase, the reflector is folded such that it is rugged and compact. A simple release and deployment mechanism is used to deploy the antenna in orbit.

The paraboloidal reflector consists of a reflecting surface formed by a copper-coated Dacron mesh supported by 48 radial thin-gauge aluminum ribs

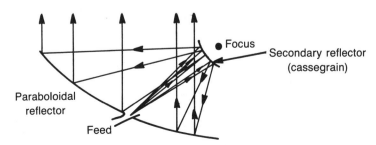

Fig. 5-11. Open cassegrain antenna.

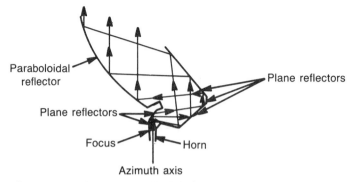

Paraboloidal reflector

Plane reflectors

Plane reflectors

Focus

Horn

Azimuth axis

Fig. 5-12. Open cassegrain antenna with triply folded horn.

Fig. 5-13. ATS-F satellite with its antenna deployed.

hinged to a central hub. The mesh is coated with silicon to protect the copper surface while on the ground and has an optical transparency of about 70% coupled with a high reflectivity at C-band frequencies. The width of the ribs is tapered approximately 2:1 from root to tip. The thickness of the ribs is also tapered. The mesh is sewn to the edge of the ribs, which have been precision machined to the required contour.

For storage during launch, the ribs are folded 90 degrees at the hinges into a tangential position. They are then wrapped around the hub into the stowage container with the mesh carefully folded between the ribs. The swing doors are folded down over the reflector and held in place by an encircling cable. The total package weighed about 64 kgs, formed a torus 0.3 meter high with an inside diameter of about 1.5 meters and an outside diameter of 2 meters. The deployed reflector has ratio of focal distance to diameter (f/D) of 0.44.

Modern Trends. The size and weight of the spacecraft has been continuously on the increase to obtain greater channel and performance capabilities. The spin-stabilized spacecraft antennas may be replaced by electronically or mechanically despun phased-array antennas or gravity-gradient stabilized antennas. Parabolic antennas may be replaced by steerable-horn antennas. Antennas that can be continuously controlled such that they are always directed towards the earth may be tried.

The usual frequency of operation for transmission from the ground is the 6 GHz band and for transmission from spacecraft is the 4 GHz band. As larger number of channels are to be accommodated in future, the use of 16 GHz and 36 GHz are being envisaged to avoid ground network interference and obtain higher bandwidths possible at higher frequencies. The atmospheric noise at these frequencies is likely to be offset by the increased antenna size.

Sometimes antenna pointing can be achieved by mutual control of spacecraft roll and control of a single-axis antenna hinge system or the antenna can have two-axis control.

5.6 MULTI-MISSION MODULAR SPACECRAFT TTC&C SYSTEM

An advanced and sophisticated TTC&C system (sometimes known as Communication and Data Handling System) was employed in the case of MMS class of satellites, wherein an onboard computer is used with Telemetry and command system. This combination enables the following functions:

☐ a) Low cost of implementation of many spacecraft on-board functions.
☐ b) Autonomous operation of spacecraft.
☐ c) The telemetry format is controlled by either the computer or a read-only memory, therefore, parameter selection and sample rates can be defined after user design is complete and modified from mission to mission at minimum cost.
☐ d) The telemetry bit rates are command selectable, 1 Kbps, 3 Kbps, 4 Kbps, 8 Kbps, 16 Kbps, 32 Kbps, or 64 Kbps.

The MMS TTC&C uses the command and data handling hardware design based on the concept of remote multiplexing of telemetry data and remote distribution of commands. Command and telemetry data are routed to and from other spacecraft and instrument subsystems via a serial digital multiplex data bus to minimize interconnect problems and to allow sizing the system to actual requirements.

The on-board computer communicates with all observatory subsystems by time sharing the multiplex data bus. This composite hardware interconnection forms an integrated electronic system that provides very flexible, yet standard, hardware. Signal conditioning of passive transducers is provided, when needed, by a constant-current source, which is applied to the devices at the time of sampling.

5.7 COMMUNICATION LINK CALCULATIONS

A point source antenna radiates uniformly (isotropically) in all directions. Thus, if an antenna radiates a power of $P(T)$ watts isotropically in all directions, then the power density, PD, at a distance R from the antenna is given by

$$PD = P(T)/[4 \times \pi \times R^2]$$

When a second antenna is used to receive the power radiated by the first (radiating) antenna, at a distance R, the total power received, Pr, by the receiving (second) antenna is given by

$$Pr = PD \times Ae$$

Where Ae is the effective area of the receiving antenna and is related to the aperture area A through an efficiency factor, n and is given by

$$Ae = \epsilon \times A$$

The efficiency factor, η, varies over a range from 0.54 to 0.81 depending upon various factors, like reflector shape, feed position, blockage, etc. If the antenna has a reflector, then all the energy is directed in only one direction. Thus radiation intensity in a particular direction has been increased compared to isotropic omnidirectional antenna. In the case of a paraboloidal antenna, a paraboloidal reflector is used to reflect all the energy in one direction and if it is directed towards a receiving antenna then the received energy is much higher than when it received from an isotropic antenna. Also, a paraboloidal reflector transforms a spherical wave from the feed at focus into a plane wave in the paraboloidal aperture. The gain of a parabolic antenna, G, is directly proportional to the area of the aperture A or the square of the diameter, D, of the reflector dish and is given by

$$G = \frac{4\pi \times A \times \eta}{\lambda^2} = \eta \times \frac{4\pi}{\lambda^2} \left(\frac{\pi D}{2} \right)^2 = \eta \left(\frac{\pi D}{\lambda} \right)^2$$

Also, the gain of an antenna with a reflector in a given direction can be defined as the ratio of the radiation intensity produced in that direction to the radiation intensity produced in the same direction by an isotropic antenna with the same power input.

The low-noise feature of the first stage of the spacecraft receiver is very important because of the very low signal power at the receiving antenna. The first stage of the receiver, namely the preamplifier, shall be very sensitive and shall not introduce any noise. To minimize the losses and noise, it is mounted directly on the antenna dish at the focus. The thermal noise power, N, generated by a matched load at a temperature of T degrees K is given by

$$N = k \times T \times B \text{ watts}$$

where k = Boltzmann's Constant, in watt-sec/degree K.
        B = Noise bandwidth

Thus, all receiving systems pick up thermal noise from the environment. Hence, the signal power intercepted by the antennas shall be higher than this thermal noise power in any practical system.

5.7.1 Link Calculations

It is assumed that the transmitting and receiving antennas employ paraboloidal reflectors. With an isotropically radiated power of $P(T)$ and transmitting antenna gain of $G(T)$, the signal power, S, intercepted by a receiving antenna (at distance R) having an effective aperture of Ae, is given by

$$S = \frac{P(T)}{4\pi \times R^2} \ G(T) \times Ae$$

As Ae can be related to receiving antenna gain, $G(R)$, as described above, the signal power can be rewritten as

$$S = \left(\frac{P(T) \times G(T)}{4\pi \times R^2} \right) \left(\frac{G(R) \times \lambda^2}{4\pi} \right) = P(T) \times G(T) \times G(R) \times \left(\frac{\lambda}{4\pi R} \right)^2$$

$P(T) \times G(T)$ is also known as equivalent isotropically radiated power (EIRP) and $(\lambda/4\pi R)^2$ is known as free space loss. It is convenient to express the above relationship in decibels by taking the common logarithm of both sides and multiplying by 10.

However, the receiving system also picks up noise power and hence the received signal power, S, shall be much higher than the thermal noise power to be of any practical use. The ratio of received signal power to the noise power is given by

$$\frac{S}{N} = \left(\frac{P(T) \times G(T) \times G(R)}{k \times T \times B} \right) \left(\frac{\lambda}{4\pi R} \right)^2$$

The required value of (S/N) depends on the information rate, the modulation scheme and the system bandwidth.

5.7.2 Example

Sample Communication link calculations for uplink and downlink are provided below with detailed explanations. Following are the Data and Assumptions:

Ground Transmitter Frequency	6 GHz
Ground Transmitter Power	300 watts
Ground Antenna Diameter	5 meters
Ground Antenna Pointing Error	0.05 degrees
Ground Receiver Noise Temperature	800 degrees K
Ground Receiver Bandwidth	40 MHz
Spacecraft Transmitter Frequency	4 GHz
Spacecraft Transmitter Power	200 watts
Spacecraft Antenna Diameter	1 meter
Spacecraft Antenna Pointing Error	0.5 degrees
Spacecraft Receiver Noise Temperature	1500 degrees K
Spacecraft Receiver Bandwidth	40 MHz
Distance between the ground station and the spacecraft	42,000 km

Tables 5-4, 5-5 and 5-6 present the uplink and downlink calculations.

Table 5-4. Uplink Budget Calculations.

Parameter	in dBs		Comments
Ground Station Related			
1. Transmitter power	24.77	dBW	10 Log (300)
2. Feed loss	−2.00	dB	
3. Antenna gain	49.94	dB	Note-1
4. Effective Isotropic Radiated power (EIRP)	72.71	dBW	(items 1 + 2 + 3)
5. Antenna Pointing error	−0.26	dB	
6. Margin	−3.00	dB	Note-2
7. Propagation/Space loss	−200.47	dB	Note-3
8. Atmospheric/Rain loss	−2.23	dB	
9. Polarization loss	−0.25	dB	
Spacecraft Related			
10. Feed loss	−0.00	dB	
11. Antenna gain	35.96	dB	Note-1
12. Antenna pointing error	−0.31	dB	
13. Received/input carrier power	−97.85	dB	Sum of items 4 to 12)
14. Noise power density	−196.84	dBW/Hz	Note-4
15. Receiver Bandwidth	76.02	dB(Hz)	Note-5
16. Receiver Noise Power	−120.82	dBW	Item 14 minus item 15
17. Carrier Power/Receiver Noise Ratio	22.97		Item 13 minus item 16

Table 5-5. Downlink Budget Calculations.

Parameter	in dBs		Comments
Spacecraft/Transmission Related			
1. Output Power	23.01	dBW	10 × Log (200)
2. Feed Loss	−0.50	dB	
3. Antenna Gain	32.44	dB	Note-1
4. Effective Isotropic Radiated Power (EIRP)	54.95	dBW	items 1 + 2 + 3
5. Antenna Pointing Error	−0.22	dB	
6. Margin	−3.00	dB	Note-2
7. Propagation/Space Loss	−196.94	dB	Note-3
8. Atmospheric/Rain Loss	−1.52	dB	
9. Polarization Loss	−0.25	dB	
Ground Station/Reception Related			
10. Feed Loss	−1.00	dB	
11. Antenna Gain	46.42	dB	Note-1
12. Antenna Pointing Error	−0.18	dB	
13. Received/Input Carrier Power	−101.74	dB	Sum of items 4 thru 14
14. Noise Power Density	−199.57	dB	Note-4
15. Bandwidth	76.02	dB	Note-5
16. Receiver Noise Power or Ground Station Net Noise Power	−123.55	dBW	Item 14 minus item 15
17. Carrier Power/Receiver Noise Ratio	26.92	dB	Item 13 minus item 16

Table 5-6. Notes for Link Budget Calculations.

Note-1

Antenna gain is given by the formula

$$= [4 \times \pi \times A_{EA}]/[\lambda^2]$$

Where A_{EA} is the antenna effective area, in square meter
λ is the transmission wavelength in meters

Note-2

Usually a margin of 3 dB is included in the link calculations due to various reasons.

Note-3

Propagation loss is also known as space loss or path loss and is given by the formula

$$= [\lambda^2]/[4 \times \pi \times r)^2]$$

where r is the distance between transmitter and receiver in meters.

Note-4

Noise power Density is given by

$$= K \times T$$

Where K is the Boltzman Constant and its value is 1.38×10^{-23} watt-sec/degrees K.
T is the noise temperature, in degrees K.

Note-5

Bandwidth in dB is given by 10 × Log (BW) where BW is the bandwidth in Hertz.

General Note

Sometimes other losses like coax cable loss, etc. are also included.

Chapter 6

Propulsion System

THE PROPULSION SYSTEM PROVIDES A CONTROLLED IMPULSE FOR ORIENTING the spacecraft spin axis during spinning phases; adjusting the orbit; controlling the attitude, the spin rate, the nutation, and the speed of the momentum wheels; and providing three-axis attitude control for a spacecraft in conjunction with the attitude control system. Some of the above functions, such as momentum wheel unloading, reorientation, etc., can be carried out using other means, namely magnetic torquers.

6.1 DESCRIPTION

The propulsion system, as shown in Fig. 6-1, generally comprises of a set of propellant tanks and a set of thruster groups suitably interconnected and isolated by latching valves to provide redundancy for all in orbit control functions. The system as mounted on the rectangular deckplate or at an appropriate location on the spacecraft structure usually contains the following components:

Fuel tanks
Oxidizer tanks (in case of bi-propellants)
High-thrust engines
Low-thrust engines
Pressurant storage tanks
Pressure regulators

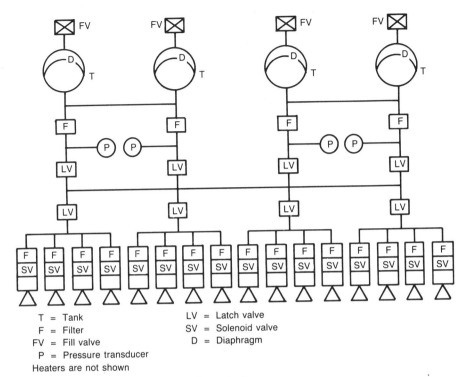

T = Tank
F = Filter
FV = Fill valve
P = Pressure transducer
Heaters are not shown

LV = Latch valve
SV = Solenoid valve
D = Diaphragm

Fig. 6-1. Propulsion system typical schematic diagram.

Isolation valves
Filters
Fill and vent valves
Fill and drain valves
Pressure transducers
Temperature transducers
Propulsion system electronics
Etc.

The Propulsion system shown in Fig. 6-1 uses monopropellant. Four spherical fuel tanks are mounted on four corners of the deckplate. A diaphragm in each tank separates the nitrogen pressurizing gas from the hydrazine fuel. Opposite tanks are connected in pairs to minimize imbalance as fuel is consumed. Fuel flows through individual filters and two latch valves between the tanks and thruster assemblies. Pressure transducers are located between the tanks and thruster assemblies and these transducers together with readings from temperature sensors, and knowledge of tank size and initial loading allow the measurement of fuel remaining in each pair of tanks.

All lines are interconnected between latch valves to minimize the effect of a valve failure. Fuel lines from the spacecraft body to the suspended clusters are heated to prevent freezing. The thrusters are also provided with heaters to maintain the proper operating temperature. In operation, latch valves are open and the system is controlled with the solenoid valves on selected thrusters.

Rocket engine modules are mounted on the sides with the nozzles pointing in different directions. Sometimes, additional thrusters are also employed and mounted at the top middle sides of the cuboidal spacecraft. The locations of the thrusters are so selected that x, y, z (or roll, yaw, pitch) coupling can be achieved even after a thruster failure.

The thruster orientation for the propulsion system is shown in Fig. 6-2. This orientation provides three-axis spacecraft motion, orbit adjust and full

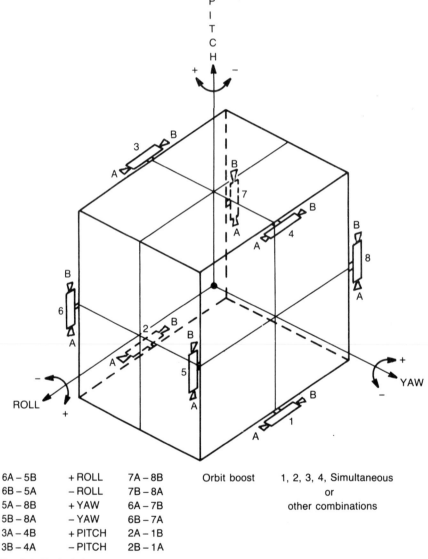

6A – 5B	+ ROLL	7A – 8B	Orbit boost	1, 2, 3, 4, Simultaneous
6B – 5A	– ROLL	7B – 8A		or
5A – 8B	+ YAW	6A – 7B		other combinations
5B – 8A	– YAW	6B – 7A		
3A – 4B	+ PITCH	2A – 1B		
3B – 4A	– PITCH	2B – 1A		

Fig. 6-2. Typical thruster arrangement and orientation on a spacecraft.

redundancy with sixteen thrusters. The primary and backup thruster firing functions is shown in Table 6-1. Alternately, twelve small thrusters are employed for three-axis stabilization control/station keeping and four large thrusters for orbit adjust.

Figure 6-3 is a photograph of a propulsion system integrated to the structure of a satellite and it uses a conventional monopropellant hydrazine blowdown configuration with four surface tension tanks manifolded to two redundant sets of catalytic thrusters. Two propellant tanks, its plumbing, some of the thrusters, electrical wiring, etc., can be seen in this photo.

6.1.1 Thrusters

Thrusters produce thrust by expelling propellant in the opposite direction with the energy deriving from a chemical reaction or thermodynamic expansion. Thrusters can use hot or cold gas. Hot gas thrusters derive energy from a chemical reaction whereas cold gas thrusters derive it from the latent heat of a phase change of the propellant. Hot gas thrusters generally produce a higher thrust and a large total impulse compared to the cold gas thrusters. However, cold gas thrusters facilitate more precise control due to their low, but constant thrust.

Types of Thrusters. Normally two types of thrusters are used on any spacecraft for more efficient operation of the propulsion system, one with high thrust and the other with low thrust. The high-thrust thruster is used for orbit changes and the low-thrust thrusters are used for orbit adjustment, orientation, momentum dumping, etc.

Injection errors have to be corrected in minutes; hence appropriate high thrust thrusters have to be selected. Typical hydrazine thruster ratings are in the range of 0.2 to 100 lb and hydrazine thrusters are used specially for station acquisition control, orbit injection correction, and for three-axis control.

For easiness, a set of thrusters (perhaps with different thrust capabilities) are assembled or fabricated into a single module with each thruster pointing toward a different direction and it is commonly known as a *rocket engine module*

Table 6-1. Thrusters and Their Functions.

Function	Thrusters
+ Pitch	3A-4B or 2A-1B
− Pitch	3B-4A or 2B-1A
+ Yaw	5A-8B or 6B-7A
− Yaw	5B-8A or 6A-7B
+ Roll	6A-5B or 7A-8B
− Roll	6B-5A or 7B-8A
Orbit Adjust	1, 2, 3, 4 simultaneous or other combinations

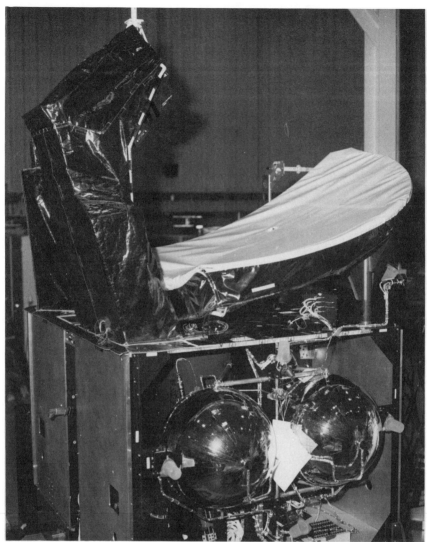

Fig. 6-3. This photograph shows the propulsion system integrated on to the structure of a satellite and it uses a conventional monopropellant hydrazine blowdown configuration with four surface tension tanks manifolded to two redundant sets of catalytic thrusters. Two propellant tanks, its plumbing, some of the thrusters, electrical wiring, etc. can be seen in this photo (courtesy of RCA).

(REM). This facilitates easy mounting onto the spacecraft and easy alignment with the spacecraft axes. Normally, the high-thrust thrusters are so mounted that their axes pass through the spacecraft center of mass. In some spacecraft, the center of mass axis changes when the spacecraft appendages are deployed from its original axis compared to the case when the appendages are not deployed. Proper care has to be taken in mounting and aligning the thruster's axis depending upon when the thruster will be used. In a sophisticated system, the thruster axis misalignment is easily taken care of by gimballing the thruster.

6.1.2 Low and High Thrust Engines

Depending upon the spacecraft mission, its orbit, etc., the propulsion system might contain thrust engines of more than one rating. A typical communication satellite in a geosynchronous orbit might contain two types, a high-thrust engine and a low-thrust engine. The high-thrust engine is used especially in the spinning mode. Pulsed mode of operation is usually employed for most spinning maneuvers. Electrical ON time is fixed and OFF time is varied depending on spin rate.

The low-thrust engine is used for both attitude acquisition (despin) and in orbit three-axis stabilized modes, including a variety of spacecraft maneuvers. The low-thrust engine is operated to produce repeatable small impulse bits at low duty cycle during the attitude acquisition and in-orbit phases of the mission. The low-thrust engine catalyst bed is normally maintained or heated to an appropriate temperature to maximize specific impulse and minimize impulse bit size for the above operations and the firing pulse width is reduced to a fraction of a second.

The thruster sizes are so chosen that, over the complete range tank pressurization (typically 3 to 1 ratio), the thrusters provide a thrust at low operating pressures high enough to provide orbit changes within a reasonable time and provide a thrust at high operating pressures low enough to allow thruster firings with appendages deployed and momentum dumping without disturbing the spacecraft attitude. Figure 6-4 shows a photograph of an electro-hydrazine thruster.

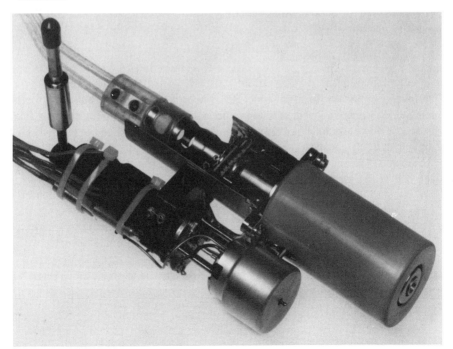

Fig. 6-4. An Electro-Hydrazine thruster (courtesy of RCA).

6.1.3 Propellant Tanks

Usually, the propellant tanks are housed inside the spacecraft such that the propellant does not get very cold and freeze. This also reduces the heater power requirement. Each propellant tank contains a positive expulsion elastomeric diaphragm separating the pressurant (nitrogen or helium) from the propellant. Each side is loaded through a separate fill and drain valve. Each tank has a certain capacity of hydrazine and single blowdown pressurization approach is normally employed for maximum reliability.

Instead of a diaphragm, other propellant management devices (PMD), such as a surface tension system can be used to force the propellant towards the tank exit port. In a spinning satellite, the centrifugal force generated by the spinning actions is sufficient to keep the propellant in the desired position with no other devices.

The propulsion system is typically designed as a single-stage "blowdown" system with a blowdown ratio of 3 to 1. This minimizes the system complexity and allows the propellant movement from the tanks to the engines by tank pressurization rather than pumping. The systems are initially pressurized, say at 300 psia. As propellant is consumed, the pressurant ullage volume increases and the pressure of the fixed mass of gas (nitrogen) decreases to an end of life value of 100 psia. This is accompanied by a decrease in thrust and specific impulse and an increase in burn time for the same delivered total impulse.

On the other hand, to provide a consistently high specific impulse, as well as to minimize variations in the delivered thrust, pressure regulators can be employed and the series redundant configuration can be selected as it provides a more precise range of operating pressure than two regulators in series.

6.1.4 Latching Valves

The latching valves, filters and pressure transducers are mounted on a common bracket to form an easily accessible module. Each propellant tank has a latching valve and hence it can be isolated. Each group of engines also has a latching valve to provide isolation. This facilitates closing the group latching valve, if any thruster valve fails to close or to seal tightly; thereby isolating that group of thrusters and enabling continuation of mission operation through the use of the backup thrusters.

All components, namely, tank, fill and drain valves, filters, and tank-to-filter interconnecting tubes are made of titanium, which offers lower weight for equal strength compared to steel. However, wherever thermal conductivity is needed stainless steel is used.

6.1.5 Filter

The filter is located just upstream of all control components in the feed lines from each tank to provide contamination protection. It will remove all particles larger than a predetermined size (typically 18 microns) in the propellant flow. Particles smaller than this can be tolerated by the valves. The filter element has certain dirt holding capacity with certain pressure drop. The entire filter is made of corrosion resistant stainless steel and has an all welded construction.

6.1.6 Pressure Transducers

The pressure transducers are used to monitor the propellant supply pressures. The transducer features a vacuum-deposited thin-film strain gauge that works over wide operating temperature range.

6.1.7 Pressure Regulators

The pressure regulator maintains precise pressurization of the propellant tanks. This is very essential as the performance of the thrusters depend upon the proper mixture ratio. If the pressurization is not proper then the mixture ratio will be different and difficult to control.

The typical failure mode of a pressure regulator is open and hence series redundancy is employed. If the blow-down approach is used then pressure regulators are not employed.

6.1.8 Heaters

If a spacecraft mission scenario contains spinning and non-spinning phases, the temperature of the spacecraft might vary over a large range. Even in the case of a three-axis stabilized spacecraft the heat fluxes will change depending upon various factors including spacecraft orientation with respect to the sun and earth. Therefore, groups of temperature control heaters are required to maintain the propulsion system above the hydrazine freezing point throughout the mission. The thermal components are selected to be compatible with the maximum spacecraft temperature. Usually, catalyst bed heaters are provided on the low-thrust engines to achieve the performance and pulse life requirements.

Heaters are provided on the catalyst beds to maintain the temperature at the appropriate range to prevent catalyst bed poisoning and to achieve high impulse repeatability.

6.2 PROPELLANTS

Cold gas propellant is stored as a liquid above the critical pressure and is self pressurizing. Heaters are not required because the propellant is a gas at low pressure at any temperature likely to be encountered in operation. Cold gas systems may have a pressure regulator between the tank and the thruster to control the propellant flow rate.

Hot gas propellant can be monopropellant or bi-propellant. The most commonly used monopropellant is hydrazine (N_2H_4) or hydrogen peroxide (H_2O_2). Monopropellant systems use tank diaphragms to separate the propellant from the pressurizing agent and also a catalyst or high temperature to promote decomposition of a single component. Hydrazine freezes at about 10 degrees centigrade and may require heaters if lower temperatures are

Table 6-2. Comparison of Monopropellants.

Parameter	Hydrazine	Hydrogen Peroxide
Symbol	N_2H_4	H_2O_2
Catalyst	Silver-Base	Shell-type
Specific Impulse	220 sec	180 sec
Long term decomposition problem	No	Yes

encountered during the mission. Table 6-2 presents a comparison of hydrazine and hydrogen peroxide monopropellants.

Advantages of simple injector design, no mixture ratio control, lack of residuals, and reduced operating temperatures, all tend to make the selection of hydrazine for most of the spacecraft.

Bi-propellant produces higher levels of specific impulse. In a bi-propellant system the fuel and oxidizer are stored separately. A commonly used bi-propellant is composed of a hypergolic mixture (the propellants burn spontaneously upon mixing) of mono-methyl-hydrazine (MMH) and nitrogen-tetroxide (NTO). Variation in propellant and oxidizer feed pressure results in variation in mixture ratio, which in turn results in a variation in thrust.

6.3 PROPULSION SYSTEM ELECTRONICS

The propulsion system operation is controlled by the attitude control system through the propulsion system electronics. The commands are received by the propulsion system electronics, processed and appropriate drive signals are applied to command designated thrusters for commanded durations.

Thus, the propulsion system electronics contain driver circuits for engine valves, latching valves, and heater groups; solid-state latches for thermal environment control heaters; and signal conditioning for all instrumentation (both telemetry and ground testing). The propulsion system electronics also contains power converters and regulators to obtain different voltages than the bus voltage and better regulation than the regulation of the bus voltage. Propulsion system electronics also houses control logic to prevent the simultaneous firing of opposite thrusters.

The electrical interface and functions performed within the propulsion system electronics are illustrated in Fig. 6-5. It interfaces with the different spacecraft support systems like, the power system, attitude control system, telemetry system, and telecommand system to receive power, commands, and to feed the instrumented signals back to processing or for housekeeping. Proper instrumentation is incorporated to estimate the remaining fuel.

Propellant tank temperature and pressure are monitored for each tank in order to predict engine performance and estimate fuel usage during the mission.

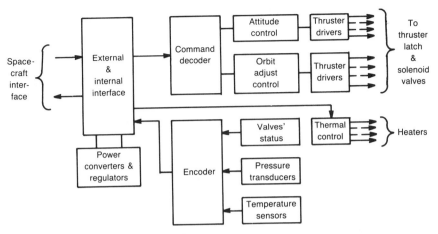

Fig. 6-5. Propulsion system electronics.

Catalyst bed temperatures are measured over the full operating temperature range of the engines to an accuracy that allows it to distinguish between heater OFF and heater ON conditions. Engine performance can be predicted by using these temperature data and engine flight performance prediction data. Voltage, current, status of the heater group circuits, the position of latching valves, etc., are also instrumented.

6.4 ALTERNATE APPROACHES

6.4.1 Magnetic Coils

In near-earth orbits, magnetic coils can be used in place of thrusters (and accompanied fuel). The control of thrusters is straightforward and simpler than those of coils, because the magnetic torque produced by a coil depends on the local magnetic field, which varies as the spacecraft moves in its orbit. As magnetic coils exert lower torques, their ON duration extends over a large fraction of an orbit or over several orbits to achieve the desired correction. Sometimes the use of magnetic coils may not be acceptable due to magnetic cleanliness requirements imposed by some experiments.

Some spacecraft combine a cold gas system with magnetic coils. The coils are usually used for attitude control. The gas jets are used for occasional but large thrust maneuvers that cannot be performed with the coils. Also, these coils cannot be used for correcting secular angular momentum changes caused by gravity gradient and residual magnetic torques.

In high altitude orbits, however, due to lower local magnetic fields, jets are the only means for generating torques.

6.4.2 Ion Jets

Ion thrusters are fine thrusters with which micro-pound thrust can be generated. Ion jets accelerate individual ionized molecules electrodynamically, with the energy ultimately coming from solar cells or self-contained electric generators. Ion thrusters are valveless and they rely upon evaporation of liquid cesium to control propellant flow.

Cesium Ion Thruster. The unit comprises the propellant storage and feed system, ionization chamber, and accelerating and neutralizing system. Cesium propellant is contained in an annular chamber in which thin nickel vanes are arranged radially to provide passages tapering toward a porous nickel wick. Surface tension forces carry liquid cesium (cesium melts at about 28 degrees C) into the wick and along to a vaporizer where it can be evaporated by applied heat. When the vaporizer heater is de-energized, the cesium cools rapidly and the mass flow falls essentially to zero (namely, it is reduced by six or more orders of magnitude). There is no valve in the system; vaporizer sealing for launch is achieved by utilizing an acrylic cement or cadmium, which is removed readily by heating.

From the vaporizer, the cesium passes over a hot tungsten cathode into a discharge chamber whose wall is a cylindrical magnet. A cylindrical anode attracts electrons from the cathode. In spiralling through the applied magnetic field, the electrons collide with many cesium atoms, giving rise to many secondary electrons and cesium ions, thus producing a plasma.

Typical ratings = 250 to 500 μlb
Typical specific impulse = 5000 sec
Typical weight of a 500 μlb thruster = 5 lb

These thrusters are good for attitude control, namely for roll, yaw, and pitch control and NS-EW stationkeeping. 11000 lb-sec of impulse is obtained using 2 pounds of cesium. This will last for 6 months of geosynchronous satellite for attitude control (6000 lb-sec) and for stationkeeping (5000 lb-sec).

In the specific impulse range of 2000 to 3000 sec, the competing technologies are the bombardment ion engine, colloid thruster, sublimation plasma-jet, and Hall accelerator.

6.4.3 Resistance Jet Thruster

Gas expulsion systems in which thermal energy is added by I^2R heating of part of the thruster hardware are known as resistance jets or resistojets. There are two types and one is known as a thermal storage resistojet, which contains a refractory heating element with a high thermal heat capacity. Low-level power is continuously supplied to the heater element, and the propellant flow is pulsed when required. The thermal heat capacity of the thruster is sufficiently great so that the heater element temperature remains essentially constant for short pulses, thereby producing constant specific impulse. The second type known as the fast heatup resistojet contains a lightweight heater element with a low thermal heat capacity. This thruster consists of a thin-walled metallic tube with an integral contoured expansion nozzle. The propellant vapor is passed through the heated tube and expanded through the nozzle, which produces thrust. In this type of thruster, both power and propellant flow are pulsed. In contrast to the thermal storage device, the heat capacity of the fast-heatup unit is held to a minimum. The primary advantage of this type of thruster is that electric power does not have to be supplied continuously.

Ammonia and hydrogen are used as propellants and have specific impulses of 560 sec at a pressure of three atmospheres and a temperature of 2500 degrees K. Typical rating of an ammonia resistance thruster is 2 mlb and is used for attitude control and as a back-up for ion thrusters.

6.5 PROPELLANT BUDGET AND TANKAGE

As described above, the propulsion system has to meet a number of mission requirements, some of which are injection orbit correction, initial and subsequent attitude acquisitions, orbit maintenance, momentum unloading, and stabilization after large thrust maneuvers. These requirements have to be met from the beginning of the mission (namely immediately after the spacecraft is launched and injected into the orbit) to the end of the mission. Depending upon the spacecraft ballistic coefficient, orbit altitude and inclination, the drag varies and the spacecraft relocates itself to a lower altitude. This process continues and the error builds up over a period. Whenever the error or deviation (new altitude, perhaps new inclination) reaches a non-acceptable limit, then it is corrected by firing appropriate thrusters for calculated durations or until the error becomes zero. Such corrections have to be carried out many times over the spacecraft mission life. Thus, each correction requires a certain velocity requirement. The velocity requirements can be calculated using the formulas given in Chapter 1. The injection error correction velocity requirement is calculated as an example.

Example. Let us assume that a spacecraft is launched into an orbit altitude of 500 nm with a 10 nm altitude error (desired altitude being 510 nm) and 0.05 degrees inclination error. Let us attain the desired orbit through Hohmann ellipse and then circularize at the apogee and then change the plane by 0.05 degrees.

The characteristic velocity of 500 nm orbit is calculated using the following formula

$$V = [\mu/a]^{\frac{1}{2}}$$

where V is in feet per second, μ is 1.40766×10^{16} ft^3/sec^2 and a is the semi-major axis of the elliptical orbit or radius of the circular orbit and the characteristic velocity of 500 nm circular orbit is calculated to be 24236 ft/sec. Now the perigee velocity V_p, of the Hohmann ellipse with 500 nm perigee and 510 nm apogee is calculated using the following formula

$$V_p = [(2\mu/R_p) - (\mu/a)]^{\frac{1}{2}}$$

where R_p is the perigee radius (earth radius plus the perigee altitude, namely 3444 nm + 500 nm) and the perigee velocity calculated to be 24251 ft/sec.

Thus, to establish an Hohmann ellipse, an additional velocity equal to [24251 − 24236] = 15 ft/sec shall be imparted to the spacecraft at the perigee (as the original orbit is circular, wherever this velocity is imparted that point becomes the perigee of the Hohmann ellipse. Now, the spacecraft is circularized when it reaches its agogee. The velocity increment needed is calculated now. First the apogee velocity, V_a, is calculated using the following formula.

$$V_a = (V_p \times R_p)/R_a$$

where R_a is the apogee radius of the Hohmann ellipse and is equal to the earth radius plus the apogee altitude and is 3954 nm. The velocity calculated is 24190 ft/sec. The characteristic velocity of a spacecraft in a 510 nm circular orbit is calculated as before and is 24206 ft/sec. Thus, the velocity to be imparted to the spacecraft at the apogee point is equal to [24206 − 24190] = 16 ft/sec.

Now the inclination is corrected and the required velocity, V_i, is calculated using the following formula

$$V_i = 2 \times V(510 \text{ nm}) \times \sin(\theta/2)$$

where V (510 nm) is the characteristic velocity of the 510 nm circular orbit and the required velocity is calculated to be 21 ft/sec.

Thus, the total velocity required for the orbit injection error correction is 52 ft/sec. However, if the inclination correction is performed simultaneously with perigee raise, then the V required is equal to the square root of [16^2 + 21^2] or 26.4 ft/sec. The total V required is therefore 15 + 26.4 = 41.4 ft/sec.

Propellant Budget. Once the mission requirements are defined, then as mentioned above, velocity requirements for each task or operation can be calculated and typical velocity requirements are summarized in Table 6-3. For

Table 6-3. Velocity Requirements Budget.

Assumptions:	Monopropellant System Blow-down Approach N_2H_4 monopropellant	
Parameter		Velocity (ft/sec)
Orbit Injection Error Correction		52
Orbit Acquisition (initial plus subsequent)		34
Station Keeping		190
Stabilization after Large Maneuvers		54
Reaction Wheel Unloading		24
Contingencies		20
Subtotal		374
Margin (10%)		38
Total		412

contingencies, some margin is allowed. From this total velocity requirement, the propellant, P_r, requirement can be calculated using the following formula.

$$P_r = W[exp(\Delta V/(g \times I_{sp})) - 1]$$

where W is the weight of the spacecraft in orbit at the end of the mission (without the propellant weight added), g is the acceleration due to gravity and I_{sp} is the specific impulse of the propellant selected for the mission. Table 6-4 gives the specific impulses for some of the propellants. Assuming a monopropellant system has been selected and hydrazine is selected as the propellant, the propellant requirement to impart a spacecraft weighting about, say 3000 lbs, a velocity of 412 ft/sec is calculated to be 180 lbs. To carry 180 lbs of hydrazine, two or four tanks of appropriate sizes are chosen. Typical propellant tank pressures are, 1600 kpa of operating pressure, 2400 kpa of proof pressure and 3200 kpa of burst pressure.

Table 6-4. Propellants and Specific Impulses.

Monopropellants:

Hydrazine	220-230 seconds
Hydrogen Peroxide	180 seconds

Bi-propellants:

Oxidizer	Fuel	Specific Impulse (sec)
Chlorine Trifluoride	Hydrazine	240
Liquid Fluorine	Hydrogen	364
90% Hydrogen Peroxide	Hydrazine	242
Nitrogen Tetroxide	Ammonia	238
Nitrogen Tetroxide	Hydrazine	254
Nitrogen Tetroxide	Hydrogen	279
Nitrogen Tetroxide	Mono-Methyl-hydrazine	310
Liquid Oxygen	Hydrogen	347
Ozone	Hydrogen	375

6.6 SOLID PROPELLANT MOTORS

Sometimes the orbit circularization function is also carried out as a part of the on-board propulsion system. Most often this is so with respect to geosynchronous satellites that the launcher usually places the satellite in the geo-transfer orbit (also known as geo-transfer elliptical orbit) whose apogee altitude is the same as the geosynchronous altitude. Such a satellite usually carries a solid-propellant motor to provide the necessary total impulse at the apogee of the geo-transfer orbit to circularize the orbit (sometimes to achieve zero degree inclination simultaneously). Figure 6-6 shows such a solid propellant motor. Velocity requirements required for circularizing and correcting the inclination can be calculated as in Section 6-5. Using the solid propellant specific impulse, the solid motor propellant requirement or total impulse can be calculated and the appropriate motor can be selected from one of the many space qualified motors.

Fig. 6-6. Photograph of a solid-propellant motor (courtesy of RCA).

Chapter 7

Thermal Control System

THROUGHOUT VARIOUS PHASES OF EVERY SPACECRAFT MISSION, THERE ARE variations in the heat dissipated internally by components and also in fluxes incident on the outer surfaces. During the spin-stabilized transfer orbit phase, the power dissipation will be relatively low, compared to the on-orbit phase. In on-orbit phase, the spacecraft is either three-axis stabilized, spin stabilized or gravity-gradient stabilized depending upon the requirement by the instruments it is carrying. The power dissipation varies over a wide range depending upon the stabilization and mission modes of operation. Provision is usually made for survival of components through eclipse periods and for operation of the spacecraft with any or all of the experiments/instruments switched off. Each piece of electronic equipment operates best over a particular range of temperature. The thermal design becomes a challenge if the spacecraft carries an apogee motor and a dimensionally stable platform is needed for the antennas and attitude sensors due to their accurate pointing requirements. The thermal control system shall maintain the temperature of subsystems within their best operating range against above constraints or power dissipations.

7.1 CONSTRAINTS AND REQUIREMENTS

The spacecraft thermal design is largely dictated by absolute temperature requirements and temperature gradients. Spacecraft thermal system typical requirements are listed below:

Temperature. Table 7-1 presents subsystem design temperature levels for a spacecraft. Although these levels change from spacecraft to spacecraft, in general, electronics mounting interface temperatures are; (i) 0 to 40 degrees centigrade for all systems in general, (ii) 0 to 20 degrees centigrade for storage batteries and (iii) – 10 to + 50 degrees centigrade for propulsion system. Batteries require a small operating temperature range and almost identical temperatures for parallel operating units. High temperatures limit the reliability and lifetime of transistors due to increased electro-migration effects. Shunts and thickfilm devices, which are high-power density elements, usually create hot spots. Some experiments require high accuracy pointing or optical requirements, which are most likely to be affected by thermally induced structural deformation.

Most of the missions require as much power at the beginning of the mission as at the end. However, the solar array produces more power at the beginning of life. Ideally, constant dissipation makes thermal design easy. On the other hand, this necessitates large radiation capability at the beginning of life and a smaller radiation at the end of life. However, modern power systems control the solar array electrically such that it provides constant power to the spacecraft over the entire lifetime.

Usually suitably sized energy storage batteries are required if the spacecraft's full operation is required. During eclipse, thermal problems become easy if there is sufficient heater power. The whole spacecraft cools down as

Table 7-1. Subsystem Design Temperature Levels.

Subsystem	Operating Design Temperature (degrees centigrade)	
	Maximum	Minimum
Communications	65	– 5
Becon	50	– 0
Transmitter	60	10
Antenna	70	– 5
Telemetry, Tracking, & Command	50	– 5
Solar Array Mechanical Assembly	90	– 80
Power System	60	0
Attitude Control System	60	0
Reaction Control System	50	5
Batteries	20	0

Note:-   The design temperatures shown represent the typical range within the subsystem. Individual components may have other design temperatures within the design range shown. Acceptance temperatures are defined as design maximum plus 5 degrees centigrade and design minimum minus 5 degrees centigrade. Qualification temperatures are defined as design maximum plus 10 degrees centigrade and design minimum minus 10 degrees centigrade.

it enters into the eclipse and the thermal control system shall keep the operating units above the lower operating temperature limit and the non-operating units above the minimum non-operating temperature limits.

Heater Power. This depends upon various factors and are typically in the range of 5% to 10% of the spacecraft power.

7.2 PRIMARY FUNCTION OF THERMAL CONTROL SYSTEM

The primary function of the thermal control system is to maintain nominal temperatures for all components on-board a spacecraft in all external environments and under all operational modes. Table 7-2 defines the temperature ranges and limits. As there is no or negligible air at spacecraft orbital altitudes, heat is transferred by conduction through solids and radiation through space. Temperature can be decreased by radiating the heat energy via radiator into deep space. Temperature can be increased by using absorbers, which absorb solar or albedo energy with a high absorption or electrically

Table 7-2. Definitions of Temperature Ranges and Limits.

Name	Definition
Operating Temperature Range	Operating Temperature Range for an equipment is that range which is guaranteed by the spacecraft thermal design in all operating cases.
Non-operating Temperature Range	Non-operating Temperature Limits are those temperatures which equipment may experience in the spacecraft while in the non-operating mode. These temperature limits are normally the same as storage temperature limits of the equipment.
Storage Temperature Range	Storage Temperature Limits of an equipment are defined on the basis of the maximum and minimum expected manufacturing, storage, transport, integration and handling temperatures to be experienced by the equipment, plus a margin, and refer to the unpowered or non operating state.
Switch on/off Temperature Range	Switch on/off Temperature Limits are those limits at which the equipment can be switched on or off at any temperature within this range without permanent damage or start-up problems.
Survival Temperature Range	Survival Temperature Range are the maximum survival temperature limits which the equipment might experience for a certain time without damage under non-energizing conditions.
Qualification Temperature Range	Qualification Temperature Limits are outside the operating temperature range by a significant margin of at least 10 degrees centigrade.
Flight Acceptance Temperature Range	Flight Acceptance Temperature Limits are outside the operating temperature range by a small margin of at least 5 degrees centigrade.

generated heat via a solar array and electronic boxes or even through energy previously stored in batteries and heaters. However, in practice radiation, absorptivity and electrical energy consumption have to be balanced.

Heat dissipated by internal equipment is conducted to radiator surfaces, which then reject the heat to space. Radiator surfaces are finished with selected coatings that minimize the external flux absorbed while maximizing the radiation to deep space. External surfaces not needed as radiators are covered with multilayer insulation blankets, and heaters are implemented to maintain acceptable temperatures during nonoperating modes or eclipse periods.

Temperature control is achieved by two methods, i.e., passive and active methods. Multilayer insulation blankets, second surface mirrors, paints, tapes, and louvers are used in the passive control method external to the spacecraft to control the temperature by radiation. Flat mounting surfaces for good thermal contact, interface filters, namely, silicon grease to improve the interface conduction, doublers, Teflon standoffs for conductive insulation and heat pipes are used internal to the spacecraft. Heater wire or mats and peltier elements are some of the components used for active thermal control. To compensate for the internal spacecraft power dissipation variations, commandable thermal control heaters are used. In addition, commandable heaters are also used to control temperatures within the propulsion system.

7.3 COMPONENT LOCATION

Each side of a spacecraft receives different levels of solar, albedo and earth fluxes, depending upon the spacecraft orbit and orientation. The most effective radiating surface (anti-sun side) on a spacecraft absorbs the least flux and sees minimal variation. In locating internal equipment, high dissipators should be mounted on the best radiating surface. Equipment which must operate within a narrow range should be located on a surface that sees little variation in external flux or within the spacecraft if possible.

Mounting equipment directly to the radiating surface of a spacecraft is desirable. Good heat conduction paths can be provided by flush mounting and should be provided for equipment with high power dissipations, such as batteries and transmitters. Mounting equipment on top of angles, channels, and brackets restricts the heat flow capability from the equipment to the spacecraft structure and should be minimized.

7.4 ANALYSIS

The temperature of an area or a spacecraft depends upon the heat input and output:

$$H_r = H_a + H_i$$

where H_r = heat radiated from the spacecraft
H_a = heat absorbed by the spacecraft
H_i = heat generated internally

The above equation can also be written as

$$[Q(\text{diss})/A(\text{rad})] + a(S+A) + e \times E = e \times F \times \sigma \times T^4$$

where e = emittance
    σ = Stephan Boltzmann constant, 5.67×10^{-8} watt/(m²) (K¼)
    a = absorptance
    F = Graybody radiation factor
    T = Temperature, degrees K
    A = Area of the spacecraft or surface of interest, m²
    S = Solar flux, watts/m²
    A = Albedo flux, watts/m²
    E = Earth flux, watts/m²
    Q = Power dissipated, watts

For detailed analysis, a spacecraft or piece of equipment is divided into a number of discrete nodes. The nodes are coupled conductively to analyze the steady state and transient conditions under different environments and in various operating modes.

The temperature in any spacecraft that experiences eclipse in its orbit varies over a wide range. The amount of temperature variation depends upon the mass of the spacecraft. The heavier the spacecraft the lower the temperature variations and vice-versa.

7.5 THERMAL CONTROL HARDWARE

External radiator surfaces are provided on spacecraft for rejecting internally generated heat to space. These surfaces are covered with thermal control coatings that minimize heat flux input absorbed and maximize radiation of heat to space. These coatings have a low solar absorptance and a high infrared emittance. Table 7-3 defines the surface properties of some common thermal control coatings. End of life properties must be considered in designing thermal control because degradation of some materials is significant. The amount of degradation varies as a function of time as well as the orbit.

External surfaces not needed as radiators are covered with multilayer insulation blankets to isolate the spacecraft from fluctuations in the external environment, minimize cold case heater power requirements, and reduce gradients across structures. Blankets are also used to isolate internal components when necessary. The basic blanket consists of several layers of aluminized kapton and dacron net separators. All blankets are electrically grounded and provide for venting during launch.

Heaters and Thermostats. To insure that minimum allowable temperatures are maintained during cold case conditions, thermostatically controlled thin-film heaters are bonded to equipment. The heaters are strips of kapton with etched foil heating elements and welded power leads. Heaters are also bonded on structures to maintain temperature levels and gradients

Table 7-3. Thermal Properties of Some Materials.

Material	Absorptivity	Emissivity
Aluminum (polished)	0.10	0.05
Aluminum silicone paint	0.25	0.28
Silvered Teflon	0.14	0.76
Silicon Oxide on polished metal	0.10	0.90
Titanium	0.80	0.18
White Paint (epoxy base)	0.22	0.81
Black Paint*	0.84	0.80
Gold	0.40	0.06
Stainless steel, polished*	0.40	0.05
Ablative material	0.90	0.90
Second Surface Mirrors, 0.15-mm silvered fused silica	0.07	0.78

*very low degradation

consistent with interface and alignment requirements. In all applications, primary and backup redundant sets of heaters should be implemented and controlled by redundant mechanical thermostats with predetermined set-points.

Louvers. Radiator surfaces can be covered with thermal louvers that dampen the temperature variations caused by fluctuations in internal heat dissipation and orbital fluxes. Louvers are designed to operate continually for an extended period of time, modulating baseplate heat rejection capability as a function of baseplate temperature. Depending on the incident environmental flux levels and heat rejection requirements, sun-shields can be added to reduce absorbed solar irradiation.

Louvers consist of highly polished aluminum blades suspended on their ends along their length onto a frame, driven by bimetallic temperature sensors. The blades close at low temperatures to reduce heat lost to space. As the temperature increases, bimetallic temperature sensors apply torques that cause the blades to rotate to the open position increasing the heat rejected to space. Since the effective emittance to space changes with temperature, utilizing louvers greatly reduces any heater power requirements.

Heat Pipes. Heat pipe exhibits a thermal conductivity greatly in excess of metals such as copper, silver, and aluminum and transports thermal energy at efficiencies greater than 90 percent. Thus, heat pipes can be used to transport heat to a radiator surface located away from a piece of internal equipment (without a radiator) and help distribute heat dissipation to minimize hot spots and provide heat radiation at moderate temperatures. The heat pipe system includes heat pipes, evaporator and condenser sections, radiating fin, and support structure. The principal control function is supplied by the heat pipe, which operates as a thermal switch to provide heat rejection when needed and minimizes supplemental heater power.

A heat pipe usually contains two channels internally, one much smaller than the other, with many small holes between the two. The heat pipe, as mentioned above, contains a working fluid (usually methanol or ammonia) and is closed at both ends. For easy understanding, it can be divided into three sections along its length, namely, the evaporator section, the transport section, and the condenser section. Heat pipe is usually mounted such that the evaporator section is over the base plate of the equipment generating considerable amount of heat over a small area and the condenser section is exposed to space or anti-sun side of the spacecraft. When the equipment generates a considerable amount of heat, the working fluid inside vaporizes by absorbing that heat. Because of the pressure buildup in that section, the working fluid flows carrying the heat through the transport section to the condenser section. Since the condenser section of the heat pipe is exposed to space or over a large radiator, the vapor condenses and gives up heat to the walls of the pipe. The heat conducts through the pipe walls into the surrounding environment or to the radiator attached to the condenser section. The condensed liquid then flows through the small holes by capillary action back to the evaporator section and no pump is required.

In addition to these components, isolators, conductive interface fillers, doublers, saddles and/or conductive straps are also utilized.

Isolators. Isolators reduce the conductive heat transfer between components. Typical materials used as isolators are titanium and inconel.

Doublers. Doublers are thick conductive plates to improve the heat transfer within the mounting platform. Thermal doubler plates are employed to distribu·e highly concentrated heat dissipation by some units over a larger area than the base of the unit itself.

7.6 EXAMPLE

Let us assume an electronic box dissipating 50 watts must operate between zero degrees C and 40 degrees C and the spacecraft is earth-oriented in an orbit with an inclination of 28.5 degrees at an altitude of 450 km.

The orbital average flux levels incident on the various sides of the spacecraft are given in Table 7-4. Since surface-1 absorbs the least amount of external flux, the box should be located on that side of the spacecraft. The view factor to space for this surface is 1.0. For silvered Teflon, the surface properties at the beginning of life are a/e = 0.08/0.76. The absorptance at end of life for this mission degrades to 0.17.

Table 7-4. Orbital Average Fluxes Incident on Spacecraft.

Surface	Solar (W/m2)	Albedo (W/m2)	Earth (W/m2)
1	0	40	59
2	424	0	0
3	5	40	59
4	28	126	186
5	289	41	59
6	288	40	59

The first step is to size the radiator area needed to maintain the temperature below 40 degrees. A hot case energy balance is performed using end of life properties to size the radiator, using the following relationship:

$$[Q(\text{diss})/A(\text{rad})] + a(S+A) + e \times E = e \times F \times \sigma \times T^4$$

The radiator size is calculated to be 0.14 m². If the box is mounted to 0.14 m² of silvered Teflon radiator, the temperature will not exceed 40 degrees C.

Now, let us calculate the power required to maintain the box at or above zero degrees C in the cold case, when only earth flux is incident, using the following relationship.

$$[Q(\text{diss})/A(\text{rad})] + e \times E = e \times F \times \sigma \times T^4$$

Power required is calculated to be 27 watts. Thus, a minimum of 27 watts must be available in the cold case to keep the temperature above zero degrees C. If the box is not operating, the power should be supplied by heaters.

If the spacecraft is limited in power, utilizing thermal louvers would reduce the heater power required.

Chapter 8

Mechanisms

V ARIOUS MECHANICAL DEVICES KNOWN AS MECHANISMS ARE EMPLOYED ON-
board the spacecraft to carry out various important functions, like antenna
deployment, antenna pointing, solar array deployment, solar array drive to
continuously point the solar array perpendicular to the sun, despin drive, etc.
Such mechanisms include deployment hinge and latching mechanisms, rotating
drives, rotational and linear pointing mechanisms. Usually motor drives like
the solar array drive, etc., are controlled throughout the life of the mission
but some mechanisms that operate only once such as the deployment of
antennas and solar arrays are achieved by means of spring loaded or
pneumatically operated actuators. These devices are released by electrically
fired pyrotechnic mechanisms. Some of the mechanisms are described below:

8.1 ANTENNA POINTING MECHANISM

Stabilizing or controlling the attitude of a communications satellite by
means of momentum or reaction wheels results in a common alignment of all
antennas on-board the spacecraft. In some applications, however, it is desirable
to provide individual adjustment of particular antenna beam directions.

For this purpose, antenna pointing mechanisms providing one-axis or two-
axis alignment are employed. One such mechanism developed and
manufactured by TELDIX is shown in Fig. 8-1 and stepper motors are used
for tilting the antenna mounting fixture. The latter is suspended in frictionless
flexural pivots which, due to their spring force, provide fail safe operation by

Fig. 8-1. Antenna pointing mechanism (courtesy of TELDIX).

returning the antenna to zero position in case of electric power failure. Microsyn pick offs can be applied if the angular position of the mechanism is to be measured for telemetry. During launch, the pointing mechanism is rigidly clamped to zero position by two strong caging brackets held in place by a steel cable. Upon telecommand, the cable is cut by a pyrotechnic guillotine so that the caging is released. Table 8-1 presents TELDIX Antenna Pointing Mechanism Technical Data.

8.2 SOLAR ARRAY DRIVE AND POWER TRANSFER ASSEMBLY

The need for the solar array drive and power transfer assembly is explained in Chapter 3 and is also known as *bearing and power transfer assembly* (BAPTA). Conventionally, slip rings are employed for power transfer from the rotating solar array to the fixed spacecraft. However, with future high-power communications satellites requiring a primary power in the kilowatt range, it will be difficult to transfer the electric power from the solar generators to the satellite body via slip rings. In order to create a more favorable alternative, TELDIX has developed a BAPTA that is described here. It performs the rotation of the solar array without mechanical contact by means of a rotary transformer.

Obviously, there is a basic difference as compared to the dc power supply systems used in satellites up to now since the inductive principle used in a

Table 8-1. Antenna Pointing Mechanism Technical Data (Courtesy of TELDIX).

Pointing Mechanism, Type		AM 1	AM 2	AM 2
Number of axes		1	2	2
Stepper motor, type		SMZ	SMZ	SMR
Range of rotation	deg.	± 3	± 3	± 3
Step angle				
toothed motor (SMZ)	arc sec.	112	112	–
friction motor (SMR)	arc sec.	–	–	12
Max. step frequency				
at 1 Nm load	Hz	250	250	–
at 0.5 Nm load	Hz	–	–	130
Max. slew rate	deg/sec.	7.7	7.7	0.43
Position sensing		step counting	step counting	mocrosyn voltage
Power consumption at 1 Hz	W	3	3	2.4
Max. weight of antenna	kg	15	15	15
Max. weight of AM	kg	1.1	2.9	3.0
Dimensions of AM				
length	mm	145	225	225
width	mm	140	150	150
hight	mm	120	125	125

transformer requires alternating current. For this reason, a static dc/ac converter is to be installed on the solar array. The ac bus in the satellite provides the additional advantage of simplified generation of a multitude of supply voltages and, at the same time, of positive dc isolation of all subsystems (see Fig. 8-2).

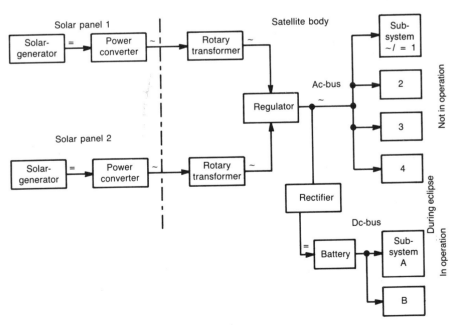

Fig. 8-2. Block diagram of a power system with ac bus (courtesy of TELDIX).

In addition, rotary transformers can be employed on earth observation and research satellites that carry high resolution instruments. On such satellites, statistical variations of the friction torque caused by slip rings during solar panel rotation can result in satellite attitude changes since the angular momentum of the overall system remains constant. This phenomenon may considerably affect the functions of the various instruments. Such a source for disturbances is eliminated by using rotary transformers. For the types of satellites stated above, the power consumption is in most cases lower, but the electrical power must also be available during the eclipse period of the orbit. Therefore, such a satellite would operate on the buffer battery and not on an ac bus as shown in Fig. 8-2.

Design. The transformer windings (see Fig. 8-3) are accommodated in an almost cylindrical housing. Adapters are used for connecting the housing (stator) with the satellite body and the rotor with the supporting tube of the solar panel. The adapters permit a universal adaptation to the satellite structure. Unlimited rotor rotation is possible in both directions.

To obtain an increase in reliability or in power transmission capacity, the windings are split into two partial transformers. The two primary windings are installed on the pivotable inner tube connected to the solar panel while the secondary windings are fixed to the housing. The electrical leads are provided with DIN terminals. Each partial transformer contains its own ferrite core with a concentric radial air gap.

Bearing. The rotor of the transformer is suspended in two axially preloaded angular contact ball bearings mounted in a back-to-back arrangement. The preload is chosen as a function of the launch conditions. There is no caging mechanism required for the launch phase.

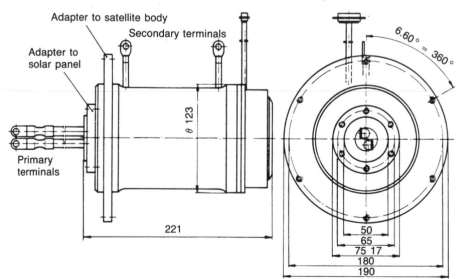

Fig. 8-3. Side and cross sectional views of cylindrical housing for transformer windings of BAPTA (courtesy of TELDIX).

The bearings have been designed for high vacuum operation and exhibit extremely low friction torques. Because of their high accuracy, the bearings ensure a steady transfer function versus the angle of rotation of the transformers.

Drive System. For rotating the solar array, the rotor of the stepper motor is used. The stator and rotor of this motor are flanged to the face of the housing and to the interior tube of the motor, respectively. Thus the motor does not require bearings of its own. The motor offers minute stepping and is particularly suited for the accurate and steady slaving of the solar panels. For communication satellites, the nominal speed of the motor is, for instance, one revolution per 24 hours. For research satellites that may require faster slewing operations, various motor versions are available to meet the specified step angle, accuracy and stepping frequency.

Signal Transmission Slip Ring Unit. For the transmission of measuring data from the solar panel to the satellite body, a small slip ring unit can be mounted inside the interior tube (rotor) of the motor. The stator of that unit is to be connected to the motor housing.

Figure 8-4 shows the photograph of a bearing and power transfer assembly (BAPTA) manufactured by TELDIX and its technical data is presented in Table 8-2.

Fig. 8-4. Bearing and power transfer assembly (courtesy of TELDIX).

8.3 GIMBALS

Gimbals are employed when angular freedom is required and gimbals can be of single-axis, two-axis or three-axis type. Various satellite components like the solar array drive, momentum wheels, or experiments can be mounted on

Table 8-2. Bearing and Power Transfer
Assembly (BAPTA) Technical Data (Courtesy of TELDIX).

Power rating	3 kVA (2 × 1.5 kVA)
Frequency	10 khz (sinusoidal)
Voltage	
Primary	35 VRMS
Secondary	163 VRMS
Turns ratio	4.75
Efficiency	>98.6%
Nominal speed	
Geostationary satellites	1 revolution per 24 hours
Earth research satellites	1 revolution per 1.5 hours
Speed for initial alignment	1 revolution per 30 minutes
Dimensions	
Housing diameter	123 mm
Housing length	221 mm (without electrical leads)
Rotor inner tube diameter	45 mm
Bore dia. for electrical leads	40 mm
Weight including stepper motor SMR 1-0	7.3 kg
Bearings	
Type	Angular contact ball bearings
Preload, axial	Optional, typical 40 to 60 N
Tolerances	ABEC 7
Lubrication	For high vacuum conditions
Drive system	
Motor	Stepper motor TELDISTEP® Type SMR 1-0
Step angle	18 arc sec
Step tolerance	±25%
Stepping frequency	$\leq$ 130 Hz at $\leq$ 0.5 Nm load
Speed	$\leq$ 1 revolution per $\leq$ 10 minutes at $\leq$ 0.5 Nm load
Dynamic torque	0.7 Nm
Holding torque	0.5 Nm
Supply voltage	15 Vdc
Power consumption	$\leq$ 2.4 W at 1 Hz
Environmental conditions	
Temperature	$-20°C$ to $+50°C$
Pressure	10^{-5} mbars (typical)
Vibration	Suitable for launchers like the
Linear acceleration	ARIANE or the Space Shuttle
Life	7 to 10 years

these gimbals to give them angular freedom. A double gimbal system developed and manufactured by TELDIX is described here.

The double gimbal systems (types SDR and KDR) have been designed for satellite attitude control systems using momentum wheels for three-axis stabilization. If a momentum wheel is fixed directly to the satellite structure, it stabilizes the satellite simply by its gyroscopic forces. A momentum wheel mounted on a double gimbal system offers the additional possibility of tilting the satellite by a few degrees with respect to the wheel. Thus, the angular position of the satellite can be adjusted with higher accuracy.

The inner gimbal system is especially suited for applications requiring redundant momentum wheels, which can conveniently be suspended on both

Table 8-3. Double Gimbal Technical Data (Courtesy of TELDIX).

Weight	< 5.5 kg, with 2 redundant stepper motors and analog pick-offs
	17.3 kg with 2 Momentum Wheels DR 25
	20.5 kg with 2 Momentum Wheels DR 50
Dimensions (envelope)	350 mm diameter, 380 mm height with 2 Momentum Wheels DR 25 or DR 50
Mechanical Interface	at customer's specifications
Gimbal Freedom	± 7.5° each axis
Actuator	stepper motors (two per axis)
Resolution	0.2 to 2.0 minutes of arc per step (optional)
Torque output	0.7 to 1.5 Nm
Max. step frequency	130 to 250 Hz
Pick-off	Two Microsyn analog pick-offs per axis
Bearings	Ball Bearings with Ceramic Balls 42 x 20 x 12 mm, 2 pairs per axis
Environmental conditions	
Linear acceleration	19 g along all axes
Vibration	
a. Sinusoidal	5 g from 5 Hz to 400 Hz 7.5 g from 400 Hz to 2000 Hz
b. Random	Increasing at 3 dB/octave from 10 HZ to 0.2 g^2/Hz at 70 Hz
	Constant at 0.2 g^2/Hz from 70 Hz to 1000 Hz
	Decreasing at 7 dB/octave from 1000 Hz to 2000 Hz
Shock	30 g, 8 ms half sine
Temperature	−20° C to + 60° C (Qualification level)

ends of the inner gimbal. The gimbal system as such is not redundant but its rugged construction meets the specified reliability requirements.

The gimbals may be suspended either in ball bearings or in flexural pivots. The former offers a relatively high load capacity but produce friction forces whereas the latter operate without friction but have a limited load capacity when low restoring torques and center shifts are required. Controlled deflection of the momentum wheel is provided by actuators and pickoffs mounted on both gimbal axes. Brushless torquers or stepper motors (friction or toothed version) may be used as actuators. The optimal selection depends on the overall altitude control concept of the satellite (e.g., analog, digital, or pulse-train operation). The same applies to the selection of the pickoffs, which may be synchro resolvers or other inductive transducers. Table 8-3 presents the TELDIX double gimbal system technical data.

Figure 8-5 shows two views of the *biaxial drive assembly,* Fig. 8-6 shows the photograph of a biaxial drive assembly manufactured by Schaeffer Magnetics Inc., and Table 8-4 presents its specifications.

8.4 DESPIN DRIVE MECHANISMS

Despin Drives are used for antenna pointing, and experiment pointing, etc. Figure 8-7 shows the Orbiting Solar Observatory Platform Despin Drive developed and manufactured by Ball Brothers Research Corporation and Table 8-5 presents its design specifications. Table 8-6 presents the specifications of the antenna despin drive used on SKYNET spacecraft.

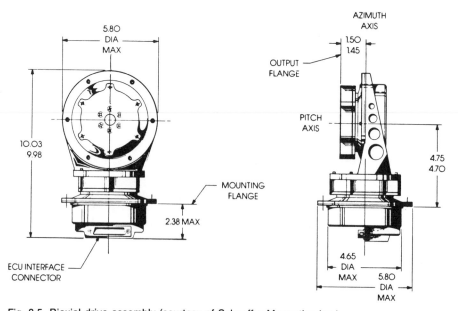

Fig. 8-5. Biaxial drive assembly (courtesy of Schaeffer Magnetics Inc.).

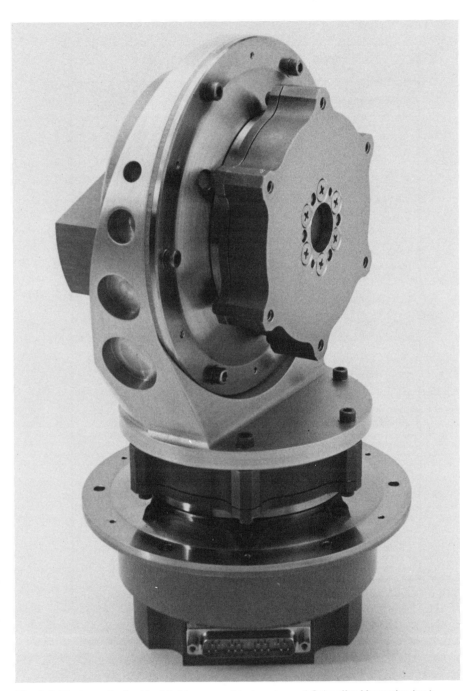

Fig. 8-6. Photograph of a biaxial drive assembly (courtesy of Schaeffer Magnetics Inc.).

Table 8-4. Biaxial Drive Assembly Specifications (Courtesy of Schaeffer Magnetics Inc.).

TYPE 55 SPECIFICATIONS

SPECIFICATION	UNITS	PITCH AXIS	AZIMUTH AXIS
OUTPUT STEP ANGLE:	DEG.	.0075	.0075
STEPS / REVOLUTION:	–	48,000	48,000
MAXIMUM STEP RATE: SLEW RATE:	STEPS/SEC (DEG/SEC)	300 (2.25) 1200 (9.00)	300 (2.25) 1200 (9.00)
POWER: (NOMINAL)	WATTS	12	12
OUTPUT CAPABILITY: INERTIAL: FRICTIONAL:	SLUG-FT² IN-LBS	350 650	350 650
HOLDING TORQUE: POWERED: UNPOWERED:	IN-LBS IN-LBS	960 250	960 250
DRIVE TRANSVERSE STIFFNESS:	IN-LBS/RAD	PERPENDICULAR TO PITCH-AZIMUTH PLANE 250,000	IN THE PITCH-AZIMUTH PLANE 85,000
DRIVE ROTATIONAL STIFFNESS:	IN-LBS/RAD	90,000 NOMINAL	
OUTPUT FLANGE LOAD CAPABILITY: AXIAL: TRANSVERSE: MOMENT:	LBS LBS FT-LBS	810 1820 220	
DRIVE WEIGHT:	LBS	9.5	

• Drive transverse stiffness is resistance to bending of the output flange.
• Drive rotational stiffness is resistance to rotation of the output flange.
• Indicated values are estimated minimums. Higher capability exists.
• Optional brushless DC motors are available for continuous drive.

Table 8-5. Despin Platform Drive Specifications (Courtesy of Ball Aerospace Systems Division).

Specification Range*

WEIGHT
Less than 21.5 lbs.

SPEED
0 to 120 RPM

POWER CONSUMPTION
Less than 9 watts

TORQUE AVAILABLE
More than 180 oz. in. stall

SLIP RING
27 power circuits at 3.0 amps ea.
58 signal circuits at 0.1 amps ea.
5 grounding circuits

*Range may be extended by design to specific requirements.

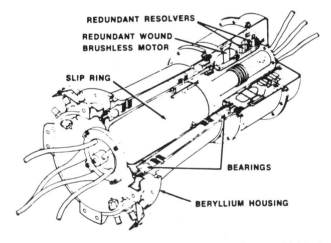

REDUNDANT RESOLVERS
REDUNDANT WOUND BRUSHLESS MOTOR
SLIP RING
BEARINGS
BERYLLIUM HOUSING

Fig. 8-7. Platform despin drive (courtesy of Ball Aerospace Systems Division).

Specification Range*
LIFE
Greater than 5 years
WEIGHT
Less than 4 lbs.
SPEED
0 to 200 RPM
POWER CONSUMPTION
Less than 6 watts
TORQUE
More than 28 oz. in.
POINTING ACCURACY
Better than 0.5 degrees

*Range may be extended by design to specific requirements.

Table 8-6. Antenna Despin Drive Specifications (Courtesy of Ball Aerospace Systems Division).

Chapter 9

Spacecraft Testing

TO MAKE SURE THAT A SATELLITE IS READY FOR LAUNCH, A SATELLITE IS repeatedly tested under actual operating conditions, namely the environment, mission phases, and diverse operating modes. Then from the test results, it can be assured that the satellite will work properly for the duration of the mission.

As satellites are quite expensive to replace, it is essential to make sure by detailed testing that it works reliably during its entire mission. Although, the component statistical failure rate data is available, as discussed in the chapter on Reliability, the design reliability is equally important in achieving the overall reliability of the satellite and it can only be verified by testing. Also the environment under which the component failure rate is evaluated is different when it is integrated and becomes part of an unit or a system.

9.1 SPACE ENVIRONMENT

The spacecraft and its building blocks are extensively tested to ensure adequate performance under space environment, which is much different from that in which they are built. Maintenance and repair is not anticipated as it is many times not practicable and when it is practicable it would be a costly affair. Thus, space environment under which a spacecraft has to work successfully is described briefly here.

Space environment features zero gravity, high vacuum, solar radiation, particle radiation, etc. For example, gravity affects the deployment of solar arrays and antennas, and liquid motion in fuel tanks, heat pipes, and batteries.

The Solar System. The solar system consists of the sun as a central body, nine planets, their satellites, numerous asteroids, and countless smaller objects such as comets and meteors. All of these bodies revolve around the sun. The asteroids are small irregular shaped bodies of 1 to 500 miles in size. Comets have an atmosphere which is formed and lost during passage through the inner solar system and is a few miles in diameter.

Space. Space is filled with a low density gas mixture, consisting primarily of hydrogen, helium, protons, and alpha particles. The estimated gas pressure in interplanetary space is approximately 10^{-18} N $\times$ m^{-2} (10^{-16} mm Hg) is lower than 10^{-27} N $\times$ m^{-2} in interstellar space. The vacuum chambers used for testing spacecraft and its building blocks have typically a pressure of 10^{-8} N $\times$ m^{-2} (10^{-6} mm Hg).

The high vacuum in space vaporizes the volatile materials. This might cause electrical short circuits, change of surface emissivities, or degrade mirrors and solar cell covers. So proper care should be taken when using such materials in space. Also the metallic vapors might condense on solar cell cover glasses and result in solar array degradation.

The Space Plasma. At geosynchronous altitudes, the magnetosphere is filled with an energy of 1 eV plasma with a typical density of 10 particles per cm^3. During magnetic substorms, large electric fields develop across the magnetotail, and the electric and magnetic field convection drives a new plasma with an initial energy of 100 eV toward the earth. As magnetic fields are strong near earth, it heats up considerably.

Also at times at geosynchronous altitudes low energy plasma is replaced by high energy (10 to 20 keV) plasma. During these times, electrostatic charging of exposed surfaces of any orbiting bodies takes place. The charging develops large differential voltages between various surfaces, especially between illuminated and shadowed dielectric surfaces. When these voltages exceed the breakdown potentials of the materials involved, arc discharges are produced. Such discharges may interfere with on-board electronics or communications and it can cause some malfunction or damage to the spacecraft. Proper care is taken to avoid any problems due to such charging by adequate shielding, and proper grounding, etc.

Meteoroids. Small particles in interplanetary space are called meteoroids. Even though the size of these particles are so small (of the order of one millimeter in diameter) due to their higher velocities (20 to 70 km per second, their impact causes damage to spacecraft components. They erode the solar cell cover glasses, and thermal coating, etc.

Gravity. The gravitational pull of the earth decreases approximately with the square of the distance from the center of the earth and is close to zero in space. Thus depending upon the spacecraft orbital altitudes, the gravitational pull can be milli, micro, or zero. Zero gravity does not cause any problem to spacecraft. However, absolute zero gravity rarely exists on the complete spacecraft as gravitational force may not pass through the spacecraft mass center, which results in a torque. This torque can degrade attitude accuracy. However, it can also provide desirable attitude control if the spacecraft is properly configured. Gravity gradient stabilized spacecraft utilizes these torques.

Magnetic Fields. The earth's magnetic field strength varies from approximately 0.30 to 0.35 gauss at the equator to approximately 0.65 to 0.70 gauss at the magnetic poles. With increasing altitude the field strength decreases approximately with the cube of the distance from the center of the hypothetical earth's magnet. A solar cell array also produces magnetic fields. The presence of residual magnetism or current loops on-board the spacecraft will result in a torque on the spacecraft due to the earth's magnetic field interaction. This can degrade the attitude accuracy. To minimize this effect, solar cell circuits are laid out such that the current loops produce no net torque. Just as with gravity gradient torque, this torque can be utilized for desirable attitude control.

Solar Radiation. The sun emits radiation in the wavelength range between 1×10^{-10} m (x-rays) and 30 m (radio frequency) and its black-body surface temperature is approximately 6000 degrees K. The space radiation environment is composed of cosmic rays, electromagnetic radiation, Van Allen belt radiation, auroral particles, and solar flare particles. Electromagnetic radiation includes ultraviolet lights, x-rays, and gamma rays. Particulate radiation consists of electrons, protons, neutrons, alpha particles, and others.

The number of particles, photons, or energy passing through a given area in a specified time, usually given in particles per square centimeter per second is known as the *radiation flux* and the total particles per square centimeter in any given time is known as the *fluence*. Radiation dosage can be expressed either in terms of the exposure dose, which is a measure of the radiation field to which a material is exposed, or in terms of the absorbed dose, which is a measure of the energy absorbed.

Bombardment of the various surfaces of a spacecraft by space radiation causes small but significant forces. Because the center of the pressure is not generally coincident with the center of mass, disturbance torques will result.

Figure 3-51(A) shows damage equivalent 1-MeV fluence in circular Earth orbits due to trapped electrons for I_{sc} and P_{mp} of silicon cells protected by 0.15-mm thick fused silica covers and infinitely thick back shields. Figure 3-51(B) multiplication factors for damage equivalent 1-MeV fluence shown in Fig. 3-51(A) for four different cover thicknesses for (a) 0-degree inclined orbits and (b) 90-degree inclined orbits.

Figure 3-52 shows damage equivalent 1-MeV fluence in circular Earth orbits due to trapped protons for I_{sc} and P_{mp} of silicon cells protected by 0.15-mm thick fused silica covers and infinitely thick back shields. Figure 3-53 shows multiplication factors for damage equivalent 1-MeV fluence shown in Fig. 3-52 for four different cover thicknesses for (A) 0-degree inclined orbits and (B) 90-degree inclined orbits.

Figure 9-1 shows average absorbed dose in 0.15-mm thick covers in circular Earth orbits due to trapped electrons. Figure 9-2 gives the average absorbed

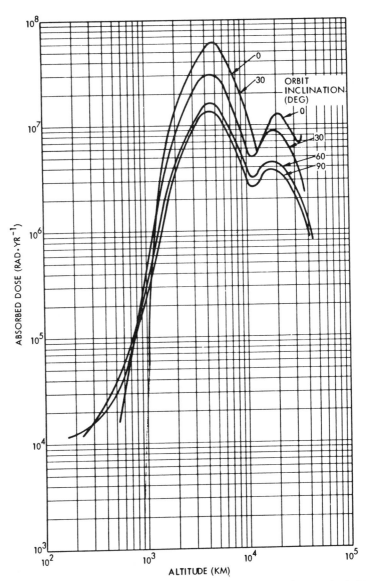

Fig. 9-1. Average absorbed dose in 0.15-mm thick covers in circular earth orbits due to trapped electrons (courtesy of Jet Propulsion Laboratory).

dose in 0.15-mm thick covers in circular Earth orbits due to trapped protons.

Albedo. The albedo of a body in space is the ratio of the amount of electromagnetic radiation reflected by the body to the amount incident upon it. The total reflected energy contains components due to reflections from clouds and scattering by the atmosphere.

Thus, the objective of spacecraft testing is to subject the entire satellite to a series of simulated environmental stresses for a period of time reasonable

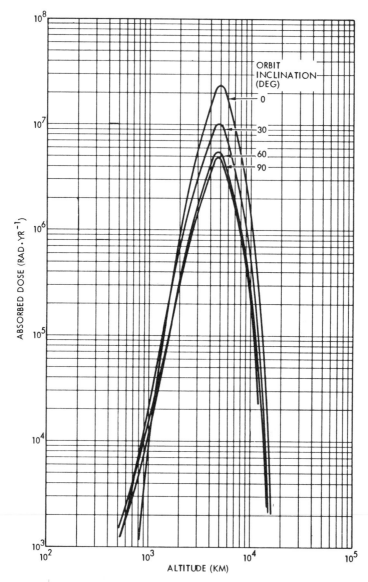

Fig. 9-2. Average absorbed dose in 0.15-mm thick covers in circular earth orbits due to trapped proton (courtesy of Jet Propulsion Laboratory).

enough to identify and eliminate failures due to improper designs, defects in workmanship, and defects in materials, etc. This follows the classical *bath tub curve*, as shown in Fig. 9-3. Early failures or infant mortalities are eliminated by conducting appropriate functional tests on the spacecraft and its components under various environmental conditions that a spacecraft will undergo from ground readiness, shipping, handling, launch, orbital operations, until it completes its mission in orbit. The environment can be mechanical (mainly

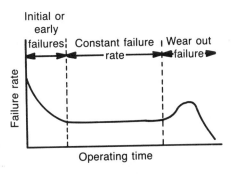

Fig. 9-3. Bath tub curve—failure rate predictor.

due to launcher and micrometeoroidal impacts), thermal (mainly due to orbital conditions, sun radiation, earth shadow, internal heat generation, etc.), electromagnetic (mainly due to other subsystems), radiative (mainly due to space particles, photons, electrons, etc.), vacuum, aging, etc. Functional tests include center of gravity test, moment of inertia test, boom extension, solar array deployment, antenna deployment, attitude stabilization, turning-on some systems or instruments, etc. The mechanical tests include shaking, vibration, noise, shock, center of gravity, dynamic balance, moment of inertia, etc., whereas the thermal tests include temperature soaking, temperature cycling, thermal-vacuum tests, etc. The electromagnetic tests include the EMI susceptibility, emission, etc.

9.2 TESTS

Tests are carried out to check some certain performance and some of the tests are described below:

Alignment Verification Tests. A detailed set of measurements are taken and the mechanical alignment of the critical surfaces of the spacecraft is determined to assure pointing accuracies of such items as the sensors, the momentum wheel, the reaction control nozzles, and the gyros.

Acceleration Test. The spacecraft structure is tested to demonstrate adequate structural design under the most severe acceleration loads expected during the launch phase.

Acoustic Tests. Capability to perform within acceptable limits under conditions of acoustic stress is verified.

Vibration Test. A vibration test is carried out to check the capability of the spacecraft structure to survive the qualification level sine and random vibration tests. Figures 9-4 and 9-5 show ULYSSES and EXOSAT spacecraft undergoing the spin balance test.

Shock Test. The spacecraft and components must survive the shock of detonating the spacecraft pyrotechnic devices.

Static and Dynamic Balance Test. To assure the balance of the satellite without nutation and coning motion, whenever it is in spin-stabilized mode, this test is carried out.

Mass Properties Measurements Test. The weight, center of gravity, and moments of inertia of the spacecraft are determined.

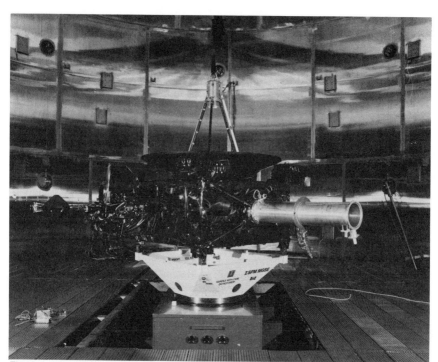

Fig. 9-4. Ulysses under spin balance test (courtesy of European Space Agency).

Fig. 9-5. EXOSAT under spin balance test (courtesy of European Space Agency).

Center of Gravity Test. It is carried out to ensure that the attitude control and orbit (altitude or plane) change operational fuel and/or energy requirements are proper or adequate.

Moment of Inertia Test. It is carried out such that proper measurements can be made for attitude control and despin operations.

Appendage Test. It is carried to make sure that the appendages, namely, antennas, solar arrays, booms, etc., deploy properly under space environment.

Antenna Pattern Test. Satellite antenna patterns are tested and checked to make sure that the ground station can receive the signals properly from the spacecraft.

Electrical Performance Tests. These tests are conducted to make sure that all connections are made properly and current and voltages are within their limits and all the equipment are performing as expected within required performance limits.

Magnetic Moment Measurement Test. The residual magnetic moment of the spacecraft is measured to permit calculating the magnetic moment disturbance torque prevailing in orbit.

Electromagnetic Compatibility Test. The spacecraft must have no spurious rf emissions that are likely to compromise the performance of the launch vehicle or support equipment. Figure 9-6 shows the SKYNET-4 in the EMC chamber.

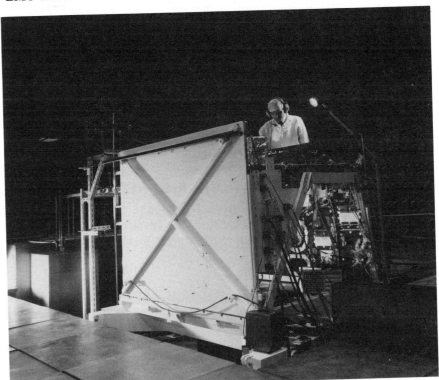

Fig. 9-6. Near-field, far-field antenna test range (courtesy of Marconi Space Systems).

Thermal Vacuum Test. This test is carried out to establish the capability of the spacecraft thermal control system to maintain component temperatures within the required envelope during on-orbit and transfer-orbit phases of the mission and to check spacecraft performance in vacuum under all probable operating configurations over the entire specified temperature range.

Solar Simulation Test. The capability of the spacecraft's thermal control system to maintain component temperatures within the required envelope during on-orbit and transfer-orbit phases of the mission is tested.

Corona Checks. Outgassing tests are conducted to demonstrate the minimum time to turn on/off all high-voltage and high frequency components when they are exposed to vacuum.

Leak Tests. Various spacecraft subsystems are tested for leakage subsequent to each spacecraft's exposure to vibration or vacuum conditions.

Ground Station Compatibility Test. Compatibility of spacecraft hardware and software signals related to tracking, telemetry, and command with ground station equipment is checked.

Combined Solar Simulation and Ground Station Compatibility Test. The capability of the spacecraft to perform as required within allowable temperature limits in orbit and its compatibility with ground station equipment is verified.

Integration Checkout and Electrical Compatibility Tests. The spacecraft subsystems are verified as they are mounted on the spacecraft for proper operation both individually and in combination with other systems.

Range Operations Tests. The spacecraft must have successfully survived shipment to the range and be ready for mating to the vehicle and for subsequent launch.

Performance Verification Tests. Functional tests are performed at the beginning, during, and at the end of each test to demonstrate the spacecraft's capability to operate within acceptable performance limits under test conditions.

9.3 TEST SPECIFICATIONS AND REQUIREMENTS

Test specifications will ensure that a satellite functions successfully throughout its mission. Specifications are generated based upon the environment experienced by the spacecraft from the place of manufacture to the end of life in orbit.

9.4 SPACECRAFT MODELS AND TESTING

Usually spacecraft testing is carried out using three spacecraft models, namely a dynamic/thermal model, an engineering or qualification model, and a flight model. The engineering model is qualification tested whereas the flight model is acceptance tested. Figure 9-7 shows the L-Sat/Olympus Spacecraft in Aeritalia Integration Room.

Fig. 9-7. LSAT/Olympus Spacecraft (courtesy of Marconi Space Systems).

Spacecraft Dynamic/Thermal Model Testing. The dynamic/thermal model is a full scale spacecraft structure made using mass dummy components and is tested to demonstrate the structural integrity of the spacecraft design. Usually after completion of the structural tests, the spacecraft is converted to a thermal model simulating thermal paths, component heat capacity, and heat dissipations and is tested for the thermal balance of the spacecraft.

Spacecraft Engineering or Qualification Model Testing. This model contains all the electrical systems and thermal control system and is identical to the flight configuration. Components are functionally identical to the flight components. Spacecraft engineering testing is carried out with qualification environment levels (which are more severe than those expected in flight) to make sure that complete operational capability, intersubsystem and intrasystem compatibility of the spacecraft exists and is capable of meeting the design requirements. The tests are conducted on an electrically functional spacecraft.

Spacecraft Flight Model Testing. These flight model qualification tests are performed with acceptance environmental levels on the spacecraft to make sure that the spacecraft performance meets the design requirements and is ready for launch.

9.5 TEST FACILITIES

To carry out all these types of tests, different types of test facilities are required and for better understanding, some of Marconi's test facilities are briefly presented here.

9.5.1 Clean Rooms

Dedicated, environmentally controlled areas, designed to the highest standards, are a prime requirement for satellite assembly, integration and test (AIT) procedures. Spacecraft are becoming increasingly complex and larger in size. In addition to four medium-sized clean areas, Marconi's newest AIT hi-bay clean room provides over 1000 square meters of floor space, sufficient for simultaneous construction and test activities for a number of satellites. This large, advanced clean room, conveniently situated adjacent to the near-field far-field antenna test range, includes an area in which specialized procedures may be isolated from other activities. A seismic block, 10 meters square, provides an area for optical measurements on a spacecraft's critical reference axes and other precise mechanical dimensions. The entire working area is serviced by an overhead crane and an adjacent clean storage bay.

The large AIT building complex has been built around a common reception and preparation area equipped with a traversing crane of 5000 kg capacity, having a hook height of 10 m. All medium-sized clean rooms provide class 100,000 clean area conditions and are designed for spacecraft system and subsystem test procedures. Figure 9-8 shows the MARECS spacecraft in the EMC/clean room. Figure 9-9 shows the Swedish Viking Satellite being integrated and Fig. 9-10 shows the ECS satellite solar array deployment under zero-G simulation.

The clean area has the following general data:

Operations area:	32 × 16 m (12 m high)
Floor loading:	1000 kg/m2
Main access door:	5 × 6 m high
Temperature control:	20 ±2 degrees C
Humidity:	35% – 60% relative humidity
Environmental	Class 3 to BS 5295 (class 100,000 to FED-STD-209B)
Crane capacity:	3000 kg
Maximum hook height:	10 m

Fig. 9-8. SKYNET-4 in the EMC chamber at Marconi Space Systems, Portsmouth, England (courtesy of Marconi Space Systems).

9.5.2 Vibration Test

Marconi Space Systems major vibration systems are housed in temperature and humidity controlled rooms and are equipped with hydrostatic bearing slip tables for operation in the horizontal mode. All vibrators are served by appropriate lifting and handling equipment.

FAR ULTRA-VIOLET SPACEBORNE
IMAGING SYSTEM DEVELOPED FOR
THE SWEDISH VIKING SATELLITE

Fig. 9-9. MARECS spacecraft in clean room (courtesy of Marconi Space Systems).

Cable booms associated with each vibrator enable up to 36 derived test signals to be routed to a central control room that conveniently overlooks the test areas. Intercommunication facilities are also included. Three modern digital control systems, capable of performing sine, random, and shock tests, are complemented by a number of sine vibration control instruments and by associated protection instruments. A wide range of piezoelectric accelerometers and associated signal conditioning, recording and analysis equipment forms part of the test instrumentation. Table 9-1 presents typical spacecraft vibration test requirements.

Electromagnetic Vibrators (7)

Max Sine Thrust (peak)	13000 N to 160000 N (36000 lbf)
Max Random Thrust (rms)	8000 N to 130000 N (30000 lbf)
Displacement	± 12.7 mm (0.5 in)
Frequency Range	5 - 5000 Hz

Hydrostatic Slip Tables (4)

Max Size	1020 × 1200 mm
Displacement	± 12.7 mm (0.5 in)
Frequency Range	5 - 2000 Hz

Fig. 9-10. ECS satellite solar array deployment under zero-G simulation (courtesy of European Space Agency).

Digital Controller (3)

Scope	Sine/Random/Shock (control & analysis)
Dynamic Range	up to 72 dB
Accuracy	± 1 dB

9.5.3 Acoustic Test

Spacecraft are subjected, during launch phases, to high intensity noise arising from rocket motors and from air turbulence. Acoustic test procedures on qualification and flight models are therefore often necessary, particularly for lightweight structures having large area-mass ratios: these tests simulate high frequency acoustic excitation during the launch phases.

9.5.4 Shock, Bump, Drop, and Topple

Marconi Space Systems can cater for all the environments in accordance with BS 2011, DEF STAN 07-55, and other specifications.

Item	Max Test Item Wt, Kg	g Range	Pulse Duration milliseconds
Bump Machine	BT 250	114	8 - 40	6 - 16
Shock Machine	Free fall Shocktest m/c 1 m	1.0	upto 2000	0.5 - 12
	Free fall shock test m/c 5 m	136	50 - 200	3.0 - 18
Drop Test	5m drop onto 25 mm thick steel plate 2400 × 2400 mm	2000		

9.5.5 Centrifuge

The Marconi Space Systems Environmental Engineering and Test Laboratory also contains a 1.2 m radius centrifuge. This unit will achieve 60 g on test items weighing up to 50 kg and is equipped with 60 slip rings to enable functional testing during acceleration. Test items of a size up to 0.5 × 0.5 × 0.5 m may be accommodated.

A smaller centrifuge is also available with a 0.3 m radius. This equipment is able to achieve 300 g with test items weighing up to 1.8 kg. In this case, the centrifuge is fitted with 50 slip rings for test item monitoring.

9.5.6 Optical Alignment Equipment

During launch phases and apogee motor burn periods a spacecraft is subjected to extreme vibration. Alignment measurements are essential to determine the effects of these conditions. This particularly applies to spacecraft attitude sensor viewing angles, individual experiment viewing angles and accuracies, and reaction control thruster axis variations. Alignment variations can also occur due to ground testing of spacecraft structures.

Marconi's optical equipment, comprising precision theodolites and seismic optical alignment blocks, are located within the AIT clean room facilities. The theodolites enable alignment checks to be made on the vertical and longitudinal planes about the spacecraft centerline to a high degree of accuracy. Measurement and adjustment of angular position is achieved by theodolite and rotary table to within 0.5 minutes of arc. Overall system accuracy alignment can be measured using an autocollimator to better than 5 seconds of arc.

9.5.7 Mass Properties Measurement Equipment

Spacecraft engineering demands highly accurate measurement processes

Table 9-1. Typical Vibration Test Requirements.

	Sinusoidal	Random	Resonance
Spacecraft			
Design Qualification	10-2000 Hz 21 g peak in thrust axis, 2 octaves per min	20-2000 Hz 0.08 g²/Hz 12 g rms acceleration 4 min per axis	500-650 Hz 0.5 octave per min, 50 g peak
Flight Acceptance	10-2000 Hz 14 g peak in thrust axis, 4 octaves per min	20-2000 Hz 0.04 g²/Hz 8 g rms acceleration 2 min per axis	500-650 Hz 1.0 octave per min, 30 g peak
Subsystems & Instruments			
Design Qualification	10-2000 Hz 32 g peak in thrust axis, 2 octaves per min	20-2000 Hz 0.08 g²/Hz 12 g rms acceleration 4 min per axis	500-650 Hz 0.5 octave per min, 40 g peak
Flight Acceptance	Not required	20-2000 Hz 0.04 g²/Hz 8 g rms acceleration 2 min per axis	Not required

to ensure that design specifications are successfully achieved. Close control of mass properties is essential. Deviations in total mass, center of gravity, moments of inertia, and products of inertia will lead to serious malfunctions in spacecraft performance and orbit control.

Mass Properties Measurement

Mass: Weight measurement - up to 500 kg (increments of 100gm)
Weight measurement - up to 5000 kg

Measuring machines

Moment of Inertia machines: Schenck M4 (MOI about roll axis)
Schenck M1
Schenck M6 + horizontal adaptor
(max. MOI 3000 kg)

Center of Gravity machines: Schenck WS 12
Schenck WM 50/6 (nominal
capacity 4000 kg)

Spin Balance machines: Schenck E5 (capacity 500 kg;
max. speed 300 rpm)
Schenck E6 (capacity 4000 kg;
max. speed 300 rpm)

9.5.8 Climatic Tests

Marconi Space Systems have a range of climatic chambers for hot, cold, damp, thermal, altitude, and reliability testing. All chambers are equipped with automatic control and dedicated monitoring equipment. In addition, data from most of the chambers are recorded on a computer based logging system, for back up purposes.

Chamber	Temp Range C		Working Volume			Special Features
	Min	Max	Width mm	Depth mm	Height mm	
Dry Heat	Ambient	+85	1800	1800	1800	Solar Radiation
Hot/Cold (small)	−40	+150	1200	600	1200	
Hot/Cold (medium)	−40	+100	1200	1300	1200	
Hot/Cold (large)	−40	+100	1800	1800	1800	
Hot/Cold	−70	+199	900	750	600	Fits over vibrator (portable) CO_2 cooled
Humidity (small)	−40	+100	900	900	900	RH 20% - 100%
Humidity (large)	−40	+100	1200	1300	1200	RH 10% - 100%
Reliability	−60	+100	1200	1200	1200	Vibrator ±3 g, 20-60 Hz (MIL-STD-781) >5C/min/ ROC

Thermal Shock	− 50	+ 120	300	250	250	BS 2011 Pt2.1 Na
Stratosphere	− 50	+ 110	530	750	600	Max ALT 100,000 ft (8 mm Hg) (BS 3G100) Water injection during pressure recovery

9.5.9 Thermal Vacuum Chambers

The Marconi Space Systems Laboratory houses two identical thermal vacuum test chambers, each having a "clean" pumping system comprising a turbomolecular pump with liquid nitrogen, capable of achieving a vacuum better than 1×10^{-5} torr.

Each chamber has a horizontal heat sink mounting plate totally surrounded by a thermal shroud. The temperature of the shroud and the heat sink can be achieved in the presence of 100 watts test item heat dissipation. Test data is recorded on a computer based data logging system and may be printed directly and stored on disk for later recall.

Pressure	Better than $1 \times 10^{-}$ torr
Temperature Range	Heat Sink − 65 to + 100 degrees C Shroud − 65 to + 100 degrees C The shroud may alternatively be cooled cryogenically to − 196 degrees C
Internal Dimensions	
Horizontal Heat Sink	700 mm × 1000 mm accurately machined flat surface, fitted with 3 inch rectangular array of 10/32 UNF mounting studs 10 mm long
Working Volume	800 mm high × 1250 mm long × 1250 mm wide (max)
Lead Throughs	4 Flanges on 150 mm ports 2 Flanges for miscellaneous use 2 Flanges, each with 4 × 55 Amphenol connectors of the following type

Inside Chamber	Outside Chamber
62 GB 12E22 55 S	162 GB 16F22 55 P
62 GB 12E22 55 SB	162 GB 16F22 55 PB
62 GB 12E22 55 SC	162 GB 16F22 55 PC
62 GB 12E22 55 SE	162 GB 16F22 55 PE

Data Logging

Chamber pressure, heat sink temperature, and shroud temperature are normally recorded. There are up to 18 thermocouple channels for unit monitoring.

9.5.10 Antenna Test Range

The large Marconi test range (comprising test and source antenna positioners) is situated within an environmentally controlled anechoic chamber. A monolithic foundation minimizes relative movement along the axes of the range to ensure a high degree of stability. This foundation supports both the elevation/roll-over-azimuth antenna and test positioner and the source/probe tower, track mounted for variable range length. Full screening of the chamber creates a stable electromagnetic environment. The enclosure also offers a high degree of physical security.

Precision optical equipment is provided to perform alignment of the range and the subsequent location of antennas under test within the chamber. Heavy duty lifting gear caters for loads up to 4000 kg to allow for the handling of complete spacecraft.

A separate and screened control room houses the scientific Atlanta Model 2022A fully computerized antenna analyzer system-equipment, which provides data acquisition and transformation software together with comprehensive data manipulation and presentation capabilities.

Operational Capability. Marconi's large test range operates primarily in the spherical near-field mode, near-field radiation pattern acquisition techniques offering important advantages over out-of-doors far-field test methods. The radiated field is sampled in the near-field of the antenna and then data is processed to derive the far-field patterns. The required levels of measurement accuracy and repeatability are achieved by this method. Environmental problems associated with external test facilities, like wind, rain, frost, electromagnetic interference, are totally eliminated. Test program time scales are shortened dramatically and benefits are gained in terms of accurate scheduling and implementation.

Marconi Space Systems' advanced test range is designed to meet all current and foreseen spacecraft antenna requirements: broad beam antennas, agile beam antennas, narrow and multi-beam antennas. In all cases, antenna systems can be tested both individually and collectively to achieve desired measurement accuracies and repeatabilities. The facility allows for the determination of gain, radiation patterns, cross polar response and boresight location. It also provides a powerful tool for the determination of antenna coupling for EMC purposes.

Large Antenna Test Range Data.

Chamber Dimensions:	15 × 15 × 31 meters long
Measurement Modes:	Spherical near-field (Extended for planar near-field and cylindrical near-field)
Quiet Zone:	5 meters diameter
Frequency Range:	0.25 - 60 GHz
Reflectivity:	Better than – 40 dB at 1 GHz rising to better than – 55 dB at 7 GHz
Cross Polar Performance	Better than – 40 dB at 1 GHz rising to – 55 dB at 7 GHz
Screening Magnetic Field:	100 dB at 200 kHz 110 dB at 1 MHz
Electric Field	130 dB from 1 kHz to 50 MHz 110 dB from 50 MHz to 1 GHz 100 dB from 16 GHz to 18 GHz 80 dB from 18 GHz to 60 GHz
Access Door:	5 m × 6 m high
Positioner Tower Load Capacity	1600 kg
Positional Accuracy	±1 mm
Pointing Accuracy	±0.03 degrees
Temperature Control	20 ±2 degrees C
Humidity Control	35% - 60% Relative Humidity
Cleanliness	Filtration to 0.5 micron

9.5.11 Electromagnetic Compatibility (EMC)

The Marconi Space Systems EMC Laboratory possesses EMC test facilities in modern clean rooms with qualified and experienced personnel. The facilities consist of two suites, comprising 5 interconnecting rf screened rooms in the main area and two in the second. All surfaces of the measurement chambers are lined with anechoic material of a depth of 18″ and 3″ respectively.

EMC measurements are achieved in accordance with a wide variety of military and commercial specifications. Detection and electromagnetic field generation is carried out over a 20 Hz to 40 GHz frequency range.

Dimensions.

Main Chamber Area	13 × 11 × 7.3 m (high)
Access Door	5 × 5 m

Screening.

Magnetic Field	>80 dB between 14 kHz and 10 MHz

Electric Field	>120 dB between 14 kHz and 10 MHz
	>115 dB between 10 MHz and 1 GHz
	>126 dB between 1 GHz and 40 GHz

Temperature.

Temperature Control	20 ±2 degrees C
Humidity	35% to 60% Relative Humidity
Environmental Cleanliness Specification	Filtration to 0.5 micron BS 5295 Class 3

Emission and Susceptibility Measurements.

| Conducted emissions and susceptibility | 20 Hz to 600 MHz |
| Radiated emissions and susceptibility | 20 Hz to 40 GHz |

Chapter 10

Reliability

RELIABILITY IS THE MAIN REQUIREMENT OF ANY SYSTEM OR EQUIPMENT. IT is defined as the probability of performing a specified function under specified conditions for a specified time. For applications, where the system or equipment becomes expensive and there is no possibility of repair as in the case of satellites, the necessity of built-in reliability becomes a rule rather than exception. Therefore continuous efforts are being made to realize better and better and higher reliability of systems or equipment used in space.

10.1 DESIGN AND DEVELOPMENT FOR RELIABILITY

There are many ways for improvement of the reliability of any electronic equipment or system, the most important of which include the component improvement, stress-level reduction, circuit or system design simplification, conservative design, and the selection of proper components and manufacturing techniques. Only after these methods have been fully exploited to increase inherent circuit or system reliability, should redundancy be considered to take care of chance failures. When judiciously employed, higher levels of reliability can be accomplished by redundancy either on a piece part or circuit basis.

Thus, the following sections deal with various methods or techniques to be considered in designing the systems or circuits for reliability.

10.1.1 Reliable Circuit Design

After considering the individual component reliability, namely, improvement by proper derating, its interaction with supply voltages, its mechanical arrangements and the environment under which it has to operate, etc., then the designer shall examine the circuit and assess its reliability. In fact, the circuit design is another aspect of the environmental design as the circuit is the immediate electrical environment of each component. It is convenient to deal with circuit reliability separately because of the wide variety of requirements that the circuit design must meet.

First, the circuit must fulfill its function; secondly, the design of the circuit has a bearing on the reliability of its components. Following the initial design of an electronic system, a mathematical analysis must be made of the individual circuits to determine whether they will permit the initially set system reliability goal to be met. Good design is universally regarded as a quality that can be equated to good reliability.

The three factors that contribute to the reliability of a system or equipment as it leaves the manufacturer are the reliability of the design, the reliability of the component parts employed, and the reliability involved in the manner in which the components and the design are fabricated. The reliability of these factors, each of which is equally important, is expressed in the following equation.

$$P_a = [P_d \times P_c \times P_f]$$

where P_a =   the overall reliability
          P_d =   the design reliability
          P_c =   the components reliability
          P_f =   the fabrication reliability

10.1.2 Design for Reliability

As explained above, the design reliability is equally important as the component or fabrication reliability. Although the component and fabrication reliability was very good in the case of a regulator, because of its poor stability in a particular operating mode the circuit did not deliver the expected output. Thus, design reliability is equally important and is dealt with in some illustrative examples.

Power Conditioning and Control System. Sometimes a system-level approach is needed in the case of a large system, like a satellite, to improve the reliability, efficiency, and to avoid unnecessary waste of power. It is no wonder if one finds several voltage regulators in series between the source of raw power and the user. Each of these regulators has inefficiency and unreliability and hence much of the available raw power is wasted as heat and the reliability of the overall system has decreased. Such a situation arises when individual units are developed in isolation and can be avoided with good control at the system level.

□ **Common Regulator.** When a single regulator powers two or more subsystems and if there is no provision to cut off power to each individual

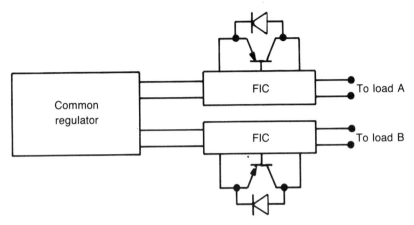

Fig. 10-1. Input short circuit protection for fault isolation circuit.

subsystem, in such cases to avoid the propagation of the malfunctions at one such subsystem to other subsystems, a fault isolation circuit (FIC) is placed between such subsystem load and the common regulator as shown in Fig. 10-1. In such cases if the load fails in short circuit suddenly, the control circuit of FIC should respond faster and isolate the failed load: otherwise the common regulator might go into current limit or foldback condition temporarily cutting off the power to all loads. Thus FIC shall respond not only to any bad impedance but also to any rate of change of load impedance. Normally, the series-pass transistor of the FIC will be failing because of two reasons, (a) any reverse voltage that may be delivered by an active load, (b) any fault in any other user that short circuits the regulator for a short time, during which input voltage to FIC drops to zero. Then the voltage across the output capacitor of FIC is applied between the collector and emitter, first zener breakdown could have occurred and the transistor would probably have been destroyed before the fault is isolated. This problem can be avoided by connecting a diode across the series-pass transistor. This allows the output capacitor to discharge rapidly.

☐ **Input Filtering.** For certain applications, L-C filters are required on power supply interfaces. In such cases, transient analysis becomes important on switch-on, both the current and voltage may overshoot to levels which could cause damage to components. Large negative voltage transients would occur at the power subsystem/user interface when the relay contact was opened (Fig. 10-2A) which can result in component damage. This can be avoided by adding a diode as shown in Fig. 10-2B.

Driving. The simplest drive source is a resistor with a pair of clipping diodes (Fig. 10-3A) or a voltage reference diode (Fig. 10-3B). If the driving source voltage levels are unsuited to either of these methods then a transistor driver circuit can be employed. Two such circuits are shown in Fig. 10-4A and 10-4B.

Inputs Driven From Switches. When a gate is driven by a mechanical

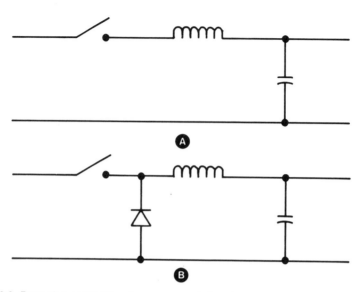

Fig. 10-2. Protection of LC filters in power supply interfaces.

contact, the circuit arrangement shown in Fig. 10-5A may be employed. The resistor reduces the possibility of noise pick-up when the switch is open. However, contact bounce in mechanical switches causes the generation of a train of spurious pulses at each operation. It is better to use the switch to trigger a monostable, which can provide a single pulse if required, or delay the application of the signal until switch bounce is ended. Alternately, if the change-over switch is available then a pair of cross-coupled gates may be used, as shown in Fig. 10-5B. Also bounce-free buffer gates can be employed for such applications.

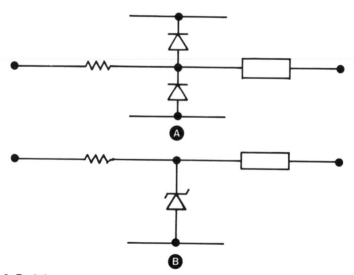

Fig. 10-3. Resistive source (A) with clipping diodes (B) with voltage reference diode.

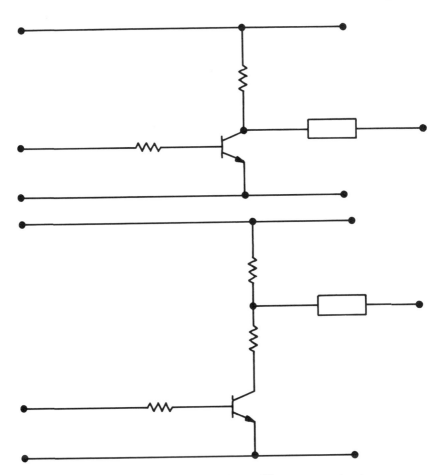

Fig. 10-4. Drive source interface (A) positive inputs (B) +ve or −ve inputs.

Interface with Relays. Relays and relay drivers are the peak power consuming circuits operating for short duration on command. Most often they respond to noise if the immunity of the circuit is not good. So, foolproof operation of relay drivers plays an important role in the successful operation of a system or equipment.

To discriminate between a true command and noise, the user interface circuit should make use of the distinct amplitude and width of the command pulse. A commonly used command interface employs an RC filter at the input (Fig. 10-6A). This circuit lacks hystersis, which provides the noise immunity. Sometimes an RC network tends to integrate any repetitive waveforms at the input. Depending upon the mark-space ratio of the noise, and the relative charge and discharge time constants, the voltage on the capacitor may build up and eventually reach a sufficient level to make the transistor conduct.

A simple circuit shown in Fig. 10-6B, overcomes all of these drawbacks. This circuit discriminates the command signal by its amplitude as well as its width and has high noise immunity because of its hystersis.

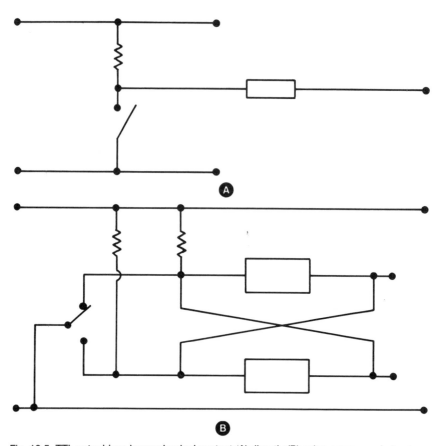

Fig. 10-5. TTL gate driven by mechanical contact (A) directly (B) using cross-coupled gates.

Digital Interface. In the case of data commands, normally 8 bits or 16 bits of information is transferred to a user subsystem. A data command user receives a clock line, a serial data line, and an address line. Each user is allocated a separate address and therefore requires a decoding circuit to recognize when the transfer of serial data is intended for him. When the correct address is decoded, a simple pulse is generated to enable the clock and data inputs to the shift register. The output of such shift register will be random when the power is applied or restored following a temporary failure, which can cause a malfunction. For example, random output of such register can turn-on/off critical relays. For example, in a spacecraft, this could result in the loss of attitude with subsequent loss of solar array power, thermal problems, etc. This can be avoided by designing the circuit such that when power is applied or restored, the shift register outputs are set automatically to a predetermined desirable safe state. Until this setting is carried out, the execution of any commands can be deferred.

☐ **TTL Integrated Circuits.** The design of reliable systems using ICs necessitates the application of certain precautions and special techniques to

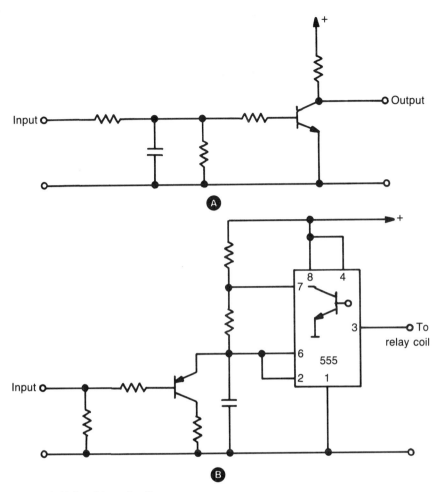

Fig. 10-6. Relay driver circuits.

the functional design as well as to the power supply decoupling, etc. The unused inputs of all OR and NOR gates and of the unused sections of AND, OR, INVERT gates must be grounded. For the best noise immunity, clock rise and fall times should be appropriate. Preset and clear pulses should be present for longer than are the clock pulses. Unused preset, clear, and J and K inputs should be treated in the same manner as are unused gate inputs.

Level Shifting. Many currently marketed COSMOS devices require the conversion of standard TTL logic levels to higher voltages. Level shifting circuits must nave a low output impedance during the edge transistion, to drive the capacitive COSMOS inputs. A high impedance output in one direction is possible in either the HIGH or the LOW state: for example, an npn emitter follower output will have a high output impedance for positive signals. Thus, if a signal of the correct polarity is induced at the output because of say, cross-talk in the load, then the driven COSMOS output may be taken out of its specified limits. A diode connected from the driver output to the power rail as shown in Fig. 10-7 will protect the device and normally only one of the logic levels requires this protection. The resistance R_s in series with the output also shown in the figure has the effect of reducing any ringing in the lines that might be caused by the series combination of the printed-wiring inductance and the COSMOS output load and stray capacitances. Nevertheless, initial precautions against such ringing are recommended and the level-shifting circuits should be located as closely as possible to the COSMOS array they drive. Excessive ringing can seriously impair the operating margin of COSMOS devices.

Operating Conditions. Manufacturers' data sheets provide information about the operating conditions of the components and it is the responsibility of the designer to evaluate the conditions, i.e., voltage, current, power dissipation, safe operating temperature (forward and reverse) to which the component will be subjected and to ensure that in no case are the limits exceeded.

For an example, the rectifier and filter of the dc-dc converter secondary was designed assuming that the load will be connected permanently and no bleeder resistor has been used. This poses a condition that the power supply shall not be turned on when there is no load. It is noticed that sometimes the power supply is turned on without the load being connected. It is human to err. In this case the capacitors have failed as the selection of the capacitor is improper. When there is no load, the output voltage builds up to a higher level than with a load.

For another example, when two capacitors are connected in series to meet the requirement of capacitors with high voltage ratings, proper care should be taken to avoid the failure of capacitors due to unequal potential division between the capacitors. Unequal potential division is possible because the tolerance of capacitors is of the order of 20%. The unequal potential division can easily be avoided by using two resistors with five percent tolerance in parallel with capacitors, as shown in Fig. 10-8.

Component Tolerances. The worst case design philosophy is commonly followed for designing reliability into an electronic circuit. A circuit design is to be checked for its performance as a function of component tolerances. The

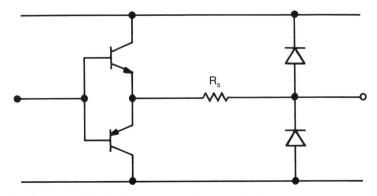

Fig. 10-7. Diode protection for COSMOS level shifting circuit.

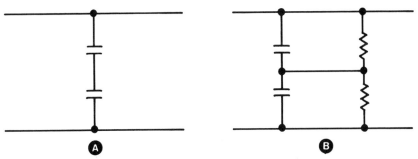

Fig. 10-8. Capacitor's protection when connected in series.

circuits that work well when they are new, may not work later because of aging. Some amplifiers may oscillate as the component values change. Hence, worst case design analysis has to be carried out taking into account the component tolerances.

10.1.3 Component Reliability

Component reliability plays an equally important role in the overall reliability of the system or equipment. The reliable performance of a particular component can be considerably improved by operating it at reduced stress levels. Typical derating factors are as follows: carbon composition (RCR) type resistor has to be derated to 50% of the rated power, maximum voltage shall not exceed 80% of the maximum rated voltage. In the case of digital IC's 80% of the manufacturer's maximum fan-out is allowed. In the case of linear ICs, the manufacturer's suggested bias voltages shall not be derated unless unless precautions are taken to ensure that such action does not cause possible malfunction. Tantalum (wet) capacitors shall be derated to 50% of rated voltage. These are only some examples.

Also depending upon the application, namely the environmental and circuit conditions, the components have to be selected. For example, solid types should not be used in low impedance circuits (power supply filters) unless protected by a series current-limiting resistor.

10.1.4 Fabrication Reliability

Fabrication reliability is equally important as illustrated in some of the examples below. A circuit or system may operate satisfactorily and reliably when alone but may exhibit unexpected behavior when assembled or integrated into a complete system. The disturbance observed often affects not only the performance of the system, but also component operating conditions and therefore life. If fabrication is poor, then the system or equipment will not function as expected even though the design and component reliabilities are very high. Thus, equal consideration has to be given to the fabrication aspect of an equipment or a system.

Testing. To ensure reliability, the testing of systems has to be done at various stages of production. Besides the production of systems or equipment always involves a number of *DO's* and *DON'Ts* that each manufacturer or contractor must adhere to. Certain materials create corrosion problems during long storage and their use must be avoided. Certain procedures like soldering, if not conducted in the prescribed manner, may cause hardening of terminals and result in subsequent failures.

If the operation of an IC is in doubt, it is commonly removed from its socket and replaced by another, without first switching off the power supply. In such a case, excessive inrush currents can occur and it is possible for a complete batch of new ICs to be destroyed, one after another, by an inadequately briefed operator.

In some cases, the supply voltage to a module containing ICs can, by handling errors, be made higher than the rated voltages of the IC. Figure 10-9 shows an example of a supply arrangement where the LT supply voltage is dropped to a value appropriate to the IC by a series resistor and a decoupling capacitor. However, a hazard can arise if the IC is removed from the socket and replaced without switching off the supply. A particularly severe hazard is represented by the charge potential developed across C1 while the socket is vacant, which will reach the full voltage. The safeguards are, (i) clearly place warning notices on the equipment and in the service manual, (ii) solder ICs directly to the panel.

Fabrication/Assembly. The reliability of any system is highly dependent upon the techniques used in the assembly of both the system and individual components. The effect here is very profound in that poor workmanship can result in poor reliability while good workmanship can be effective in improving the reliability. To cite an example of the impact of adopting good practices, tapped winding leads from toroids have been observed to lead to problems when both heavy gauge and fine gauge wires are wound over each other. The fine wire if wound over the heavy wire, has a tendency to sink down into the interstices of the large winding. Thermally induced movement of the large size wire will pinch the small size wire and cause open circuits. This is usually prevented by wrapping a layer of Teflon tape over the heavy winding before installing the fine wire winding.

10.1.5 Redundancy

Though the inherent circuit reliability could be maximized as illustrated in the previous sections by reliable circuit design, judicious selection of

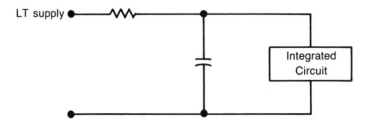

Fig. 10-9. High voltage hazards.

components, proper derating and good workmanship, the chance failure that could partially or totally jeopardize a mission can be taken care of only by adopting proper redundancy to ensure the overall mission reliability at the required level.

Reliability is the important characteristic of the system or equipment. For applications, where the systems or equipment becomes expensive and that there is no possibility of repair as in the case of satellites, the necessity of built-in reliability becomes a rule rather than an exception. Thus, the various techniques for the improvement of the reliability with particular reference to the reliable circuit design are presented.

10.2 RELIABILITY AND REDUNDANCY

Though the inherent circuit reliability could be maximized by optimized circuit design, judicious selection of components and good workmanship, the chance failure that could partially or totally jeopardize a mission can be avoided only by adopting redundancy. When judiciously employed, higher levels of reliability can be accomplished by redundancy either on a piece part or circuit basis. This ensures the overall mission reliability at the required level. Thus various redundancy approaches are presented here.

10.2.1 Redundancy Approaches

There are different approaches to redundancy such as standby redundancy, task sharing redundancy, majority logic redundancy, shared mode of standby redundancy, partial redundancy, etc., which are described below in detail.

Standby Redundancy. In this case only one of the two or more similar units is in operation at any instant of time. This configuration is shown in Fig. 10-10. Two independent failure detectors are needed to monitor continuously the performance of the units. Thus, the failure detectors are powered continuously to make the system more reliable. An alternative is to switch ON the particular failure detector of the working unit only. In that case it is assumed that the failure detector and change over network are highly reliable and the redundant units do not degrade while inoperative. An important point

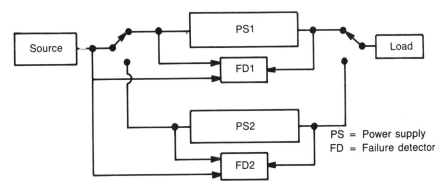

Fig. 10-10. Standby redundancy.

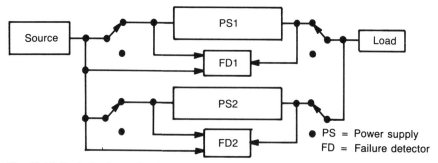

Fig. 10-11. Task sharing redundancy.

to be noted with this standby redundancy is the existence of a momentary outage of unit output during change-over without some special provision. This exists for a maximum period of 3 milliseconds, which is the relay contact change-over period. If this is not acceptable then standby redundancy should be excluded from such systems.

Task Sharing Redundancy. In this case, two or more units operate at the same time, sharing the total task. If a unit fails, that unit is isolated and the remaining units continue to operate, sharing the task in larger proportions. This configuration is shown in Fig. 10-11. There will not be any outage of output during change over. In this type the units operate at other than the maximum efficiency design point. Therefore the total system efficiency is lower.

Majority Logic Redundancy. In this case, three or more units operate at the same time and in parallel. The failed unit is isolated and the remaining units share the task in larger proportions. Though individual failure detectors are not required, an integrated failure detector is required to perform the comparison of functions. This configuration is shown in Fig. 10-12.

Shared Mode of Standby Redundancy. In this case, two tasks (A and B) are performed by two units and there is only one common redundant unit. Three separate failure detectors are employed. This configuration is shown in Fig. 10-13. The redundant unit is capable of taking the sum of the tasks A and B. In case of failure of unit 1, the redundant unit performs task A,

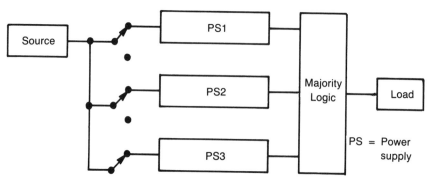

Fig. 10-12. Majority logic redundancy.

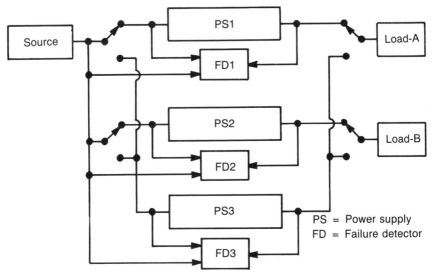

Fig. 10-13. Shared mode of standby redundancy.

isolating the unit 1 from the source and task A. This system operates as a shared mode of standby redundancy, initially working as a standby redundancy having a single redundant unit for two units. If a unit fails then the redundant unit works at lower efficiency as it is designed for the sum of the tasks A and B. Unit 2 together with the redundant unit work as if they are sharing the task, though they are performing independent tasks. If the unit B also fails, then the redundant unit itself performs the task A and B.

Partial Redundancy. This type of redundancy, as shown in Fig. 10-14, is based on the fact that the failure rates of passive components are less than those of active components. Hence instead of employing a complete circuit as a redundant unit, judiciously selected parts or sub-circuits primarily containing active components are only duplicated as redundant elements.

10.2.2 Optimization

In selecting a redundancy approach, the following main factors shall be considered, i.e., reliability, weight, circuit complexity, efficiency, cost, etc. One has to select the redundancy approach depending upon the mission requirements and by carrying out a trade-off analysis giving appropriate weightages for different performance parameters of interest to that particular mission.

10.2.3 Failure Detector

The detection of failure at the first instant is very important and the failure detector shall continuously monitor unit performance. Single or particular combination of deviations in unit performance parameters will be interpreted as a failure. Different parameters, like temperature, frequency, current or voltage, and/or other parameters at some point internal to the unit may be used to detect a failure. Need for a highly reliable failure detector is very essential.

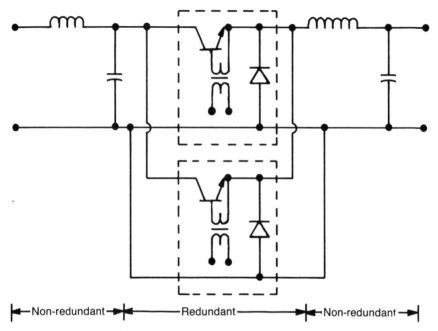

|←—Non-redundant—→|←————Redundant————→|←—Non-redundant—→|

Fig. 10-14. Partial redundancy.

10.3 RELIABILITY AND
FAILURE MODE AND EFFECTS ANALYSES

The importance of reliability has been illustrated and various methods and approaches to the improvement of reliability have been presented in the previous sections. High performance, high levels of reliability, and lower costs are the primary considerations in the design of any system. A reliability analysis provides a measure of inherent reliability. Thus, the inherent reliability of a typical regulator is calculated as an example. Also included is the *failure mode and effects analysis* (FMEA) for a spacecraft's power system. The primary purpose of FMEA is to identify and eliminate, where possible, critical single-point failures. Where critical single-point failures cannot be eliminated, the main goal is to reduce the probability of occurrence of such failures and to minimize their failure effects.

10.3.1 Description of a Spacecraft Power System

As a spacecraft power system is selected as an example to carry out FMEA, a brief description of the same is presented here to facilitate better understanding of FMEA. Figure 10-15 shows a typical power system for a communication satellite at geosynchronous altitude. This power system is known as a nondissipative regulated bus power system. A common control unit drives the power stages of shunt, charge, and discharge regulators. The charge regulator

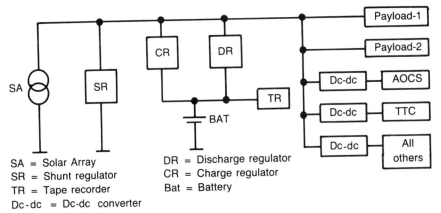

SA = Solar Array   DR = Discharge regulator
SR = Shunt regulator   CR = Charge regulator
TR = Tape recorder   Bat = Battery
Dc-dc = Dc-dc converter

Fig. 10-15. Typical spacecraft regulated bus power system.

itself maintains the bus at a fixed voltage and charges the battery at a variable current. When the battery is completely charged, the shunt regulator is turned on and the charge regulator is switched from the charge mode to the trickle charge mode. Now the shunt regulator maintains the bus at a fixed voltage. During eclipse period, the battery discharge regulator which is of boost type, boosts the battery voltage to bus level and maintains it at a fixed voltage.

The bus voltage is supplied to the payloads directly, whereas through dc-dc converter-regulators or dc-ac inverters to the bus loads, namely, the telemetry, tracking, command and communication (TTC&C) system, attitude and orbit control system (AOCS), propulsion system, etc. Some of the loads like the tape-recorder, etc. are supplied from the storage battery through commandable redundant fuses.

10.3.2 Description of a Charge Regulator

Because the charge regulator is selected as an example to show the reliability calculations, a brief description of it is presented here. A detailed schematic of the charge regulator, part of the spacecraft power system (Fig. 10-15), is shown in Fig. 10-16. The control is achieved by comparing the scaled-down bus voltage to a reference voltage. The error voltage is amplified, compensated, and fed to a pulse-width modulator through an attenuator. The modulator is synchronized at a clock frequency and its output is a pulse width proportional to the compensated-amplified error voltage but limited by the attenuator. This signal is employed to switch Q into saturation or cut-off. Thus, the current flow into the battery is controlled such that the bus voltage is maintained at V_o. When Q is ON, energy is stored in the inductor (L) and this maintains a continuous current flow into the battery when Q is OFF.

10.3.3 Charge Regulator Reliability Calculations

To assess the inherent reliability of the charge regulator shown in Fig. 10-16, a reliability analysis is performed. The general assumptions are (a) all

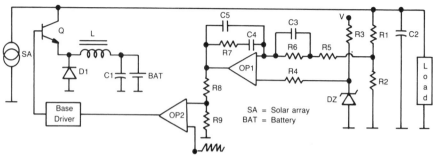

Fig. 10-16. Detailed schematic of a charge regulator.

components are considered to be equally important for proper functioning, (b) the effect of one component failure on the other component is not considered. The MIL-SPECS of the components used in the calculations are given in Table 10-1 as extracted from MIL-HDBK-217A. The constant failure rate is calculated by

$$R = e^{-\lambda t}$$

Where R = the reliability
e = natural log base
λ = failure rate
t = duration in hours

Here a mission duration (t) of one year is assumed for reliability calculations. Standard reliability calculation techniques are used to calculate the

Table 10-1. Failure Rates of Components Employed in the Charge Regulator.

Components	Technology	Style	MIL-SPEC	Quality Factor	Failure Rate xE-9/Hr
Resistors:					
-Low Power	Carbon Composition	RCR	MIL-R-39008	S	0.45
-Low Power	Metal Film	RNR	MIL-R-55182	S	2.80
Capacitors:					
-Non-Polarized	Silver Mica	CMR	MIL-C-39001	S	0.30
-Polarized	Tantalum, solid	CSR	MIL-C-39003	S	11.00
Semiconductors:					
-Transistor	Silicon, NPN		MIL-S-19500	JANTXV	28.00
-Diode	Silicon		MIL-S-19500	JANTXV	12.00
-Zener			MIL-S-19500	JANTXV	17.00
-IC	Linear		MIL-M-38510	JAN CLASS A	12.00
Inductor:					
-Power					7.50
Filter					

NOTE:-Quality Factors, S = 0.01; JANTXV = 0.1; JANTX = 0.2; JAN = 1.0;
　　　　JAN CLASS A = 0.5; JAN CLASS B = 1.0

reliability of different series and parallel components/circuits/units if there are any in the charge regulator. The reliability of the entire regulator is obtained by taking the reliabilities of all the components/circuits/units in series.

For series calculations

$$R_{series} = (R1)\ (R2)\ (R3)......(R_n)$$

For parallel calculations

$$R_{parallel} = [1 - (1 - R1)\ (1 - R2)\ (1 - R3).....(1 - R_n)]$$

where $R1$, $R2$, $R3$,.....R_n are the reliabilities of the components/circuits/units that are in series or parallel, respectively. Thus, the calculations along with the reliability calculated for a mission duration of one year is given in Table 10-2. The quality factors and the failure rates for various components of the charge regulator, as mentioned above, are given in Table 10-1. Table 10-2 shows five columns, namely, component description, failure rate, derating factor, number of components, effective failure rate. The failure rate shown in the second column includes the effect of the quality factor. The derating factor is the ratio of the applied electrical stress to the rating of that particular component. The last column is the product of the columns 2, 3, and 4. Charge

Table 10-2. Charge Regulator Reliability Calculations.

Component Description	Failure Rate	Derating Factor	Number of Components	Effective Failure Rate
Resistors				
Carbon				
Composition	0.0045	0.20	2	0.001800
Metal Film	0.0280	0.10	1	0.002800
	0.0280	0.20	5	0.028000
	0.0280	0.50	1	0.014000
Capacitors				
Tantalum, Solid	0.1100	0.30	2	0.066000
Silver Mica	0.0030	0.25	3	0.002250
Semiconductors				
Diode	1.2000	0.50	1	0.600000
Zener	1.7000	0.20	1	0.340000
Transistor				
NPN, Silicon	2.8000	0.40	1	1.120000
IC (op)	6.0000	0.60	2	7.200000
IC (Driver)	6.0000	0.75	1	4.500000
Inductor	7.5000	0.60	1	4.500000
Total Failure Rate				18.374850E-9/Hr

Reliability, R = e $\quad \dfrac{(- 18.37485E-9)\ (8760)}{= 0.999839}$

regulator total failure rate is obtained by adding the last column, which is 18.374850E-9/hour. Finally, the probability of the charge regulator functioning as per the design, meeting expected performance over a period of one year is 0.999839.

10.3.4 Failure Mode and Effects Analysis

As mentioned above, the primary purpose of FMEA is to identify and eliminate, where possible, critical single-point failures. However, if critical single-point failures cannot be eliminated, the main goal is at least to reduce the probability of occurrence of such failures and to minimize their failure effects. Critical single-point failures are those failures occurring singly that disable the power system from providing power to the spacecraft's critical loads.

To realize the complete mission goal, all the subunits are essential. One can equally envision cases where certain equipment will fail without affecting the complete system and hence it can partially fulfill the mission goals. FMEA is carried out on subsystem/unit level and for each unit analyzed, all failure modes are included. Structure and passive elements such as power busses and wiring are excluded from the analysis. The basic failure modes considered are open, short, and degradation.

Thus, the FMEA carried out on spacecraft power system is presented in Table 10-3. From this table, it is clear that to avoid single-point failures, the charge, discharge, and shunt regulators have to be redundant. The storage battery shall have cell open-failure bypass circuitry and shall be overdesigned with respect to the storage capacity to avoid any effect due to cell shorts. In addition to connecting the tape-recorder to the battery directly, some switching mechanism shall be incorporated such that it can also be powered from the regulated bus.

The importance of reliability and various approaches to the improvement of reliability has been presented. The inherent reliability of a typical regulator is calculated as an example. Also included is the *failure mode* and *effects analysis* for a spacecraft power system. FMEA helps to identify and eliminate critical single-point failures.

Table 10-3. Spacecraft Power System Failure Mode and Effects Analysis.

Subsystem/ Unit	Failure Mode	Effects on Other Systems
Solar Array	Open	As solar array consists of n sections and are connected using isolation diodes, opening of a section only reduces the solar array output by 1/n.
	Short or degradation	Same as above.
Shunt Regulator	Open	Bus voltage will not be maintained. Loads connected to the bus may not accept the variation in the bus voltage, which is the solar array voltage.
	Short	Shorts the solar array and some component may fuse open. Bus voltage will not be maintained.

Subsystem/ Unit	Failure Mode	Effects on Other Systems
Charge Regulator	Open or Degradation	Battery cannot be charged. Effect is the same as the loss of the battery. Tape recorder will not function. Hence satellite can work only in real time mode.
	Short	Bus voltage clamps to battery voltage. Battery charge rate is not controlled. Once the battery is charged up, the battery has to be protected by opening the battery. Whenever the battery is charged, the bus loads cannot work. However time sharing is possible.
Discharge Regulator	Open	Battery cannot be discharged. Loads cannot be supplied power during eclipse period.
	Short or Degradation	Battery voltage cannot be boosted to bus voltage and bus voltage clamps to battery voltage. Hence, battery has to be opened.
Battery	Open	Tape recorder cannot operate. Eclipse operation of the satellite is not possible.
	Short	Power will be shunted to ground. Battery has to be opened. Tape recorder cannot operate. Eclipse operation of the satellite is not possible.
Dc-Dc Converters	Open	Automatically changes to the redundant unit.
	Short	Automatically changes to the redundant unit.
Power Control Unit		
Charge/ Discharge Relay	Charge Mode	Satellite will work only during oribital day.
Emergency Relay	Discharge Mode	Equal to battery opening
	Normal Mode	Much care has to be taken in draining the battery which can be affected by commanding.
	Emergency Mode	Equal to battery getting isolated/open

Part II
Applications of Satellites

Chapter 11

Communications Satellites

S ATELLITE COMMUNICATIONS OPENED A NEW ERA IN GLOBAL COMMUNICA-
tions by providing reliable and cheaper communications to areas where
earlier methods could not provide acceptable service and the entire globe
including the ocean regions can be covered by communications satellites.

11.1 SATELLITES FOR COMMUNICATION

Even before the advent of satellites, conventional communication
techniques had advanced quite well and enabled dialing almost anyone,
anywhere in the country directly. Also, there are TV networks and cables across
the oceans.

The first transatlantic telegraph cable was opened for service in 1866. But
it took 61 years more to carry the human voice across the ocean, through high
frequency radio. Since radio waves like light travel in straight lines, the high
frequency radio systems depend on the ionosphere to reflect the radio waves
around the curvature of the earth. But the ionosphere is an unreliable reflector
and it took another 29 years to establish a more reliable system. In 1956, the
first submarine telephone line was laid across the Atlantic Ocean. This line
originally had 36 telephone circuits, which were later raised to 96 channels
by using a high speed switching technique. Even though in 1964, there were
4 cables across the Atlantic Ocean, these systems were limited in their
bandwidth and capacity.

To reproduce voice reasonably, a bandwidth of about 4 kHz is required.

Thus a 2-way telephone channel requires 8 kHz bandwidth. Similarly, a TV requires about 4.5 MHz to 6 MHz, which is equivalent to 600 telephone channels or 13,200 telegraph channels. This is where switching over to microwaves takes place. High frequency radio waves that use the ionosphere as a reflector extend from 3 to 30 MHz only. On the other hand, microwaves extend from 1000 to 10,000 MHz and hence microwave circuits can carry several TV channels or several thousands of telephone or telegraph channels. However, microwaves can only be used for straight-line paths between points that are within sight of each other as they cannot use the ionosphere as a reflector. For distances that extend beyond the horizon, several repeaters are needed to relay the signals from point to point. The normal distance between two repeaters is 50 to 80 km and to cover longer distances many repeaters have to be employed in tandem. This is possible over land, but not across the oceans. Though theoretically one microwave tower of 650 km high above the Atlantic Ocean, can link America and Europe, it is not practicable. A "fixed" satellite positioned high over the Atlantic can be considered as a "virtual" tower that overcomes this limitation.

A satellite at 35,786 km altitude can see more than one third of the globe as shown in Fig. 11-1 and three properly positioned spacecraft can cover the entire earth as shown in Fig. 11-2. A satellite can be accessed by several points in this coverage area simultaneously. Also the satellite can be moved over to a new location if required.

Communications Satellites. A glorious era of space exploration started when the first earth satellite was launched in 1957. The first communications satellite, named SCORE, was launched on 18th December, 1968. Since then, several communications satellites have been launched, namely, ECHO I & II, COURIER, TRISTAR I & II, RELAY I & II, the SYNCOM series, MOLNIYA series, the ATS series and the INTELSAT series.

Communications satellites are of two basic types, (a) the passive and (b) the active. The passive satellite acts merely as a reflector of signals. These need very powerful ground transmitters. These satellites have no electronic parts. Passive communications satellites can have large discrete structures like the ECHO satellite or a large volume of space filled with a large number of tiny passive dipoles like the WEST FORD. The large structure usually takes the form of a sphere and when a radio signal is beamed at the sphere much larger than the wavelength of the signal, it gets reflected in all directions. This type of sphere is known as an isotropic reflector because the power in the signal is equally distributed in all directions. The sphere, a large balloon, can be made of a thin plastic material. The reflector surface can be coated with a thin layer of metal. The sphere is folded into the nose cone of the launcher and later inflated in orbit. Of course, the receiving ground system receives a fraction of the power that the reflector intercepts and reradiates.

The active satellite, on the other hand, receives the signal, amplifies it and transmits back to the earth. Just like the microwave repeater, the satellite shall be visible to both the transmitting and receiving stations. As the ground receiver receives only a fraction of the power that the passive satellite intercepts

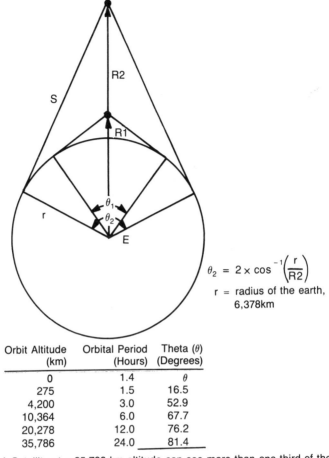

$$\theta_2 = 2 \times \cos^{-1}\left(\frac{r}{R2}\right)$$

r = radius of the earth, 6,378km

Orbit Altitude (km)	Orbital Period (Hours)	Theta (θ) (Degrees)
0	1.4	θ
275	1.5	16.5
4,200	3.0	52.9
10,364	6.0	67.7
20,278	12.0	76.2
35,786	24.0	81.4

Fig. 11-1. A Satellite at a 35,786 km altitude can see more than one third of the globe.

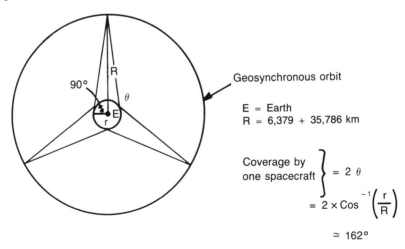

Geosynchronous orbit

E = Earth
R = 6,379 + 35,786 km

Coverage by one spacecraft $\Big\}$ = 2 θ

$$= 2 \times \cos^{-1}\left(\frac{r}{R}\right)$$

$$\simeq 162°$$

Fig. 11-2. Three properly positioned satellites cover the entire Earth.

and reflects, they are very inefficient and hence active satellites are preferred. Also in the case of active satellites, the re-radiation just covers the earth only by using directional antennas. This further enhances the efficiency of active satellites. Here onwards, only active satellites are considered and are called simply satellites or spacecraft.

Thus, a communications satellite functions as a repeater and it receives uplink carriers from earth stations, processes, and retransmits on the downlink to other earth stations. Current communications satellites function as multi-channel repeaters and these function similar to line-of-sight microwave repeaters. The characteristics of the equipment on-board the satellite depend upon various factors, namely, the distance between the satellite and the ground station, the loss in the medium, etc.

Table 11-1 presents the frequency bands commonly used for satellite communications. Generally, rain attenuation is higher at higher frequencies. Also the transmitter becomes more expensive as its transmission frequency increases. However, as lower frequencies are already allocated, new allocations are only for higher frequencies.

Different uplink and downlink frequencies are used to prevent feedback within the satellite while simultaneous transmission and reception takes place. Also, uplink frequencies are usually well separated from downlink frequencies to minimize interference between transmitted and received signals. The lower frequency is generally selected for the downlink due to lower atmospheric losses and thus requires less power from the satellite. Until 1979, for international and commercial communications the 6 GHz band (5.925 to 6.425 GHz) was used for the uplink, and the 4 GHz band (3.7 to 4.2 GHz) was used for downlink.

In 1979, 14, and 11 GHz bands were additionally allocated for the international and commercial communications. To overcome the shortage of frequencies, a number of alternatives have been thought of, namely, reuse of the same frequency at different beams using spatial isolation. Polarization diversity is used to double the effective frequency use. Reverse frequency allocation has also been thought of at some extra cost.

Synchronous Satellites.. At low altitudes, a satellite views a smaller area of earth. At lower altitudes the earth's magnetic field is relatively strong and is responsible for the trapping of high energy protons in the Van Allen radiation belt. In the inner belt, this radiation can adversely affect an active satellite due to rapid degradation of solar cell efficiency and damage to other solid-state components. The magnetic field causes precession of the spin and reduction of the spin rate of spin-stabilized satellites by interacting with the power supply currents and the eddy currents induced in the metallic parts of the satellite.

The lower the altitude the higher the fraction of the orbital period spent in the shadow of the earth as shown in Fig. 11-3. Hence, extra solar cells are needed to charge the storage batteries to have uninterrupted power supply on-board the spacecraft. The lower the altitude the lower the orbital period and the satellite passes by a ground station quickly. Hence, the ground station

Table 11-1. Frequency Designations.

Frequency Range	Designation	
Conventional Designations:		
30 to 300 Hertz	ELF	Extremely low frequency
300 to 3000 Hertz	VF	Voice frequency
3 to 30 Kilohertz	VLF	Very low frequency
30 to 300 Kilohertz	LF	Low frequency
300 to 3000 Kilohertz	MF	Medium frequency
3 to 30 Megahertz	HF	High frequency
30 to 300 Megahertz	VHF	Very high frequency
300 to 3000 Megahertz	UHF	Ultra high frequency
3 to 30 Gigahertz	SHF	Super high frequency
30 to 300 Gigahertz	EHF	Extremely high frequency
300 to 3000 Gigahertz or 3 Terahertz	VEHF	Very extremely high frequency
Alternate Designations:		
0.1 to 0.3 Gigahertz	VHF	
0.3 to 1.0 Gigahertz	UHF	
1.0 to 2.0 Gigahertz	L	
2.0 to 4.0 Gigahertz	S	
4.0 to 8.0 Gigahertz	C	
8.0 to 12.0 Gigahertz	X	
12.0 to 18.0 Gigahertz	Ku	
18.0 to 24.0 Gigahertz	K	
24.0 to 40.0 Gigahertz	Ka	
40.0 to 100.0 Gigahertz	mm	
Modern Designations:		
0.1 to 0.25 Gigahertz	A	
0.25 to 0.5 Gigahertz	B	
0.5 to 1.0 Gigahertz	C	
1.0 to 2.0 Gigahertz	D	
2.0 to 3.0 Gigahertz	E	
3.0 to 4.0 Gigahertz	F	
4.0 to 5.5 Gigahertz	G	
5.5 to 8.0 Gigahertz	H	
8.0 to 10.0 Gigahertz	I	
10.0 to 20.0 Gigahertz	J	
20.0 to 40.0 Gigahertz	K	
40.0 to 60.0 Gigahertz	L	
60.0 to 100.0 Gigahertz	M	

shall employ tracking antennas and more ground stations are needed to hand over traffic from one satellite to the other. Also lower altitude orbits also require more satellites. The cost of the ground station also goes up because of the complex tracking system. The motion of satellites at lower altitudes introduces Doppler shifts, variation in path lengths, and irregular periods of mutual visibility between stations.

According to A.C. Clarke, a British scientist, who gave the concept of a "geostationary satellite and its possible applications for communications", a satellite in 35,786 km circular orbit in the plane of the equator becomes stationary relative to earth motion and is known as a synchronous satellite.

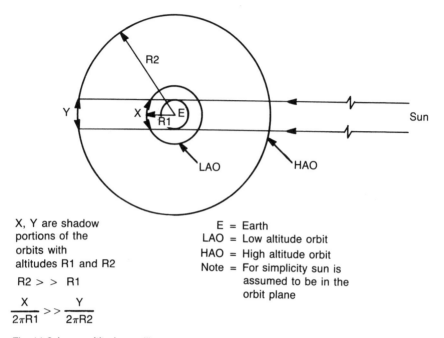

X, Y are shadow
portions of the
orbits with
altitudes R1 and R2

R2 > > R1

$$\frac{X}{2\pi R1} > > \frac{Y}{2\pi R2}$$

E = Earth
LAO = Low altitude orbit
HAO = High altitude orbit
Note = For simplicity sun is
assumed to be in the
orbit plane

Fig. 11-3. Lower altitude satellites spend a higher fraction of the orbital period in the shadow.

In a synchronous orbit, the orbital period is the same as the Earth's period for one complete rotation about its axis. Thus, the satellite remains "fixed" relative to a point on Earth. Also, at this altitude, three properly positioned satellites could cover almost the entire Earth. Compared to this we would need as many as 30 low altitude satellites to cover the entire earth. Another advantage is that a single synchronous satellite can provide uninterrupted communications service 24 hours daily. The stationary orbit is sunlit for more than 99% of the time (the infrequent eclipses occur only at midnight). This simplifies the generation and storage of power and reduces the number of temperature cycles.

Two-way propagation delays caused by the high altitude of the stationary satellite is about 0.5 seconds. Although this delay is more than the delay with submarine cable transmission, it is tolerable provided no echo exists.

Equatorial geosynchronous orbits are very useful for earth stations close to the equator. The advantage diminishes the farther the earth station is from the equator. Earth stations far away from the equator use satellites in highly elliptical polar orbits similar to the Molniya orbit shown in Fig. 11-4. In Molniya orbit, the spacecraft stays for about 6 hours at the apogee and four such spacecraft are needed for 24 hours coverage.

11.2 CHOICE OF FREQUENCY

The selection of the frequency for satellite communications depends upon technical factors, FCC allocation, and interference caused by and to others.

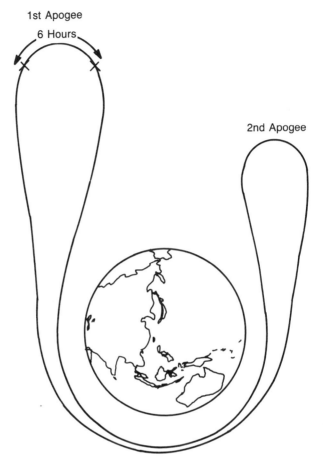

Fig. 11-4. Molniya orbit.

The main technical factor in frequency selection is the ground system noise temperature achievable at various frequencies. Ground receiver noise is influenced by (a) galactic noise and (b) atmospheric noise. An antenna pointed in the direction of the galactic center sees many noise sources and experiences high antenna temperatures. Of course, when the antenna looks out of the galactic plane, the noise temperature is much lower. For most regions of the sky, the antenna temperature can be nearer minimum value. Thus, it is important to know how the spacecraft travels so that the higher values of galactic noise can be minimized. Galactic noise is important at frequencies below 400 MHz and becomes negligible for frequencies above 1 GHz. Atmospheric noise is negligible below 8 GHz, however, oxygen and water vapor absorption increases atmospheric noise rapidly above 10 GHz. Thus, the region between 1 GHz and 10 GHz and preferably between 2 to 6 GHz is the desirable region.

To achieve large bandwidths, it is desirable to use frequencies above 10 GHz. For example, a TV bandwidth of five megahertz will then be only a fraction of a percent of the transmission frequency. Thus, at these frequencies there is room for many broadband signals.

As mentioned earlier, to avoid self-interference, the uplink frequency received at the satellite differs from the downlink frequency transmitted by it to the ground station.

11.3 MODULATION

Usually the selection of transmission frequencies depends upon various factors like atmospheric attenuation, medium of transmission, and the distance to be travelled, etc. Normal transmission frequencies are in the megahertz range. However, the information or signal to be transmitted is in a different frequency range. For example, the information of direct importance is in the audible range, namely in the range of 30 to 20,000 cycles. Also, the data to be transmitted usually falls in this range. Thus if the information is to be transmitted, then the signal shall be used to modulate the waves of higher frequency in the range of megahertz used for transmission due to obvious reasons given in the previous section. The higher frequency waves carry the low-frequency signal. Thus the higher frequency is known as carrier and is said to be modulated by the signal.

Usually the spacecraft transmitters and receivers operate at very high frequencies. Many signals may be multiplexed together and then used to modulate the carrier. When the carrier modulated by the signals is transmitted, the electromagnetic waves of this carrier frequency travel through the atmosphere and space. A receiver tuned to this carrier frequency picks up this modulated carrier. However, this frequency is not audible to the human ears and hence a reverse process of modulation is to be carried out to separate the signal from the carrier so that it is in the audible range. Such a process is known as demodulation.

The desirability of various modulation schemes depends upon their communication efficiency such as loading and frequency utilization factors. With present limitation of spacecraft power, there is need for modulation methods that conserve power.

Types of Modulation. Among the modulation types, the three basic types of modulation are a) amplitude modulation, b) frequency modulation and c) phase modulation. In amplitude modulation, the amplitude of the carrier is modulated to carry the signal. If the modulating signal is a sine wave, then the modulated carrier's amplitude varies continuously from one level to another level just to follow the modulating signal. In the case of frequency modulation, the frequency of the carrier is varied over a small range as a function of modulating signal amplitude and in the case of phase modulation, the phase of the carrier is modulated over a range to correspond the amplitude variation of the modulating signal.

In demodulating the modulated carrier, exact replica of the original carrier is usually required. This is obtained in different ways. Sometimes a reference frequency is generated by a high-precision oscillator. It is also possible to derive the carrier from the modulated carrier. Sometimes an unmodulated carrier is also transmitted along with the modulated carrier and sometimes the signal is interrupted at certain intervals to transmit the carrier.

Analog and Digital Modulation. When an analog signal, like voice or music is used to modulate the carrier, the amplitude of the carrier varies continuously in accordance with the variations of the input signal. This modulation is known as analog modulation. However, if the modulating signal is a binary on-off signal, then the amplitude of the modulated carrier will have only two levels. In this case, the modulation is known as digital modulation. The above discussion assumed amplitude modulation, although the analog or digital modulation can also be attributed to other types of modulations, i.e., frequency and phase modulations. When the modulating signal is digital, then AM, FM, and PM are known as amplitude-shift keying (ASK), frequency-shift keying (FSK), and phase-shift keying (PSK).

Modulation Process. Irrespective of the type of the modulation, when the carrier is modulated, the energy is distributed all over the carrier spectrum—upper and lower sidebands and the carrier. In the case of amplitude modulation, the sidebands will have frequencies of $f_c + f_m$ and $f_c - f_m$ (f_c is the carrier frequency, m_f is the modulating signal frequency) and are known as upper and lower sidebands respectively. The two bands are mirror images of each other and they contain the same information. Though the modulating frequency referred to above has a single frequency, in practice it contains many over a bandwidth. Thus $f_c + f_m$ is not a single frequency, but it is a band with its center frequency being $f_c + f_m$. For example, if the carrier frequency is 90 megahertz, and the signal to be transmitted has a maximum bandwidth of 20 kilohertz with a center frequency of 10 kilohertz, then $f_c + f_m$ is 9010 kilohertz.

Theoretically, as both sidebands contain the same information, it is sufficient to transmit only one sideband and this also saves transmitter power. However, to make demodulation simpler, sometimes both sidebands are transmitted. Also as mentioned in the earlier section that the carrier can be derived from the modulated signal, it is not necessary to transmit the carrier.

When the carrier is modulated, it has to meet certain requirements so that when the modulated carrier is received, it is possible to recover the signal to match the original signal. Thus, if the amplitude of the carrier is A_c, and the amplitude of the modulating signal is A_m, then the ratio A_m/A_c is known as the modulation index. If the modulation index is greater than 1, then the recovered signal upon demodulation will be distorted and is known as over modulation. On the other hand, if the modulation index is less than 1, then the recovery of the original signal will be inefficient and this is known as under modulation.

As the power is equal to the square of the voltage, the power needed for carrier transmission is proportional to A_c^2 and for single sideband transmission, the power required is proportional to $(A_m/2)^2$. This is the main reason why the carrier is usually suppressed and only one sideband is transmitted. This also reduces the bandwidth to half.

Usually two types of demodulation are employed, one needs the regeneration of the carrier and the other approach does not require the carrier and they are known as coherent demodulation and envelope demodulation, respectively. When a single sideband is transmitted, coherent demodulation has to be used. For this purpose, a pilot carrier is transmitted. The carrier is transmitted intermittently, or only just enough carrier is transmitted.

When the carrier and the two sidebands are transmitted, envelope demodulation/detection can be used. Envelope detection is simple and it simply rectifies and filters the modulated carrier to recover the original signal. The choice of method of modulation has to be decided making a trade of various factors involved, i.e., the bandwidth, the complexity of demodulation, etc.

Frequency Modulation. In amplitude modulation any corruption due to noise will mislead and may be interpreted as a signal. It is difficult to separate the noise from the signal, as the noise has a wide bandwidth. However, in frequency modulation, as the amplitude variation is not used to recover the original signal, the effect of noise is almost negligible. Thus, the amplitude of the frequency modulated waveform is almost constant and is not varied purposely.

When the carrier is frequency modulated, it produces an infinite number of sidebands (and not only two sidebands as in the case of amplitude modulation). If the frequency of the carrier is f_c and the frequency of the modulating signal is f_m, then the ratio, f_c/f_m is known as the modulation index. The bandwidth of the modulated wave changes as a function of the modulation index. The higher the modulation index, the higher the bandwidth and the energy is spread over a wide range of frequencies. (In the case of amplitude modulation, the energy of the amplitude-modulated wave results in three clusters, i.e., carrier, lower, and upper sidebands.) Thus a compromise between bandwidth and power is to be made in deciding the modulation index.

When the frequency modulated signal is received, any noise on its amplitude is removed by passing it through a voltage limiter. There are many methods to recover the original signal, one of which at low frequencies uses a scheme consisting of a differentiator, rectifier, monoshot, and a low-pass filter to recover the original waveform.

Phase Modulation. Transmission and detection of a small change in phase is complex and the accuracy involved is very poor. Hence, phase modulation is not generally used for the transmission of continuously varying or analog signals. This type of modulation is usually employed for the transmission of data. Here the modulating signal phase modulates the carrier and the phase changes are usually in steps of 90 or 180 degrees for digital data.

When the phase-modulated carrier is received, demodulation of the phase modulated wave requires the carrier for its original phase information or the carrier has to be generated from the phase-modulated carrier. Just like in the case of amplitude modulation, sometimes the carrier is also transmitted in this case continuously or intermittently.

Pulse-Code Modulation. *Pulse-code modulation* (PCM) and its variants are essentially encoding processes that convert analog signals to digital data streams. In pulse-code modulation, first the analog signal is sampled and quantized. The higher the number of quantization steps, the higher the accuracy. The analog signal is sampled at judiciously decided intervals, and each sampled value is quantized so that the inaccuracy introduced in the quantization process is acceptable. Thus, if the quantization levels are 32, then at any instant when the signal is quantized the signal value greater than 30 and less than 31 is read and 30. This introduces one bit error, namely one in 32 or about 3%. Thus practically used quantization steps are 128 resulting in a rounding error of less than one. Due to various known advantages, the steps or pulses are binary coded so that a maximum of 7 bits are required to represent any value between zero and 127. This overall process is known as pulse-code modulation.

The higher the number of samples, the higher the accuracy of the recovered signal, but higher samples result in higher bandwidth and on the other hand a very low number of samples result in inaccuracy of the recovered signal. Thus, practical and mathematical studies conducted resulted in an optimum number of samples. Thus, if the highest frequency of the signal is of Y cycles, then a sampling frequency of $2Y$ cycles is sufficient to recover the signal. For secure coding and improved communication efficiency, pulse-code modulation is desirable for some systems.

Differential Pulse-Code Modulation. In pulse code modulation, the sampling rate and quantization steps are decided as a compromise between the bandwidth and the recovery of the signal. However, if the signal can be recovered even with a lower sampling rate or with a lower number of quantization steps, then the transmission cost can be reduced. Thus, in *differential pulse-code modulation*, the difference between the amplitude of the previous sample and the current sample are encoded. This results in the requirement of 5 bits compared to 7 bits in the case of conventional PCM. Thus, when the sampling rate is the same, the bandwidth requirement reduces by about ⅖ or about 28%.

Delta Modulation. In *delta modulation*, the encoding only indicates whether the waveform amplitude increases or decreases at the sampling instant. Thus, in delta modulation, the signal difference between the previous and current sample is encoded but only one bit per sample. This type of modulation mainly depends upon the type of modulating signal. Thus, if the signal varies at a very fast rate then it requires more quantization levels or sometimes more bits-per-sample are used. On the overall, delta modulation requires less bandwidth compared to the differential PCM.

Frequency-Shift Keying (FSK). In *frequency-shift keying*, the information is represented by discrete frequency shifts. For synchronous systems the optimum frequency spacing is $1/t$ cps, where t is the bit length in seconds. Synchronous operation provides better signal-to-noise advantage over asynchronous operation. The number of discrete frequencies required is N, where N is the level of the code being used. Envelope detection is employed to demodulate frequency-shift keyed signals.

Phase-Shift Keying (PSK). In *phase-shift keying*, the information is represented by discrete shifts in phase. The optimum phase shift is $360/N$, where N is the level of the code being used. Product detector is used for demodulation, in which the received signal is multiplied by a stable reference signal which is phase locked to the received signal. The reference signal may be derived from the received signal or from a stable clock. PSK systems have an inherent requirement for synchronous operation.

Continuous Wave ON-OFF Keying (OOK). In On-Off keying, the information is represented by switching the carrier on and off. This modulation technique has given way to more sophisticated methods. This modulation is very effective even under extremely unfavorable signal to noise ratio conditions.

Spread Spectrum Technique. Whenever total interference power is large, signal to noise advantage (noise rejection) can be achieved by spreading the signal energy beyond the normal intelligence bandwidth according to a prearranged sequence. Pseudo-Noise modulation and correlation detection are effectively employed in such systems. Another method is frequency hopping, in which the energy is spread over a frequency-time plane.

Burst Transmission. This works on the principle of time compression and stores up generated traffic at its normal input rate. When propagation path conditions are proper for the support of the transmitted energy, the system releases the stored intelligence in a time compressed form at rates 10 to 100 or even more times the normal rate. This information is received and recorded at this rapid rate, and then slowed down to normal rate before being fed to the end user or customer.

11.4 MULTIPLE CARRIERS

Sometimes it may not be economical that only one carrier is transmitted through one transponder, because the transponder bandwidth is much higher than the carrier bandwidth. For example a voice channel requires a maximum of 20 kHz bandwidth (currently used voice channel bandwidth is only about 4 kHz). Thus a large number of voice channels can be transmitted simultaneously through one transponder. Thus a transponder having a bandwidth of 4 MHz, can carry 200 voice channels (assuming 20 kHz per channel or 1000 assuming 4 kHz per channel) or with some guard band it can easily carry 100 voice channels (the carriers, for two-way transmission, are allocated in pairs). On the other hand, it can only transmit, perhaps one TV carrier.

Thus, more information can be transmitted through a transponder only if one carrier bandwidth equals the transponder bandwidth and there is no need for guard band interleaving. However, if more than one channel of information is to be transmitted then as mentioned above, to avoid cross-talk some guard band has to be used; thus only 70% of the bandwidth may be effectively utilized. Also to avoid saturation of the travelling-wave-tube amplifier (as intermodulation products cause interference) it is operated below full power, resulting in underutilization of the transponder.

Frequency-Division Multiple Access. To make effective use of transponder bandwidth a large number of channels shall be transmitted. One way of making sure that all channels are kept separate is to use frequency-division multiple access. In frequency-division multiple access (FDMA) the transponder bandwidth is divided into smaller bandwidths allowing adequately for guard space. Each earth station uses one or more of these subdivided bandwidths and no two earth stations transmit on the same subdivided bandwidth at the same time. The subdivisions, if not dedicated, are reallocated from one earth station to another depending upon the traffic with prior knowledge and arrangement.

Time-Division Multiple Access. In this mode, each earth station is allowed a particular time slot or slots for transmission. The transmission time slots are properly planned such that no two earth stations overlap their transmissions. For the time slot allocated for a particular earth station, complete transponder bandwidth is available, thus that particular earth station can transmit a high speed burst of bits of information. This mode of operation allows efficient use of the transponder, bandwidth, and power.

Every earth station in a TDMA system receives the entire bit stream and extracts those bits that are addressed to it. Synchronization is very important, when TDMA is employed for high speed data transmission because channels to different earth stations are of different path lengths and hence propagation times. The propagation delay varies from its nominal value depending on how frequently the satellite position changes. Some digital equipment cannot process the data if the delay exceeds 200 microseconds. Also, satellites experience a daily oscillation in position due to the attraction of the sun and moon. Thus, depending upon the delay variation, the data rates have an upper limit or indirectly the equipment has to be synchronized at frequent intervals.

Advantages of TDMA. The TDMA needs very high speed burst modems, high speed processing of bit streams, stringent system timing, and large data buffers at each earth station. The rapid advances in TDMA technology are helping TDMA replacing FDMA systems. This is because TDMA offers significant advantages over FDMA, namely, a) as there is no intermodulation of carriers, the transponder can be used at its full power; b) use of digital speech interpolation reduces bandwidth requirements and it can easily be used with TDMA system; and c) channels of widely differing capacities can be intermixed and thus TDMA is very flexible.

Errors. Satellite carrier data channels usually have error rates better than equivalent terrestrial channels and to achieve low error rates, error correcting codes are employed; although an error rate of 1 error in 100 bits for digital telephone channels is acceptable, a lower error rate is very important in other data communication applications. Hence, efficient error detection and correction methods are employed and very low error rates of about 1 error in 10^7 are currently achieved. As the satellite link has a low signal-to-noise ratio, it is very important to make sure that no more noise is added due to improper equipment used on the ground. Transmission delays can cause further problems.

11.5 PRIVACY

Privacy is very important in satellite communications or data transmission. The signals from a spacecraft can be received anywhere within the spacecraft antenna coverage. Depending upon the information being transmitted, the degree of privacy varies. For example, a commercial TV broadcast is allowed to be received by as many people as possible, as TV station advertisement rates and income might depend on the size of the audience. On the other hand, business or technical information about a product being transmitted from one branch to another branch of the same company must be protected from unauthorized hands, as that company has invested and spent a lot of money in generating that information and it is proprietary to that company.

The interception and processing of the spacecraft signal requires expensive equipment. Extracting data from the modulated spacecraft carrier in the megahertz frequency range, though involving a complex process, is possible. Thus, to discourage any unauthorized interception, usually the signals are encrypted before modulating the carrier. When the authorized party receives the encrypted signal, it has the equipment or knows the code to decrypt or decode the signal for final use.

11.6 PROPAGATION DELAY

The distance between satellite and earth station varies with the angle of elevation, and consequently the propagation delay also varies. A satellite in a geosynchronous orbit vertically overhead is 35,786 km away and the maximum distance from a ground station can be 41,679 km, assuming the earth to be spherical and the angle of elevation is zero (from Fig. 11-1, the maximum slant range, S, is given by square-root of the term $[R2^2 - r^2]$ where $R2 = (35,786 + 6378)$ km and $r = 6378$ km).

As the transmission time delay is essentially independent of transmission frequency, the time taken by the signal to travel to the satellite in a geosynchronous orbit is given by the distance it travels divided by the velocity of light, 300,000 km per second and it varies from 119.3 to 139.0 milliseconds depending upon the elevation of the ground station. It will take two times this delay for a round trip.

11.7 ORBITAL SPACING

Spacecraft using the same frequencies must be placed far enough from each other to avoid or minimize any interference. Although, the physical spacing can be controlled so that the spacecraft can be located a half a degree apart (363.7 km apart at geosynchronous altitude of 35,786 km) due to the frequency interference requirements, the spacing used to be 5 degrees apart a few years ago; it is now 2 degrees apart.

How close two spacecraft can be located is difficult to answer as it depends on many factors related to space and ground segment, i.e., transmitting frequency, the bandwidth of the ground transmitter, its antenna size, etc. Use of larger ground antennas, as they can focus the beam into a narrow spot, allows closer spacecraft positions. Also, the use of higher frequency permits more spacecraft to share the orbit, as higher the frequency, the narrower the beamwidth for the same antenna size.

11.8 INTELSAT

INTELSAT is an international organization of 112 countries that owns and operates the global commercial communications satellite system serving the entire world. The system is used primarily for international communications and by more than 26 countries for domestic communications. INTELSAT was created in 1964 and began operations in 1965 with INTELSAT I (also called Early Bird as shown in Fig. 11-5) and recently launched the last satellite (Fig. 11-6) of the fifth generation. The first of the next generation satellites, the INTELSAT-VI (Fig. 11-7) series, is scheduled for launch in 1989. Today, through a network of 15 satellites in geosynchronous orbit over the Atlantic,

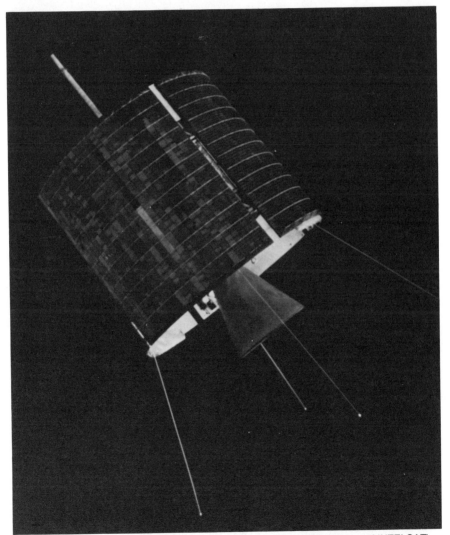

Fig. 11-5. INTELSAT I geosynchronous communications satellite (courtesy of INTELSAT).

Fig. 11-6. INTELSAT V geosynchronous communications satellite (courtesy of INTELSAT).

Indian, and Pacific ocean regions and more than 680 earth stations, INTELSAT links together more than 165 countries, territories, and dependencies around the globe. Two-thirds of the world's international telephone services and virtually all international television transmissions are carried over INTELSAT satellites.

INTELSAT has recently introduced two new programs for developing countries. Project SHARE (satellites for health and rural education) is designed to stimulate the use of telecommunications to help meet basic needs, such as education and health services, in the developing world. The INTELSAT development fund is designed to assist the developing countries in building their telecommunications infrastructures.

11.8.1 INTELSAT Services

INTELSAT services have expanded dramatically since 1964. Just ten years ago, international public-switched telephone service was 100% of INTELSAT's services. In 1986, domestic and maritime services accounted for 26.5 percent of INTELSAT's total services. INTELSAT currently provides the following services:

International Telephony Service. This was one of the first services provided through the Early Bird Satellite, launched in 1965, and includes international telephone, data, telex, and facsimile service. Table 11-2 presents

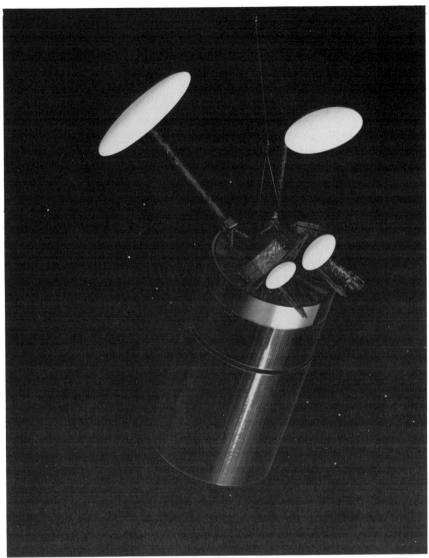

Fig. 11-7. INTELSAT VI geosynchronous communications satellite (courtesy of INTELSAT).

the evolution of INTELSAT satellites. Capacity on individual INTELSAT satellites has expanded from Early Bird's original capacity of 240 simultaneous telephone calls, to 120,000 telephone calls (plus three television channels) on the INTELSAT VI satellite, scheduled for launch in 1989. Today, more than two thirds of all international calls go through the INTELSAT system. The satellites continue to offer distinct advantages over fiber-optic cables for point-to-multipoint distribution and service to thin-route areas.

International Television Service. Virtually all international television service is provided through INTELSAT. Coverage of the Apollo moon landing in 1969 was the first time an event was telecast instantaneously around the

Table 11-2. Evolution of INTELSAT Satellites.

INTELSAT DESIGNATION	I	II	III	IV	IV-A	V	V-A	VI
Year of First Launch	1965	1967	1968	1971	1975	1980	1985	1989
Prime Contractor	Hughes	Hughes	TRW	Hughes	Hughes	Ford Aerospace	Ford Aerospace	Hughes
Width Dimensions, m. (Undeployed)	0.7	1.4	1.4	2.4	2.4	2.0	2.0	3.6
Height Dimensions, m. (Undeployed)	0.6	0.7	1.0	5.3	6.8	6.4	6.4	6.4
Launch Vehicles	Thor Delta	Thor Delta	Thor Delta	Atlas Centaur	Atlas Centaur	Atlas Centaur or Ariane 1, 2	Atlasr Centaur or Ariane 1, 2	Ariane 4 or NASA STS (Shuttle)
Design Lifetime, Years	1.5	3	5	7	7	7	7	14
Bandwidth, MHz	50	130	300	500	800	2,144	2,250	3,300
Capacity								
Voice Circuits	240	240	1,500	4,000	6,000	12,000	15,000	120,000
Television Channels	—	—	—	2	2	2	2	3

world. INTELSAT's current network of 15 operating satellites in three ocean regions offers the unique ability to serve most destinations around the world.

Domestic Telecommunications Services. Begun in 1985, Planned Domestic Service offers the purchase or long term lease of INTELSAT transponders to satisfy domestic communications requirements. The service is offered on a non-preemptible, fully protected basis, which provides the assurance and guarantee of service availability that countries require when planning domestic communications networks. Preemptible Domestic Service, which has been offered since 1974, provides service on a spare capacity and preemptible basis.

Vista Service. The VISTA Service brings domestic and international telecommunications services to rural and remote communities. The service, which is especially effective for low traffic requirements is offered on a channel-by-channel basis, on a preemptible or non preemptible basis. Super VISTA Service under the control of a Demand Assigned Multiple Access (DAMA) system, can be extremely cost effective, by passing on to the customer the more efficient utilization of the INTELSAT space segment that the DAMA equipment provides.

INTELSAT Business Service (IBS). It is a digital service tailored to meet the specific needs of the business community and designed to carry all types of telecommunication services, including voice telephony, high and low speed data, packet switching, video conferencing, electronic mail, and telex. IBS was introduced in 1983 and first used by the Bank of Montreal for communications between its corporate headquarters in Toronto and its branch office in London. The initial IBS offering provides a speed range from 64 kilobits per second to 8.4 megabits per second with various features attractive to the business user. Super IBS, introduced in 1986, provides IBS service with ISDN quality. With the addition of CARIBNET Service (a combination IBS/INTELSAT service for the Caribbean Basin), the service can now be offered on a regional basis.

INTELNET. It is a digital service designed for data collection and distribution, using small, inexpensive microterminals and a large central hub earth station. Applications for this service range from networks for sending out financial news and other data, to networks for collecting information on oil and gas exploration, gathering environmental data, and controlling and analyzing inventory data.

11.8.2 The INTELSAT Space Segment

The space segment of the INTELSAT network currently consists of 15 satellites in geosynchronous orbit. At present, global services are provided by a combination of IV-A, V, and V-A satellites. The first of the next generation of satellites, INTELSAT VI, is scheduled for launch in 1989.

INTELSAT maintains telemetry, tracking, command, and monitoring stations at eight locations around the world. These stations track the satellites orbits, check their positions and constantly relay vital information on their

Fig. 11-8. INTELSAT ground station (courtesy of INTELSAT).

operations. Information received at these stations is relayed to the INTELSAT Operations Center and Spacecraft Control Center in Washington, D.C.

11.8.3 The INTELSAT Ground Segment

The ground segment connected to the INTELSAT space segment consists of over 680 earth stations located in more than 165 countries, territories, and dependencies around the world. Figure 11-8 shows the Cayey, Puerto Rico earth station. Generally, these earth stations are owned and operated by the telecommunications organizations in the countries in which they are located. Earth station standards related to space segment access are set by INTELSAT. Several approved types of earth stations can use INTELSAT's services.

Standard-A. Currently the most widely used in the system for international communications, this standard originally employed a 30-meter antenna earth station, designed to operate with 6/4 GHz bands of frequencies. Availability of increased power from satellites has allowed this standard to be downsized to less than 18-meters in diameter while still making most efficient use of satellite capacity. New antennas of this size are now under construction.

Standard-B. This standard offers a lower cost alternative to Standard-A antennas. It has an 11-meter antenna designed to operate with 6/4 GHz bands of frequencies and is particularly suitable for areas with lower-or medium-sized traffic demands.

Standard-C. This standard originally provided for antennas of between 14- and 19-meters and is designed to operate with 14/11 GHz bands of frequencies with INTELSAT-V satellites. As in the case of the Standard-A antennas, this standard has also recently been downsized to about 13- to 15-meter size while still making efficient use of satellite capacity.

Standards-D1, -D2. These standards are designed for use with the Vista low density telephone service and offer antennas from 4.5 to 11 meters.

Standards-E1, -E2, -E3, -F1, -F2, -F3. These standards provide antennas with diameters from 3.5 to 9 meters and are designed for use with INTELSAT Business Service (IBS).

Standard-G. This standard is designed for use with INTELSAT service and other international services not covered by existing earth station standards. It allows the use of a wide range of earth station sizes including its use with microterminals as small as 0.8 meters in diameter.

Standard-Z. This standard is authorized for use with domestic services. Specific guidelines for technical performance are provided and the earth station owner may choose performance within these set parameters.

11.9 SOME IMPORTANT COMMUNICATIONS PROGRAMS AND SATELLITES

11.9.1 Advanced Communications Technology Satellite

The advanced *Communications Technology Satellite program* (ACTS) is a research and development program under the sponsorship of the National Aeronautics and Space Administration (NASA). Its major objective is to significantly increase the amount of data that can be transmitted to small aperture earth stations. NASA is encouraging users to experiment with space-based switching even to the level of a voice circuit and with new and innovative applications for these advanced communications technologies. Unique features of ACTS include operation using both fixed and scanning spot beams, in conjunction with advanced switching and processing systems, to concentrate communications capacity into narrow beams and thus automatically illuminate only those portions of the country that receive or transmit messages. These beams are rapidly formed and exist only for the time required. This results in an optimum use of the frequency spectrum and maximizes the potential data throughput. ACTS operates in the 30/20 GHz frequency range or Ka band (27.5 to 30.0 GHz uplink, 17.7 to 20.2 GHz downlink). The technology, however, is readily adaptable to the Ku-band frequency range (14/12 GHz).

Two switching techniques based on TDMA used are: (a) Matrix switch mode, which allows 220-megabits per-second burst-rate terminals capable of supporting high-rate data for trunking applications between satellite beam locations. It was three fixed beams with a total capacity 220 Mbps/beam. It has provision to compensate for rain by transmitter power management. (b) On-board baseband processor mode, which operates at a burst rate of 27.5 or 110 Mbps per beam. These techniques allow an individual user with a small terminal to access the satellite on incremental 64 Kbps circuits. The potential

ACTS family of terminals will include a 1 meter class (1 m to 1.8 m), a 3 m class and a 5 m class. The maximum burst rate for the 1 m class terminal will be 13.75 Mbps coded (forward error correction), and 27.5 Mbps uncoded. It has two scanning spot beams with a total capacity 220 Mbps/beam. Uplink burst rate is 27.5 and 110 Mbps and downlink burst rate 220 Mbps. It employs FDM/SSTDMA with SMSK modulations and has provision for compensating for rain by burst rate reduction.

ACTS Technology will have a wide range of applications and is expected to spawn and/or expedite the following emerging industries: customer premises services, flexible trunking, shared tenants services, efficient international communications, rapid data-base access and transfer, commercial video distribution, emergency communications restoration, mobile communications, electronic mail, teleconferencing, distributed computer processing, high definition television, and electronic data transfer.

11.9.2 NASA's Tracking and Data Relay Satellite System

Every year NASA launches scientific and experimental satellites that generate large volumes of valuable data. These satellites looking at the earth operate at low altitudes (200 to 1000 km) since they can see more detail by being close to the earth. Satellites that look away from the earth also operate at low altitudes since a launcher can carry a much larger payload to low earth orbit. Usually scientific and experimental data is stored in tape recorders, and dumped at high speed when the satellite is over one of NASA's Satellite Tracking and Data Network (STDN) stations. Collection of satellite data is expensive as it requires a large number of ground stations and they are to be located over the entire globe.

In 1980, the Tracking and Data Relay Satellite system (TDRSS) began its operation. Figure 11-9 shows a Tracking and Data Relay Satellite in orbit. NASA's scientific and experimental satellites send their data up to Tracking and Data Relay Satellites (TDRS) instead of down to earth stations. Also, now satellites can transmit data to TDRSS for a larger percentage of time, thus data storage requirements on-board the satellites are reduced. This also allows missions that generate volumes of data that are beyond the capacities of current tape recorders. As TDRS operates at geosynchronous altitude, they remain over fixed locations on the surface of the earth and also can see more than half of all the user satellite orbit locations.

11.9.3 Fleet Satellite Communications (FLTSATCOM)

FLTSATCOM satellites are the spaceborne portion of a worldwide Navy, Air Force, and DoD communications system, linking naval aircraft, ships, submarines, ground stations, the Strategic Air Command, and the Presidential Command Network. This consists of four satellites for continuous global communications coverage (except for polar regions).

Spacecraft Description. The main body is a hexagon structure, 7.5-ft wide and 4-ft high, containing the payload and spacecraft modules. The bottom of the structure to the top of the offset spiral antenna mast measures 21.7 ft.

Fig. 11-9. Tracking and data relay satellite (TDRS) (courtesy of TRW).

Two solar arrays, attached to booms extending from the spacecraft module, rotate to continually face the sun, and produce a minimum of 1259 watts of power for at least five years. The solar wings measure 43.4-ft tip to tip. Three 24-cell nickel-cadmium batteries provide power during solar eclipse periods. Three-axis body stabilization is employed to maintain vehicle attitude, and keep antennas pointed at Earth. Hydrazine jets alter the satellite longitudinal location. At launch, the satellite weighed 4,100 lb; in orbit it weighs 2,200 lb.

Program Objectives. FLTSATCOM provides worldwide high priority SHF and UHF communications between naval aircraft, ships, submarines military ground units, the Strategic Air Command and the Presidential Command Network. The communications subsystem provides the capability for more than 30 voice and 12 teletype channels and is designed to serve small mobile users. Operation in the UHF band permits economical use of present line-of-sight and satellite communications equipment, simple aircraft, and submarine antennas with hemispheric coverage patterns. It eliminates the need for bulky satellite tracking antennas. Design life is five years.

Payload. The payload module contains three antenna systems, transponders for 23 channels and the electronics required to support communications functions. The UHF receive antenna is an 18-turn deployable helix 12.5-ft long, offset from the satellite module; the UHF transmit antenna is a 16-ft diameter deployable dish and surrounding mesh with a central spiral post. A small separate conical helix antenna atop the central mast serves as the S-band tracking, telemetry and control antenna. A super-high-frequency antenna horn protrudes through the transmit UHF antenna mesh. The communications system includes nine fleet relay 25 kHz channels and 125 kHz Air Force narrowband channels designed to serve small mobile users; one 25 kHz fleet broadcast channel; a 500 kHz DoD wideband channel, and an SHF beacon. The UHF band operates in the 244 to 400 MHz range.

Launch History. FLTSATCOM 1 launched February 9, 1978 was positioned at 100 degrees west longitude. FLTSATCOM 2 was launched May 4, 1979 and positioned at 23 degrees west longitude and relocated to 75 degrees east longitude for FLTSATCOM 3 operation.

FLTSATCOM 3 was launched January 17, 1980, and positioned at 23 degrees west longitude. FLTSATCOM 4 was launched October 30, 1980, and positioned at 172 degrees east longitude. FLTSATCOMs 1 through 4 complete the constellation required for global (except for polar regions) communications coverage. FLTSATCOM 5 was launched August 6, 1981 to be placed in a geosynchronous orbit at 73 degrees west longitude above the Equator and was intended for use as an in-orbit spare, but was damaged during launch.

Launch Vehicle. The launch vehicle was an Atlas Centaur. The two stage, liquid fueled vehicle stands 39.9-m tall and is 3-m in diameter. With payload, it weighs 149,050 kg at liftoff. Atlas SLV-3D first stage thrust is 1,917,288 newtons at liftoff. Centaur second stage vacuum thrust is 133,400 newtons.

11.9.4 Small Business Satellite (SBS) Systems

Small Business satellite systems are designed to serve business communication needs in the United States, and will be the first domestic American commercial satellite to operate in the relatively uncrowded "K-Band" frequency range.

Satellite Description. The SBS satellites consist of two concentric cylindrical solar panels, the outer of which telescopes in space. This feature doubles the spacecraft's solar power generation capability over many previous satellite (spinner) designs. The antenna system on the satellites is folded at launch and deploys in space. The satellite is 7-ft in diameter and more than 9-ft in length in its stowed configuration. When it is fully deployed in space, it is nearly two stories tall.

Program Objectives. SBS was created to provide intercompany networks for the largest communication users, typically companies whose communications bills exceed $1 million per year. The SBS satellites expand voice, video, high speed data, and electronic mail services to business and industry throughout the U.S. Life span of the spacecraft is expected to be seven years.

Payload. The SBS spacecraft provides point-to-point K-Band (12/14 GHz) communications using a shaped beam to give higher gain over the East Central and West Coast portions of the US. The all digital system, operating in a time-division multiple-access (TDMA) mode, is capable of transmitting directly to rooftop antennas as a result of its operations in the 12/14 GHz frequency range. The satellite has 10 channels, each with 43 MHz bandwidth. Each channel uses a 20-watt multicollector traveling-wave tube. The radiation patterns, using the large shaped-beam reflector, provide a signal strength exceeding 43.7 dBW in the primary eastern coverage zone. With low-noise solid-state FET microwave integrated circuit techniques used in the redundant receivers, weighted beam receive sensitivites (G/T) range from 2.0 to 2.5 dBK over the most densely populated regions of the US.

Launch History. SBS-2 was launched from Cape Canaveral on September 24, 1981. It is now operating at an altitude of about 35,888 km at 97 degrees west longitude. Some eighty 18-ft diameter rooftop antennas receive direct transmissions from the craft. An additional 20 ground stations with 25.4-foot antennas will also be included in the system for a total of 100 ground stations.

Launch Vehicle. The launch vehicle was a Delta 3910. SBS uses conventional first and second stages of the Delta 3910s for launch. However, instead of a conventional third stage, a payload assist module is used. The module injects the spacecraft into an elliptical transfer orbit. The basic components of the module are a spin table, solid fuel motor and payload attachment system. It is a privately developed rocket vehicle that can operate either as the third stage of a NASA/McDonnell Douglas Delta booster or from the cargo bay of a Space Shuttle Orbiter to lift unmanned spacecraft to high altitude orbits.

11.9.5 India's Multi-Mission Spacecraft

The Indian National Satellite (INSAT) is basically a three-in-one spacecraft with three major objectives to provide, (a) telecommunications (voice and data communications) between major locations within India, (b) broadcasting (video and audio transmission for direct broadcasts to low-cost television, radio, and disaster warning receivers) throughout India, (c) meteorological services (VHRR observations and transmission of data to the meteorological center, and real time relay of data from remote meteorological data collection platforms (DCPs) to the meteorological center. This is an applaudable achievement as it carries the functions of three spacecraft, each with different mission requirements. It is understood that INSAT is the world's first spacecraft using and operating S band for direct broadcast.

Spacecraft. The INSAT is a three-axis stabilized spacecraft whose body looks like a rectangular box with a single wing solar array on the south side and a counter-balancing solar sail mounted on the top of a 12.19-meter boom on the north side. The main body dimensions are 2.2-m × 1.6-m × 1.4-m while north-south faces are of 2.2-m × 1.4-m an east-west faces are of 1.6-m × 1.4-m. Two-wing solar array design is purposely not used because the Very High Resolution Radiometer (VHRR) cooler (mounted on north side) needs to look continuously into deep space as it has to maintain radiometer detectors at cold temperatures. Use of a solar sail minimizes the fuel requirements for attitude control of the spacecraft. Antennas are mounted on east and west sides and they face toward the earth when deployed in orbit. Some of the spacecraft subsystems and their characteristics are listed below:

General	7 years design life 2 satellites Location—74 degrees East and 94 degrees East
Payload	C Band 12 Transponders each of 5-watt rf power and 36 MHz bandwidth Uplink 5.855 MHz to 6.425 MHz Downlink 3.71 MHz to 4.2 MHz
	S Band 2 transponders (TV) each of 50-watt rf power and 36 MHz bandwidth; Downlink 2.575 MHz, 2.615 MHz center frequencies
	DCP Uplink 402.75 MHz Downlink C Band
	VHRR C-Band downlink
Control	Three-axis stabilization
Telemetry	Dual Frequencies C Band (60-degrees wide Omni Antenna in Transfer orbit) C-Band Dish (in-Orbit)
Command	Dual frequencies C-Band Omni Antenna
Tracking	C-Band Transponder Mode Remodulation 27.7 kHz max tone

Communications and Broadcasting. INSAT uses 12 C-band transponders to provide communications and TV programs distribution via the C-band and the C/S-band antennas. The C/S-band antenna is used to transmit meteorological data, the TV downlink, and half of the telecommunications downlink channels to the ground. The C-band antenna is used to relay domestic telecommunications and direct broadcasting services. Direct TV broadcasting to community TV sets, radio program distribution, national TV networking and disaster warning are provided on S-band downlink at 2555-2635 MHz and the uplink (for the transmitting centers) is on C-band 5855-5935 MHz.

Meteorology. The meteorological payload uses VHRR and can produce a full earth picture about every 30 minutes in order to provide observations of weather systems such as cyclones, typhoons, monsoons, sea and cloud top temperatures, snow cover, etc. The radiometer has visible and infrared channels that provide resolutions of 2.75 km and 11 km respectively.

The meteorological system also collects and transmits meteorological, hydrological, and oceanic data from unattended remote data collection platforms (DCPs) and relays them to a central processing center.

Launch. INSAT-1A was launched on April 10, 1982 on a Delta 3910 using a PAM. The spacecraft failed in September of the same year as one of the antenna reflectors and the solar pressure counter-balancing solar sail did not deploy. The solar sail problem eventually led to the spacecraft failure.

To avoid such a failure, some modifications and improvements were made on INSAT-1B launched on August 31, 1983 using Shuttle STS-8. It became operational from October 15, 1983.

Chapter 12

Other Applications and Examples

12.1 MARITIME SATELLITE COMMUNICATIONS

The maritime satellite communications system consists of Space segment, ship earth stations on water, and coast earth stations on land. The space segment includes satellites and facilities for the monitoring and controlling of the satellites. The ship earth station implies an earth station installed on board the ship, on floating facilities, or on a platform to receive and or transmit. Coast earth stations are the stations on land and are generally located on or near the coast.

12.1.1 International Maritime Satellite Organization (INMARSAT)

The commercial maritime satellite communications service was first provided by the MARISAT system, which was established in 1976 by four major communications entities of the U.S. This system was originally intended for U.S. domestic communications, but the service is open to vessels of other countries, too. MARISAT satellites are now positioned one each over the Atlantic, the Pacific and the Indian Oceans. In August 1976, two earth stations were opened for maritime service in the United States, followed by the commissioning of the Yamaguchi coast earth station, in Japan in Nov 1978. As a result, the MARISAT system is now able to provide service practically on a global scale. As of 1st July 1981, there were 733 vessels registered in this system to utilize the service.

Meanwhile, with the aim of providing maritime mobile satellite communication service on a worldwide scale, INMARSAT (International Maritime Satellite Organization) came into existence in July 1979 with headquarters in the U.K. The INMARSAT convention was drafted after the agreement relating to the INTELSAT.

Figure 12-1 shows the second-generation maritime communications satellite being built for INMARSAT. Two Maritime Communications Satellites

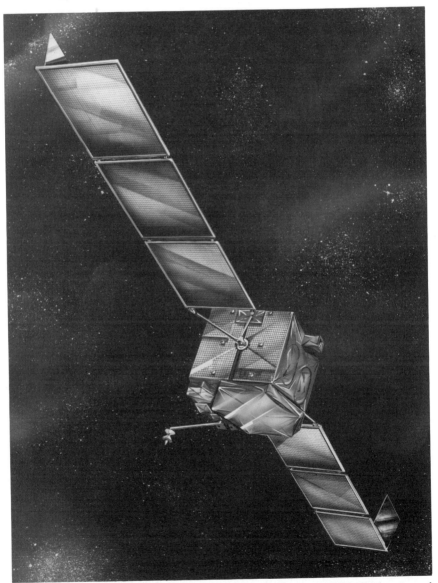

Fig. 12-1. Second generation maritime communications satellite INMARSAT-2 (courtesy of British Aerospace).

(MARECS) supplied by British Aerospace are currently operating in INMARSAT's existing global maritime satellite communications network.

12.1.2 Maritime Communications Satellite (MARECS)

The European Space Agency's MARECS program covers the development, launch and in-orbit operation of communication satellites to be integrated in a global maritime communication system. Development began in 1973 with funding from Belgium, France, Italy, United Kingdom, Spain, and West Germany, later joined by the Netherlands, Norway, and Sweden. The MARECS satellite is part of a global communications system configured to provide high quality full-duplex, reliable real-time voice, data, and teleprinter services between ship earth stations and coast earth stations with automatic connections to the terrestrial network.

The MARECS-1 was launched on December 20, 1981 by the Aerospatiale Ariane Booster into an orbit with apogee of 35,743 km, perigee of 35,620 km and inclination of 2.3 degrees. The MARECS satellite consists of two modules: a service module, which is a derivative of the European Communications Satellite (ECS) bus, and a payload module. The satellite has a design life of seven years, a three-axis attitude control system and a tracking, telemetry and command system that uses VHF during transfer orbit and C-Band, through the communications subsystem, on station. The overall satellite weighed 1006 kg at launch.

The payload consists of a C-Band to L-Band forward transponder and a L-Band to C-Band return transponder incorporating a Search and Rescue (SAR) channel. The payload is capable of operating without continuous ground control.

12.2 WEATHER AND EARTH RESOURCES

Since TIROS-1 in 1960, survey of the earth from space by optical means became possible. Since TIROS-9 in 1965, global photographic coverage has become a reality. The current operational meteorological satellites can photograph every portion of the world at least once a day and meteorologists can access this data within 24 hours. Procedures for rapid recognition of significant weather phenomena have been developed and the automated processes for disseminating information have become routine. During the first five months of operation of ESSA-1, the operational version of TIROS for the Environmental Science Services Administration, 295 storm warnings were issued to 23 countries. TIROS-3 in 1961 identified hurricane Esther at least two days before any other means of observation could have discovered it.

With the Applications Technology Satellite, meteorological observation in color became a reality from a 24 hour stationary orbit. Stationary orbits offer the advantage of continuous assessment of a large portion of the earth by means of frequent observations. This type of observation helps in detecting small storms. The improved contrast of clouds (due to color photographs) against the ground or ocean background permits a higher effective resolution, and the color helps in estimating the altitude of observed clouds. The ratio of blue to red in the cloud top photographed is affected by scattering of the light as it moves through air. The greater the amount of air through which the light moves, the redder its color. With colored photographs, the blue/red ratio can be measured and the amount of air is through which the light has passed deduced. Since the angles of the observed cloud from the spacecraft and from the sun are known, the altitude of the cloud top can be estimated. Measurement of the temperature of the cloud by IR sensors permits greater accuracy in determining the height of clouds, since the relationship between altitude and temperature is more direct and more accurately predictable. IR sensors have been used on relatively low altitude missions to provide cloud cover photographs of the night side of the earth. Airglow photometers detect cloud patterns in moonlight and, with more difficulty, with only starlight and zodiacal light.

Photography in certain spectral bands in the visible and near IR is a powerful tool in the classification of vegetation and soil types. Monochromatic and polychromatic sensing of reflected and emitted infrared can complement visual photography, enabling delineation of boundaries of vegetation, water courses, roadways concealed by vegetation, and similar features of interest. In the band from 9,000 to 55,000 Å, the energy radiated is primarily reflected solar energy, modified by absorption of water. Within this band, with multispectral data in the visible spectrum, spectral signatures permit unique identification of types of vegetation. Also, infrared radiation at a wavelength of 14,000 Å indicates the water content of the underlying terrain. IR spectrometry from 80,000 to 140,000 Å can yield data bearing on the mineralogical and chemical composition of dry surface rocks barren of vegetation. Many other applications of IR mapping of potential value have been demonstrated, including the detection of subsurface fresh water runoff into the oceans, ice-depth, and the characteristics of snow cover.

Weather. Meteorological satellites provide information in the form of photographs so that meteorologists can study and forecast the weather. These satellites also measure the speed of the wind, the temperature of the air and the amount of moisture in it. All of this information is used in predicting weather and spotting hurricanes. Satellites also help predict the amount of rain that a hurricane will bring. Temperature monitoring enables the prediction of freezes far enough ahead so that farmers can take precautions to protect the crops. These satellites can monitor pollution over cities and large industrial areas so that pollution alert and reduction measures can be taken.

Oceans. Oceanographic instruments observe the earth's oceans, monitor the height of the waves, the ocean currents, temperature of the water, and locate icebergs. These instruments also observe the color of the water, which tells about plant life, oil spills, etc. Ocean temperature affects the temperature of the air. Colder ocean temperatures in certain places mean a colder than normal winter. Water temperature charts and location of cooler and warmer currents helps fishermen as large numbers of fish are found in areas of the ocean where cool and warm water currents meet. Satellites also measure the amount of sediment flowing in the sea. As the sediment settles, it changes the crust of the earth. This helps in predicting what happens to the sediments in the future.

Land. Satellites see which crops are doing well and which are in trouble from drought or pests. They monitor the movements of animals and scan the areas where these animals graze. Land maps today are more accurate than ever before, because of these satellites.

Sun. Satellites observe the effects of sunspots and solar flares and earth's magnetic field. Disturbances on the sun can change the weather on earth. Sunspot electromagnetic waves jam, or interfere with the reception of broadcast signals. Sometimes they can cause power blackouts.

Environmental satellites photograph the earth at least once a day, every day, and all photographic data is stored, which can be retrieved later, when scientists want to study long term trends.

12.2.1 Global Weather Experiment

For the Global Weather Experiment five geostationary satellites and a number of polar orbiting satellites were deployed around the globe to take part in observations. Satellite communications ensured the availability of this observed data in time at various locations. For example, the CNES' ARGOS system mounted on the TIROS operational satellite enabled determination of the location of up to several thousand platforms and received the data from the platforms and distributed it. Figure 12-2 shows the METEOSAT linking with central control and processing station, user station, and data collecting platforms. Measurements with this technique from free floating buoys deployed in the southern oceans were a particularly important data source for the Glob-

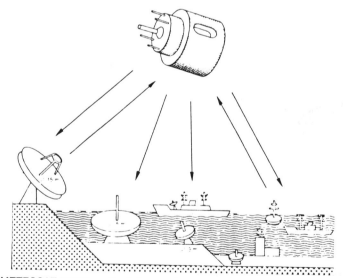

Fig. 12-2. METEOSAT linking on the left central control and digital processing station, in the middle with the user stations (main, local, shipborne), and on the right with the data collecting platforms (ground stations, hydrology stations, ships and buoys) (courtesy of European Space Agency).

al Weather Experiment. Since the development of the global meteorological observing system in 1979, it has continued through the World Weather Watch (WWW) program of the World Meteorological Organization (WMO) with a goal to achieve, on a continuous basis, observational coverage of the globe.

12.2.2 Instruments

Multi-Spectral Radiometer: Visible and infrared radiances are received by the radiometer as its field of view scans the earth as the satellite spins, and as between each satellite rotation the telescope mirror is moved in small stages so as to view different lines across the earth. Thus, full coverage of the earth is achieved.

Usually all radiometers have a few spectral channels. For example, Meteosat Radiometer has two identical visible channels between 0.4 and 11 micrometers, and in the water vapor band between 5.7 and 7.1 micrometers; and one in the infrared window between 10.5 and 12.5 micrometers. The resolution of this visible channel at the sub satellite point is 5 km for the infrared channels and 2.5 km for the visible channels when both visible channels are employed.

Temperature Remote Sensing Instruments. Satellite observations of the atmosphere between about 10 and 100 km in altitude have dramatically improved the study of the atmosphere, which is populated by planetary waves generated in the troposphere and under suitable conditions they propagate upwards. In winter they can reach very large amplitudes up to about 100 degrees k in temperature amplitude at about 45 km altitude.

The temperature of the earth's surface and the upper surfaces of clouds may be determined from the thermal infrared part of the spectrum. The wavelength employed for such thermal imaging is one in which the atmosphere is largely transparent, namely the 10-12 micrometer band. By measuring the radiance at wavelengths where the atmosphere absorbs, properties of the atmosphere including its temperature and composition can be inferred. Also clouds are transparent to microwave radiation unless they are precipitating clouds.

High Resolution Infrared Sounder. This is a twenty channel radiometer which includes channels in both the 4.3 μm and 15 μm spectral regions where carbon dioxide emits.

Stratospheric Sounding Unit. This is a three-channel radiometer for sounding the stratosphere (10-50 km altitude) with spectral selection based on the pressure modulation technique.

Microwave Sounding Unit. This is a four-channel radiometer observing emission from stratospheric oxygen near 0.5 mm in wavelength.

Advanced Microwave Sounding Unit. It consists of two separate instruments, one possesses 15 channels for sounding temperature from the surface to 50 km altitude and the other possesses 5 channels near 180 GHz in frequency for sounding the water vapor distribution.

Along Track Scanning Radiometer. The exchange of heat and momentum and the earth's surface is of great importance in determining the atmosphere's circulation. Thus, the measurement of surface temperature especially over the ocean is very important. The input of heat energy into the atmosphere is dominated by the latent heat input through evaporation, which varies as a function of surface temperature. The Along Track Scanning Radiometer (ATSR) on ERS-1 measures sea surface temperature to better than 0.5 degree K. Figure 12-3 illustrates the Along-Track Scan of Landsat-7 ALS and EOS Polar Orbiter and Cross-Track Scan of Landsat-6 Enhanced Thematic Mapper, whereas Table 12-1 presents the details of these instruments.

Electronically Scanned Microwave Radiometer. This radiometer provides continuous observation of ice cover due to very different emissivities of ice and water at 1.5 cm wavelength. Further, this instrument enables recognition of different types and ages of ice. Emissivity of the land surface at these wavelengths depends strongly on soil moisture.

Scanning Multichannel Microwave Radiometer. The combination of images at different wavelengths can help in the interpretation of satellite data. For instance, low cloud or fog can be recognized in infrared images by making use of the different spectral properties of the clouds and the land or

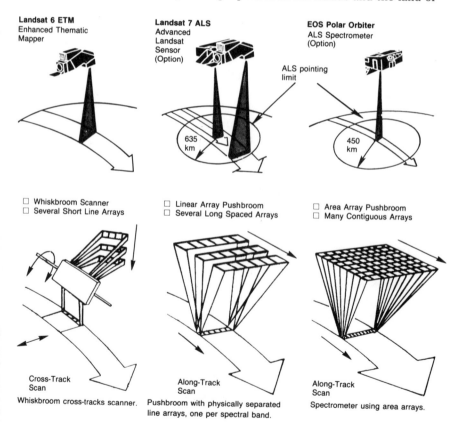

Fig. 12-3. Illustration of Along Track Scan of Landsat-7 ALS and EOS Polar Orbiter and Cross-Track Scan of Landsat-6 Enhanced Thematic Mapper (courtesy of Earth Observation Satellite (EOSAT) Company).

Table 12-1. Details of Along-Track and Cross-Track Scan Radiometers Used on Landsat-7, EOS Polar Orbiter and Landsat-6 Enhanced Thematic Mapper.

PERFORMANCE CHARACTERISTIC	INSTRUMENT/MISSION		
	ENHANCED THEMATIC MAPPER LANDSATS 6/7	MLA (ALS) LANDSAT 7 (OPTION)	ALS SPECTROMETER EOS POLAR ORBITER (OPTION)
Spatial			
☐ Orbit (km)	705	705	Variable 250-1000
☐ Swatch Width (km)	185	41	41 from 705 km orbit.
☐ Line-of-sight Pointing	Nadir view only	Nadir to 635 km off-nadir; omnidirectional pointing from 705 km orbit. Stereo and cross-track imaging.	Nadir to 450 km off-nadir; omnidirectional pointing from 500 km orbit. Stereo and cross-track imaging. 9 with 17° pointing from 500 km orbit.
☐ Revisit Cycle (days)	16	16	
☐ Resolution (m)	15 Panchromatic, 30 VNIR/SWIR, 120 Thermal	☐ 10 Visible/Near Infrared, ☐ 20 SWIR	☐ 7 VNIR (from 500 km orbit), ☐ 14 SWIR
Spectral			
☐ Coverage (μm)	0.45-2.35, 10.4-12.5	0.4-2.5	0.4-2.5
☐ Number Bands	8	32	64-128
☐ Resolution (nm)	70-270, 2100	8 (simultaneously), 20-270	20
Radiometric			
☐ Quantization	8 bits (256 levels)	10 bit internal precision, 8 bits transmitted	10 bit internal precision, 8 bits transmitted
☐ Average Sensitivity	0.5% NE$\Delta\rho$, 0.5k NEΔT	0.5% NE$\Delta\rho$ reflective	0.5% NE$\Delta\rho$ reflective
☐ Accuracy (%)	5-10	2-5	2-5

ocean surface. The extent of snow cover and its variations has been determined by combining observations of different wavelengths from the Scanning Multi-channel Microwave Radiometer on Nimbus-7.

2-CM RADAR. The NASA SEASAT satellite obtained surface winds by observing with a 2-cm radar the back scattered signal from small capillary waves on the sea surface whose amplitude and direction is a function of the instantaneous wind.

12.2.3 Landsat Applications

Landsat is an earth-oriented sun-synchronous satellite carrying remote sensing and earth resources instruments and its applications are described below.

Coastal Zone Applications. This includes tidal wetland monitoring, barrier island migration assessment, beach erosion studies, and development activity monitoring. Inland bays are shallow by nature and problems arise when tributaries carry large amounts of sediments into the bays, or when nutrients associated with agricultural runoff create extensive algae and plant growth.

Agribusiness. This includes crop conditions, assessment of soil associations, irrigation practices, and freeze damage assessment. Repeat and synoptic coverage provided by Landsat data permits analysts to assess the productivity and quantity of agricultural crops in a region. This information is necessary to plan crop storage and transportation needs and to plan on the amount and type of seed that will be required for spring planting. This Thematic Mapper scene of an agricultural area demonstrates crop maturity differences. Growing vegetated areas appear as red, harvested fields in green, and freshly plowed or recently planted fields of spring crops appear dark blue or almost black.

Geology. This includes mapping geologic structure, tectonic analysis of large areas, locating seismic and other geophysical traverses, planning field logistics, and guiding selection of land acquisitions and permits. Thematic Mapper data is extremely useful in focusing the efforts of mineral and petroleum exploration programs, especially in low vegetation areas. By analysis of the geologic structure and spectral characteristics, the remote sensing geologist can eliminate large areas of low interest and concentrate on more prospective ground.

Urban Applications. This includes, (a) monitoring the urbanization of rural/forested areas, (b) land cover and impervious surface analyses for urban watershed mapping and modelling, and (c) environmental stress analysis on urban parklands and greenways.

Thematic Mapper data has applications in a wide variety of urban and regional planning applications. The sensitivity of the Thematic Mapper sensor to spectral contrasts between developed and undeveloped land, combined with the 30-meter resolution, provides an excellent tool for monitoring the direction and magnitude of urban growth.

Forestry. This includes, identification and monitoring of insect and/or fire damage, and inventory sample design. Landsat data has a number of important applications to forestry. Forest managers have demonstrated that by using repetitive Landsat coverage, temporal changes in the forest composition associated with timber harvest, or defoliation caused by insect, disease, or fire can be monitored. Knowledge about the total forest area available for harvest and its location is very important to the forest manager. Landsat data provides an attractive alternative to traditional survey methods.

Disaster Assessment/Emergency Planning. Landsat images are useful in identifying natural buffer zones, safe haven areas and evacuation routes for the creation and updating of emergency plans. Analysis of current images for annual review of emergency plans is often required to assess the impact of land use change and population dispersal areas on evacuation routes.

The Landsat Multi-spectral scanner image collected after a spring flood was used to assess the flood damage to winter wheat. Such imagery is helpful in calculating flood damage insurance payments. It is also useful in identifying mitigating measures that could be taken to prevent similar damage in the future.

Hydrology. A large percentage of the agricultural water supply in the western United States comes from melting snow packs in the surrounding mountains. In California, approximately 75% of the water used for crop irrigation comes from the Sierra Nevada snowpack. In the past, the principle application of satellite imagery to snow hydrology measured the mountain area above the snowline. This, however, is a poor measure of the total volume of snow. The improved Thematic Mapper spatial resolution now allows the hydrologist to use textural information to estimate snow thickness.

Land Use and Regional Planning. Landsat data are widely used in developing countries for land suitability/capability analyses, land-use inventories and development planning.

Range Management. Seasonal Landsat coverage, augmented by specialized computer models, can be applied to vegetation biomass mapping to determine range conditions and carrying capacities. Using Landsat data, areas of overgrazing can be identified and surface water conditions can be monitored.

Cartography. Landsat data are appropriate for many cartographic applications such as thematic mapping, mosaicking for regional basemap purposes, and the creation of geometrically accurate photo maps.

12.3 SATELLITE SOLAR POWER STATIONS

Satellite Solar Power Stations (SSPS) are conceived to supply electrical power needed by the entire earth by employing about 100 satellite solar-power stations. It is estimated that the total cost of power generation in geosynchronous orbit including the launch could be as low as $2 per watt if it is done on a large enough scale.

These stations will be very big with many miles-long and many miles-wide solar arrays, and will be in geosynchronous orbit. This orbit is such that the station stays over the earth at a fixed point and stationary as the earth rotates. The solar cells convert the solar radiation into electrical power which is converted into microwaves and beamed to ground stations. On ground the microwaves are converted back to electrical power and sent to users.

Electricity using satellite solar power stations is clean energy. There will be no radioactive waste, or air pollution. Though they are costly to build, over a period they offer cheaper electricity as the sun's radiation is free and abundant.

On the other hand, the microwaves may be dangerous and the cost of building the stations would be enormous. New technologies of producing cheaper solar arrays and making satellites might bring down the cost of these stations.

12.4 FUTURE APPLICATIONS

In this section some of the future satellite applications are described. A study conducted for NASA by the Aerospace Corporation described a system which could permit wristwatch radio telephones. Nations are connected with telephone today. The next step is to connect them through video phones. Telecommunications with video link might become a future substitute for physical travel. Countries with no oil or minimum oil resources might not have to depend upon oil-rich countries. They could substitute satellite communications for physical travel.

Today commercial interest in processing efforts in space is focused on the following fields, (a) pharmaceuticals, (b) protein crystals and biotechnology, (c) advanced metals, alloys, glasses, and ceramics and (d) semiconductor materials. The processing in space can be done on dedicated satellites or space platforms or on space stations. Figure 12-4 shows the Columbus pressurized module attached to the U.S. Space Station.

The McDonnell Douglas/Johnson & Johnson pharmaceutical venture focuses on the use of the electrophoresis separation process in space to obtain mass quantities of a hormone that cannot be separated practically in normal gravity on Earth. In the four processor flights on the shuttle, McDonnell Douglas has been able to demonstrate that as much as 716 times more separated material can be obtained in space compared with Earth-based separation. A four- to five-fold increase in product purity also has been demonstrated with the space processing.

The Lovelace Medical Center is interested specifically in studying the production of monoclonal antibodies in space for the treatments for cancer or other diseases such as malaria. Research indicates it may be possible to obtain much more of the antibody material from space-based processing than would be possible in the Earth's gravity field, which overpowers much of electrophoresis separation capability. Battelle Columbus Laboratories, USA, is interested in zero gravity process for growth of collagen fibers to be used in the repair and replacement of human connective tissues.

Aluminum Company of America is interested in using the space environment to manufacture aluminum lithium alloys and high strength powder metallurgy alloys. Bethlehem Steel, USA, wants to determine the effect of zero gravity solidification on the structure of graphite in cast iron. Honeywell, Inc.,

Fig. 12-4. Columbus pressurized module attached to the U.S. Space Station (courtesy of Aeritalia).

USA is interested in investigating mercury cadmium telluride crystal growth in space. These are some of the large number of companies interested in space-based processes and manufacturing.

There are plans to manufacture organic crystals, thin films, glass-forming alloy systems, ultra-pure and bubble free glass, improved silicon chips, etc., in space. Production of new electronic semiconductor materials, like pure gallium-arsenide crystals, could provide a fundamental improvement in computer power. Gallium arsenide is considered extremely valuable, particularly in very pure form, for use in high-speed computers and other sophisticated electronics equipment. Space-processed spheres are more uniform compared to the ground processed ones and will be used in a variety of calibration applications.

Some analysts forecast that by the year 2000, commercial space operations could generate $65 billion in annual gross revenues, consisting of $20-$27 billions from pharmaceutical products, $10-$15 billions from other products, $15-$25 billions from advanced space communications, remote sensing, launchers, etc. This in turn could generate $13 billion in annual tax revenues.

12.5 SATELLITE APPLICATION EXAMPLES

In this chapter, some of the satellites have been described. Every satellite uses most of the subsystems described in Chapters 1 to 10.

12.5.1 TIROS-N/NOAA Weather Satellite Series

Advanced TIROS-N (ATN) weather satellites are designed and built by RCA for the National Oceanic and Atmospheric Administration (NOAA) under the management of NASAs Goddard Space Flight Center. ATN satellites provide scientists with the most comprehensive meteorological and environmental information since the start of the U.S. space program. The ATN satellites are the latest in a series of RCA weather satellites dating back to 1960.

The TIROS-N/NOAA system collects meteorological readings from several hundred data collection locations on land, in the air and at sea. It takes vertical measurements of the temperature and moisture distribution in the atmosphere. In addition, the satellites measure energy particles for solar research and radiation warning. In addition to the normal complement of meteorological sensors, these advanced satellites are equipped with instrumentation for global search and rescue missions that locate downed aircraft and ships in distress. Figure 12-5 shows the locations where 524 persons were rescued by ATN satellites. ATN will also carry instrumentation for ozone mapping and for monitoring the radiation gains and losses to and from the earth.

The spacecraft are launched from the Western Test Range by Atlas-E rockets. They are placed in sun synchronous polar orbits 450 to 470 nautical miles above the earth. Weather observations, including visible and infrared pictures, are provided to scores of ground stations world wide. Total global weather and environmental information from the satellites are received, processed and distributed by NOAA's National Earth Satellite Service, USA. In addition to weather forecasting, the information is used for hurricane tracking

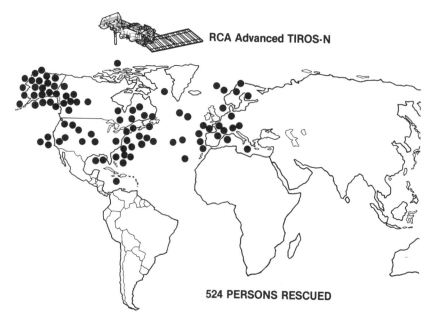

RCA Advanced TIROS-N

524 PERSONS RESCUED

Fig. 12-5. ATN satellites rescued 524 persons (courtesy of RCA).

and warnings, agriculture, commercial fishing, forestry, maritime, and other industries.

Objective. Provide the Department of Commerce, National Oceanic and Atmospheric Administration (NOAA), with the global environmental data required to support both the national operational systems and experimental programs of the National Weather Service and of the World Weather Program. The data collected by the satellite's advanced instrument complement will be processed and stored on-board for transmission to the central processing facility at Suitland, Maryland, via the Command and Data Acquisition stations. Data will also be transmitted in real time, at VHF and S-band frequencies, to globally distributed remote stations.

Launch Information.

Launch Site   Western Test Range at Vandenberg AFB, California
Launch Vehicle   Modified Atlas-E

Orbital Elements.

Circular   Near polar sun synchronous orbit, nominal altitude of 870 km, 1500 hrs ascending node orbit or 833 km, 0730 hrs descending-node orbit (Fig 12-6)
Period   101.6 minutes for 833 km and 102.4 minutes for 870 km
Inclination   98.7 degrees for 833 km and 98.9 degrees for 870 km

Spacecraft Information.

Height	491 cm
Diameter	188 cm
Liftoff Weight	1712 kg
In-Orbit Weight	1030 kg
Payload	352 kg

Power.

Array	1470 Watts
Spacecraft Bus	286 Watts orbit average
Sensor Payload	263 Watts orbit average

General Description. NOAA satellites are equipped (see Fig. 12-7) with instruments that provide data on Earth's cloud cover, surface temperature, atmospheric temperature and humidity, water-ice moisture boundaries, and electron flux in the vicinity of Earth. Additionally, a data collection system aboard the satellites receives environmental data such as temperature, pressure, altitude, etc., from fixed and floating platforms, buoys, and balloons located throughout the world.

Figure 12-8 shows the NOAA-G satellite standing two stories high in a thermal/vacuum test chamber at RCA. The spacecraft have integrated systems with the capability to control the spacecraft's injection into orbit after separation

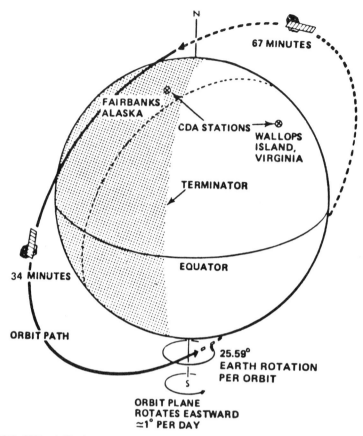

Fig. 12-6. ATN satellite in a sun-synchronous orbit (courtesy of NASA Goddard Space Flight Center).

from the Atlas-E launch vehicle; maintain proper spacecraft attitude in orbit; control the temperature of the spacecraft subsystems and instruments; process, record, and transmit the environmental data to ground stations; and receive and decode ground commands. The spacecraft structural and thermal design is modular in concept, permitting growth or modifications for each mission.

Two redundant on-board computers control spacecraft injection into orbit, control satellite operations in orbit, and issue the commands required to maintain proper spacecraft attitude. The data processing and communications subsystems are designed to satisfy all mission requirements without exceeding the limitations of the existing network of meteorological ground stations.

The advanced TIROS-N satellite series is a cooperative effort of the United States (NASA and NOAA), the United Kingdom, Canada and France. NASA, using NOAA funds, procures and launches the satellites. NOAA operates the two command and data acquisition stations, the satellite operations control center, and data processing facilities. The United Kingdom, through its Ministry of Defense meteorological office, provides the stratospheric sounding unit, one of the three atmospheric sounding instruments on each satellite. The Centre

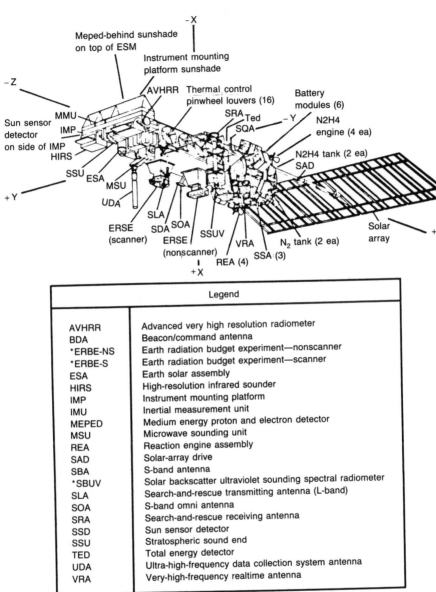

Fig. 12-7. ATN spacecraft with major features identified (courtesy of NASA Goddard Space Flight Center).

Legend	
AVHRR	Advanced very high resolution radiometer
BDA	Beacon/command antenna
*ERBE-NS	Earth radiation budget experiment—nonscanner
*ERBE-S	Earth radiation budget experiment—scanner
ESA	Earth solar assembly
HIRS	High-resolution infrared sounder
IMP	Instrument mounting platform
IMU	Inertial measurement unit
MEPED	Medium energy proton and electron detector
MSU	Microwave sounding unit
REA	Reaction engine assembly
SAD	Solar-array drive
SBA	S-band antenna
*SBUV	Solar backscatter ultraviolet sounding spectral radiometer
SLA	Search-and-rescue transmitting antenna (L-band)
SOA	S-band omni antenna
SRA	Search-and-rescue receiving antenna
SSD	Sun sensor detector
SSU	Stratospheric sound end
TED	Total energy detector
UDA	Ultra-high-frequency data collection system antenna
VRA	Very-high-frequency realtime antenna

National d'Etudes Spatiales of France supplies the data collection and location system and a search and rescue processing system for the satellite and also the ground facilities to process the data collection system data and make it available to users. Another French agency, the Centre d'Etudes de la Meteorologie Spatiale, uses its ground facilities to receive atmospheric sounder data from the satellite during orbits in which the satellite does not come within contact range of the Command and Data Acquisition stations in the United

Fig. 12-8. NOAA-G satellite stands two stories high in a thermal/vacuum test chamber at RCA (courtesy of RCA).

States. The Canadian Department of Commerce Communications Research Center Provides a search and rescue repeater system.

Instrument Payload. The primary environmental sensors are given below:

☐ **TIROS Operational Vertical Sounder.** It is a three-instrument system consisting of high resolution infrared radiation sounder, stratospheric sounding unit, and microwave sounding unit.

☐ **High Resolution Infrared Radiation Sounder (HIRS).** It is a 20 channel instrument that makes measurements primarily in the infrared region of the spectrum. The instrument provides data that permits calculations of: 1) temperature profile from the surface to 10 mb; 2) water vapor content in three layers in the atmosphere; and 3) total ozone content. The design is based on the HIRS instrument flown on the Nimbus satellite.

☐ **Stratospheric Sounding Unit.** It employs a selective absorption technique to make measurements in three channels. The spectral characteristics of each channel is determined by the pressure in a carbon dioxide gas cell in the optical path. The amount of carbon dioxide in the cells determines the height of the weighting function peaks in the atmosphere (NOAA-E only).

☐ **Microwave Sounding Unit.** It is a 4-channel Dicke radiometer that makes passive measurements in the 5.5 mm oxygen band. This instrument, unlike those making measurements in the infrared region, is little affected by clouds.

☐ **Advanced Very High Resolution Radiometer (AVHRR).** It is a 4- or 5-channel scanning radiometer sensitive in the visible, near infrared, and far infrared window regions.

☐ **Space Environment Monitor.** It is a system consisting of two separate instruments and a data processing unit. The instruments are the total energy detector, and medium energy proton and electron detector and are described below:

☐ **Total Energy Detector.** It measures a broad range of energetic particles from 0.3 keV to 20 keV in 11 bands (NOAA-E & -G only).

☐ **Medium Energy Proton and Electron Detector.** It senses protons, electrons and ions with energies from 30 keV to several tens of MeV (NOAA-E & -G only).

☐ **Data Collection System (DCS).** It is a random access system to acquire data from fixed and free floating terrestrial and atmospheric platforms. Data collected from each platform includes identification, as well as environmental measurements. Location of the platform carrier frequency as received on the satellite.

☐ **Search and Rescue.** This system is used in a joint U.S.A., Canadian, French, and Russian program to perform an experimental mission that will provide data for identifying and locating downed aircraft and ships in distress on a timely basis and dramatically reduce casualty losses by alerting potential rescue teams and guiding them to disaster scenes. The SAR system will detect signals from aircraft emergency locator transmitters and from emergency position identification radio beacons on ships.

☐ **Solar Backscatter Ultraviolet Instrument.** It is used on-board NOAA-F and -G, will measure the earth's ozone distribution.

☐ **Earth Radiation Budget Experiment.** It is used on-board NOAA-F and -G, to determine the radiation gains and losses to and from earth.

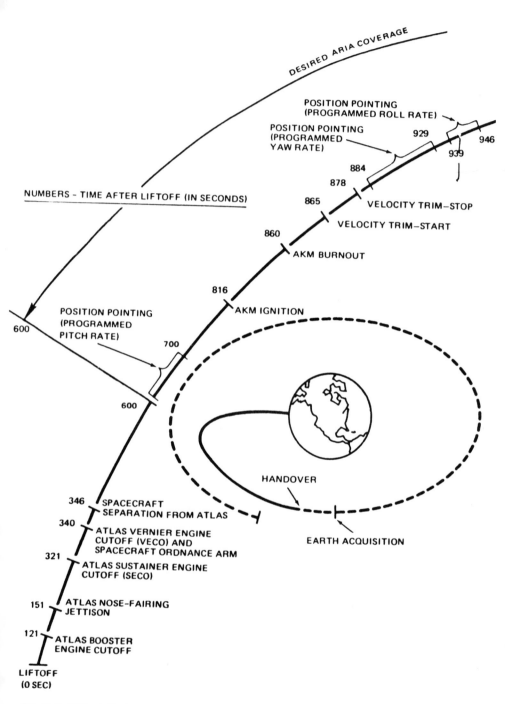

Fig. 12-9. ATN satellite launch-to-orbit injection sequence (courtesy of NASA Goddard Space Flight Center).

SAR ANTENNA DEPLOYMENT 1575

2040

START SOLAR-ARRAY BAND DEPLOYMENTS
1175 NO. 2
1375
1535
1675 1705

HYDRAZINE BLOWDOWN

T + 2040 HANDOVER

1170 NO. 1
1030

START BOOM DEPLOYMENT

START VHr ANTENNA DEPLOYMENT

T + 2039 DEACTIVE CONTROL SYSTEM

START SOLAR-ARRAY CANT

START DATA COLLECTION SYSTEM ANTENNA DEPLOYMENT

MAJOR LAUNCH EVENTS

Event	Time from Liftoff (seconds)
Liftoff (L/O)	T + 0 at ≃15:50 GMT
Booster engine cutoff (BECO)	121.5
Booster package jettison (BPJ)	124.6
Nose fairing jettison (NFJ)	151.5
Sustainer engine cutoff (SECO)	320.9
Vernier engine cutoff (VECO)	339.9
Spacecraft separation	345.9
Pitch rate—start	600.0
Pitch rate—stop	700.0
Solid motor—ignition	816.4
Solid motor—burnout	860.0
Velocity trim—start	865.0
Velocity trim—stop	878.5
Hydrazine isolation	879.1
Yaw rate—start	884.0
Yaw rate—stop	929.0
Roll rate—start	939.0
Roll rate—to orbit rate	946.0
Hydrazine blowdown—start	1030.0
Hydrazine blowdown—stop	1170.0
Array deployment	1170.0
Boom deployment	1375.0
Array cant	1535.0
SAR ant. deployment	1575.0
VRA deployment	1675.0
UDA deployment	1705.0
Handover	2040.0

MAJOR PRELAUNCH EVENTS

Work Days Before Launch	Event
8	Mate of launch vehicle
7	Spacecraft aliveness test
6	Spacecraft readiness test
5	Spacecraft readiness test
3	Fairing squib verification
2	Booster fuel load
1	Countdown begins
0	Launch at ≃15:50 GMT

*Greenwich Mean Time

Attitude Control Subsystem. All first stage launch events are controlled by the Atlas-E vehicle. Figure 12-9 shows the launch-to-orbit injection sequence. After first stage separation, control is provided by a combined hydrazine/nitrogen *Reaction Control Equipment* (RCE) system integral to the spacecraft. The RCE provides spacecraft control during the solid stage burn, provides a velocity trim burn after orbit insertion, and is used to orient the spacecraft and provide control during the deployment of the solar array and antennas. The hydrazine system is deactivated following the burn while the nitrogen (GN_2) system is maintained in a ready condition during the oribital life of the spacecraft, to provide auxiliary reaction wheel unloading capability and to keep the spacecraft on its designated flight path (the trajectory profile is stored on the on-board central processing units). This software also provides the discrete commands for the ascent guidance sequence.

After acquisition, spacecraft attitude is maintained to ± 0.2 degrees in three axes by a zero momentum system that uses: a static infrared earth horizon sensor to provide pitch and roll attitudes; gyros to provide attitude information about the yaw axis via sun sensor updates; and reaction wheels with magnetic unloading, to supply control torques. The system is autonomous requiring only weekly ephemeris updates from NOAA's Satellite Operations Control Center.

Command And Control Subsystem. The command and control subsystem provides the functions of decoding ground commands, storing commands for later executions, and issuing control signals. The *redundant crystal oscillator* provides a highly stable frequency source for all spacecraft clock related functions. The *Controls Interface Unit* (CIU) provides a bi-directional interface between the two spacecraft computers and the other spacecraft components. The CIU collects data from the attitude sensor and passes the data to the computers for processing. The CIU accepts data and control signals from the computers and outputs them to the various spacecraft components.

The computers regulate the final stages of injection into orbit by computing required thrust levels and is suing control commands to the *reaction control equipment.* They also control operation of the satellite in orbit and issue commands to maintain spacecraft attitude within predetermined limits.

Data Handling Subsystem. The data handling subsystem consists of *TIROS Information Processor* (TIP), the *Manipulated Information Rate Processor* (MIRP), five digital tape recorders and the cross strap unit.

The TIP formats all low-bit-rate instrument and telemetry data available for transmission from the satellite. It also adds synchronization, identification, and time code data before simultaneously transferring the data to the beacon transmitter, tape recorder interface (by command), and MIRP. Within the MIRP, the TIP data is combined within three output data formats, high resolution picture transmission, global area coverage, and local area coverage.

The MIRP processes data from the AVHRR to provide separate output for: (a) high resolution picture transmission in real time, (b) medium resolution APT transmission in real time, (c) tape recorded global area coverage for central processing of reduced resolution data, and (d) tape recorded local area coverage for central processing of selected areas of high resolution data. The MIRP, in addition to formatting, adds synchronization, identification, telemetry, time code, and (except for APT) the TIP output to the AVHRR. The high resolution AVHRR is reduced in resolution by averaging for both APT and global area coverage uses.

The five digital tape recorders, each with a single electronic module and dual-tape transport, provides recorded data for subsequent transmission to the central data processing facility. Each transport has an adequate tape capacity to record approximately 4.5×10^8 bits of data.

The cross strap unit provides interconnections among the MIRP tape recorders and S-band transmitters. The cross strap unit also generates clock signals used by the instruments.

Communications Links. The ATN spacecraft communications links include:

☐ (a) A real time S-band digital link (1698 or 1707 MHz) for continuous transmission of high-resolution video data to local ground stations.

☐ (b) A real time VHF link (137.5 or 137.62 MHz) for continuous analog transmission of medium resolution video data (APT) to local ground stations.

☐ (c) An S-band digital playback link (1698, 1702.5, or 1707 MHz) for recovery by the Command and Data Acquisition stations of global and local area recorded AVHRR data.

☐ (d) A beacon downlink (136.77 or 137.77 MHz) for real time transmission of telemetry and low-rate sensor data.

☐ (e) A command uplink (148.56 MHz)

☐ (f) A uhf data collection uplink (401.65 MHz)

☐ (g) An S-band digital playback link (1698, 1702.5, or 1707 MHz) for recovery or recorded TIP data by a European ground station.

☐ (h) An S-band digital ascent telemetry link (1702.5 MHz).

☐ (i) An L-band (1544.5 MHz) search and rescue downlink to ground local user terminals.

☐ (j) Search and rescue uplinks: Search and rescue repeater system (121.5, 243 and 406.05 MHz); Search and rescue processing system (406.025 MHz).

Power System. Spacecraft power is provided by a direct energy transfer system whose primary source is a 125-sq-ft planar solar array; the secondary source is three nickel cadmium batteries. The solar array is made up of eight panels of solar cells, each 61.4-cm × 237.5-cm. The array, which must be deployed from its stowed launch position, is canted at 36 degrees to the orbit normal. A solar array drive system causes the array to rotate once per orbit so that the array continuously faces the sun. Current supplied to the satellite through slip rings during daylight portions of the orbit is used to operate the satellite and to charge the three 26.5-ampere-hour nickel-cadmium batteries. These batteries supply spacecraft power during dark portions of the orbit and augment the array during daylight peak load conditions. Total orbit average load capacity for the system is about 435 watts at the end of two years in orbit at a worst case sun angle.

Thermal Control Subsystems. The thermal control subsystem provides accurate temperature control of the spacecraft subsystems and the instrument payload. Both active and passive control elements are used for this purpose.

Passive Control Components: Blankets, finishes, insulators, shades.

Active Control Components: Pinwheel and vane louvers, heaters, control electronics.

12.5.2 DSCS Defense Communications Satellites

The DSCS III program is an important part of the comprehensive plan to meet national defense communication needs. Figure 12-10 shows the typical system operation. Users of the system will range from airborne terminals with 33-inch diameter antennas to fixed installations with 60-foot diameter antennas and elaborate data processing equipment. Mobile terminals supporting ground and naval operations will communicate with each other and the command chain through the satellite with FDMA or TDMA.

Figure 12-11 shows the DSCS spacecraft, whereas Fig. 12-12 shows its principal components. Table 12-2 presents the key features of the satellite, whereas Table 12-3 gives the technical data of the spacecraft.

Figure 12-13 shows the frequency plan for six channels and Table 12-4 presents the frequency allocation for efficient spectrum use. A six-channel communications transponder with each channel operating through its own rf amplifier will serve the users. This allows compatible grouping of users for efficient use of the frequency spectrum and transponder power. Signals are

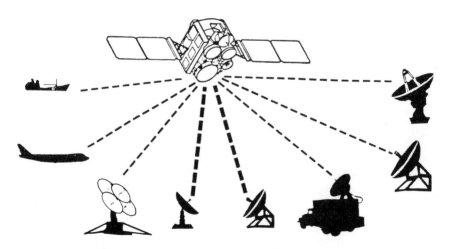

Fig. 12-10. Typical system operation (courtesy of General Electric).

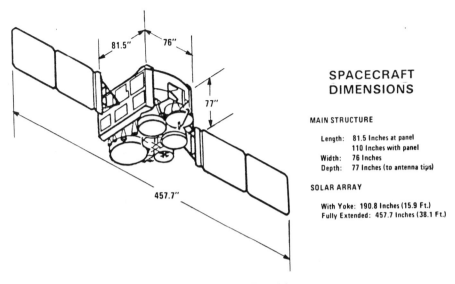

SPACECRAFT
DIMENSIONS

MAIN STRUCTURE

Length: 81.5 Inches at panel
         110 Inches with panel
Width: 76 Inches
Depth: 77 Inches (to antenna tips)

SOLAR ARRAY

With Yoke: 190.8 Inches (15.9 Ft.)
Fully Extended: 457.7 Inches (38.1 Ft.)

Fig. 12-11. DSCS spacecraft (courtesy of General Electric).

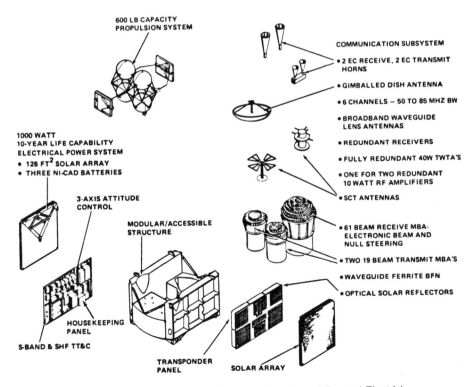

Fig. 12-12. DSCS spacecraft principal components (courtesy of General Electric).

Table 12-2. DSCS Spacecraft Key Features (Courtesy of General Electric).

KEY FEATURES
GENERAL

ITEM	KEY FEATURES
Spacecraft Summary	• 3-Axis Stabilized • Single Axis Sun Oriented • Large North/South Viewing Panels for Passive Heat Rejection • 10-Year Design Life • North/South Stationkeeping • Reliability Exceeds .7 at 7 Years • Low Life-Cycle Cost
TT&C Subsystem	• Command and Telemetry Interface with SCF, DCS Terminals, and Shuttle • Rapid MBA Reconfiguration • Incorporation of S-Band and SHF COMSEC Equipment
Single Channel Transponder	• SHF and UHF Receive Frequency Capability • Compatible with initial AFSATCOM I System • High A/J performance with AFSATCOM II Modulation • Wide bandwidth, fast response synthesizer • Digital matched filter MFSK/FSK demodulator • Integral operational command via comm channel • High power UHF P.A. at 60% power efficiency • SHF and UHF Downlink Capabilities
Attitude Control Subsystem	• Autonomous initial acquisition and operation • Time Shared Central Digital Processor for all Control Modes • Earth & Sun Sensors for Attitude Sensing • Four Skewed Reaction Wheels • $.08^{\circ}$R, $.08^{\circ}$P, 0.8°Y Control Accuracy $.09^{\circ}$R, $.09^{\circ}$P, 1.8°Y
Electrical Power Subsystem	• Regulated Bus − 28V ± 1% − 126 sq. ft. Solar Array − 1240 Watts Output (BOL) • 105 AHr NiCd Battery Capacity

COMMUNICATIONS

ITEM	KEY FEATURES
Receive Antennas	• Wideband 61 Beam Waveguide Lens Multibeam Antenna (MBA) • Full 61 Beam Control − Amplitude and Phase • Endfire Disc-On-Rod Feed Horns • Broadband Nulling • Accurate, Rapid Selective Coverage Pattern • Two Earth Coverage Horns
Transmit Antennas	• Two 19 Beam Waveguide Lens Multibeam Antennas (Efficiency 60%) • Full 19 Beam Amplitude Control • Endfire Disc-On-Rod Feed Horns • Accurate, Rapid Selective Coverage Algorithm • Two Earth Coverage Horns • High Gain Parabolic Dish Connectible to Channels 1 & 4; 2 & 4; or separately to 1, 2, or 4

KEY FEATURES
GENERAL

ITEM	KEY FEATURES
Electrical Power Subsystem (cont.)	• Fully Redundant • Rapid Response to Load Changes • Load Fault Isolation • Transient Protection
Propulsion Subsystem	• Hydrazine Propulsion System with Redundant Thrusters and Tanks • 600 Lb. Capacity • 1.0 Lb. Thrusters
Thermal Subsystem	• Passive • North/South Radiator Panels using Optical Solar Reflectors • Imposes no Satellite Operational Restrictions • Survive Failure Modes Including Attitude Loss and Total Battery Failure
Structure Subsystem	• Accessibility/Modularity • Parallel Subsystem Assembly and Test • North/South Array Through Drive Shaft • Independent Propulsion Module • Vibration Damped Equipment Panels • Lightweight/Stiff/Dimensionally Stable • Growth and Option Flexibility
Survivability	• Overall Hardening Approach is Based Upon JCS Guidelines
Launch Vehicle	• Shuttle IUS (DSCS III-B Series) • Titan 34D IUS (DSCS III-A1) • Titan 34D Transtage (DSCS III-A2)

COMMUNICATIONS

ITEM	KEY FEATURES
SHF Transponder	• High Gain for Enhanced Small Terminal Operation • 60 MHz Nominal Channel Bandwidth • 85 MHz in Channel 3 • Low Noise Figure (5.2 dB) • Passive Thermal Design for Maximum Reliability • Fully Hardened Components • Low-Loss, Light-Weight Filters • Low Phase Distortion
Single Channel Transponder	• Antennas SHF . . . . . . MBA or Earth Coverage Horn UHF . . . . . . Greater than Earth Coverage • Communications . Emergency Action Messages • Commands . . . . . 40 Discrete Commands. Serial command capability
Redundant JLE	• High Accuracy • High Dynamic Range

Table 12-3. DSCS Spacecraft Subsystems'
Technical Data (Courtesy of General Electric).

COMMUNICATIONS

Transponder	Effective Isotropic Radiated Power EIRP (dBw)	
X-Band Single Frequency Conversion — 6 Channels	Earth Coverage	Narrow Coverage
CH1 (MBA)	29.0	40.0
CH2 (MBA)	29.0	40.0
CH3 (MBA)	23.0	34.0
CH4 (MBA)	23.0	34.0
CH3 (EC Horn)	25.0	
CH4 (EC Horn)	25.0	
CH5 (EC Horn)	25.0	
CH6 (EC Horn)	25.0	
CH1,2(Gimballed Dish)	44.0	
CH4, (Gimballed Dish)	37.5	
System Noise Temperature G/T (dB/°K)	Earth Coverage	Narrow Coverage
Multi-Beam Antenna	-16	-1
Earth Coverage Horn	-13	-
Power Output	CH1,2	40 Watts
	CH3,4,5,6	10 Watts
Bandwidth	CH1,2,4,5	60 MHz
	CH3	85 MHz
	CH6	50 MHz
Gain Control Range	39 dB	
Antenna		
Earth Coverage		
MBA (Transmit)	Gain 15.0 dB Min	
Horn (Transmit)	17.0 dB	
Narrow Coverage		
MBA (Transmit)	Gain 26.0 dB Coverage — 1.0° Diameter Circle on Earth Disc	
Gimballed Dish	Gain 30.2 dB Min Coverage — 3.0° Diameter Circle	

received and transmitted through an interconnected set of multibeam antennas. These antennas have the capability to spatially distribute receiver pattern gain and transmission power according to user requirements. Transmitter power can be concentrated on small isolated terminals or distributed optimally over wide areas.

The single channel transponder on DSCS III supplements dedicated AFSATCOM spacecraft for command and control communications. The regenerative transponder receives and transmits uhf signals using AFSATCOM

SPACECRAFT

Spacecraft Weight (Dry)	Approximately 2000 lbs.
Attitude Control Subsystem	Accuracy — 0.08° Roll 0.09° Pitch 1.0° Yaw 0.2° Circular Error Radius, Overall Accuracy of RF Beam Axis Pointing ±0.1° Orbit Positioning Accuracy
TT&C Subsystem	Command Capacity —600 Discretes — 16 Messages (18'Data Bits each) Telemetry Capacity —490 Bi-Level 320 Analog 18 Serial Digital Inputs
Power	1240 Watts Array Power (Beginning of Mission) 980 Watts Array Power (10 Years) 28 V ± 1%
Propulsion Subsystem	Capacity — 600 lbs Monopropellant Hydrazine Thrust Levels — 1 lb. to 0.3 lbs. Blow-Down Ratio — 4:1 with Full Load Specific Impulse — 228 sec at initial conditions
Thermal Subsystem	Maintains all Components within Specified Tempera- ture Limits during Launch, Transfer and Synchronous Orbit
Overall Reliability	Greater than 0.7 at seven years

I modulation. AFSATCOM II signals are also processed and afford a high degree of anti-jam protection, receiving at either uhf or shf.

Transponder. Users are served by a six channel shf communications transponder with variable payload configuration responsive to user needs as shown in Fig. 12-14 and with each channel powered by its own rf amplifier to make most efficient use of the available frequency spectrum and power. Two channels at 40 watts and four channels at 10 watts are provided at bandwidths ranging from 50 to 85 MHz. The transponder amplifiers and

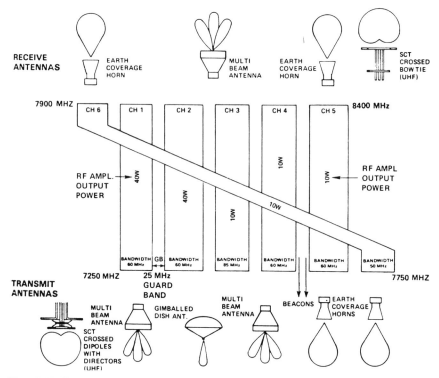

Fig. 12-13. Frequency plan (courtesy of General Electric).

associated electronic components are packaged on the North panel of the space-craft. Passive thermal control, augmented by thermostatically controlled electric heaters, maintains specified temperature control for all payload elements.

Antenna. Signals are received and transmitted through a flexibly inter-connected antenna set which includes, a) four Earth coverage horns, two each for receive and transmit; b) a 61-beam waveguide lens receive antenna, with associated beam forming network, to provide selective coverage and jamming protection; c) two 19-beam waveguide lens transmit antennas with beam-forming networks to rapidly produce selected antenna patterns that fit the network of ground receivers; d) a high-gain gimballed dish transmit antenna for spot-beam fixed coverage and 3) uhf antennas, bow-tie receive/cross dipole transmit for the single channel transponder.

Single-Channel Transponder. A single channel transponder payload is integrated into the spacecraft for secure and reliable dissemination of the *Emergency Action Message* (EAM) and single *Integrated Operational Plan* (SIOP) communications from world wide command post ground stations and aircraft.

The single channel transponder (SCT), as shown in Fig. 12-15 receives command and control communications from ground terminals and airborne command posts at shf or uhf. A high degree of security and anti-jam protection is provided for both uplink and downlink communications. Digital signal

Table 12-4. Frequency Allocation
for Efficient Spectrum Use (Courtesy of General Electric).

Spectrum	SHF from 7900 to 8400 MHz Receive from 7250 to 7750 MHz Transmit
	UHF (SCT) from 300 to 400 MHz Receive from 225 to 260 MHz Transmit
	SHF (SCT) Channel 1
	SHF (TTC) Channels 1, 5 Receive 7600, 7604 MHz Transmit
	S-BAND (TTC) SGLS Channels 12, 16 1807.764, 1823.779 MHz Receive 2257.5, 2277.5 MHz Transmit
Channels 1-5	725 MHz Up-Down Translation
Channel 6	200 MHz Up-Down Translation
Guard Bands	25 MHz

RECEIVE PLAN

ANTENNA CHANNEL	MULTI BEAM	EARTH COVERAGE HORN	UHF BOW TIE
CH 1	X	X	–
CH 2	X	X	–
CH 3	X	X	–
CH 4	X	X	–
CH 5	–	X	–
CH 6	–	X	–
SCT	X	X	X

TRANSMIT PLAN

ANTENNA CHANNEL	MULTI BEAM	EARTH COVERAGE	GIMBALLED DISH	UHF CROSS DIPOLE
CH 1 (40W)	X	–	X	–
CH 2 (40W)	X	–	X	–
CH 3 (10W)	X	X	–	–
CH 4 (10W)	X	X	X	–
CH 5 (10W)	–	X	–	–
CH 6 (10W)	–	X	–	–
SCT	X	–	X	X

processing and digitally controlled frequency synthesizers are used to give flexible, efficient uplink demodulation, and downlink modulation that are compatible with both first and second generation terminals. Transponder operation can be controlled by the command posts using the integral secure command system as well as by the DSCS III TT & C system. A high-power

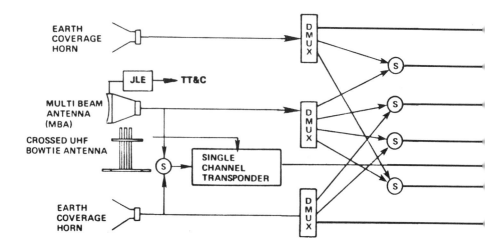

Fig. 12-14. Variable payload configuration responsive to user needs (courtesy of General Electric).

solid-state transmitter drives the transmit uhf antenna giving worldwide coverage.

Unique SHF/S-Band TT&C Subsystem. A unique dual-frequency telemetry, tracking (S-band only) and command subsystem, as shown in Fig. 12-16 was selected to meet DSCS III requirements for varied command and telemetry interfaces (SCF, DSC terminal; provide rapid MBA reconfiguration; operation in various jamming environments; long life; and hardened design). The subsystem provides tracking and command capabilities via both S-Band and shf links. The primary function of the shf link is to control channelization and antenna system configuration through its links with satellite configuration control elements under jurisdiction of the DCA. It uses the earth coverage horns and the multibeam antennas for receive and transmit.

The S-Band subsystem is primarily for spacecraft control, tracking and housekeeping functions through its links with the remote tracking stations of the AF Satellite Control Facility. It uses a dedicated S-Band antenna. Although separate functions are assigned to the two frequency channels during normal operation, their functions are interchangeable whenever conditions may warrant.

Ground Control Configuration Provides Operational Flexibility. The satellite Configuration Control Elements (SCCEs) command and control the satellite payloads to satisfy real time user requirements. Using the SHF TT&C channel, the SCCEs generate commands and command sequences that reconfigure channel and antenna-beam allocations, and control COMSEC equipment. The SCCE's are linked for control and priority assignment by ei-

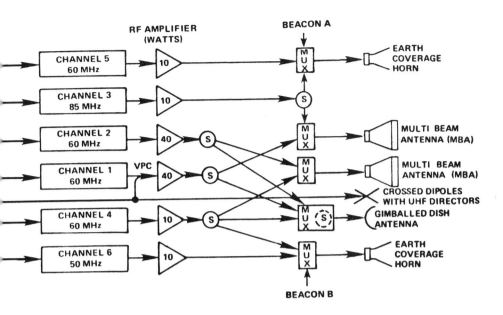

ther hard lines or satellite command channels. They are compatible with automatically routing configuration demand data to a central processor for integration.

12.5.3 European Communications Satellite

The following material is from the European Space Agency's Publication—*ESA BR-08: ECS Data Book.*

General Description. The general configuration of the satellite is shown in Fig. 12-17. The satellite has a central body maintained in a stable Earth-pointing mode, and solar array wings, mounted on the north and south faces, which are sun tracking.

The central body contains all the satellite subsystems (with the exception of the solar array), and is of a modular construction. The modular construction allows:

—The separate integration of the satellite payload (repeater) on one module and the complete service module (containing all the other subsystems) on the other;
—Other payloads as yet to be defined, to be carried by the same basic unmodified service module.

The Earth-pointing face of the body carries the communications antennas and the north and south faces carry the main power dissipating elements of the repeater.

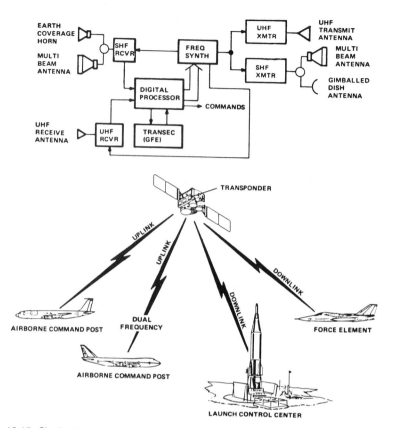

Fig. 12-15. Single channel transponder system and its links with various users (courtesy of General Electric).

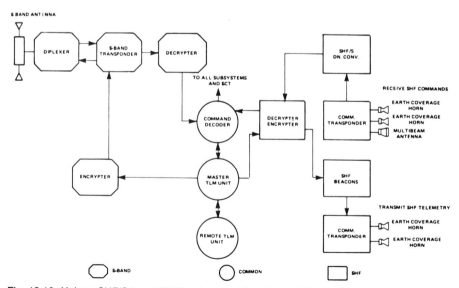

Fig. 12-16. Unique SHF/S-band TT&C subsystem (courtesy of General Electric).

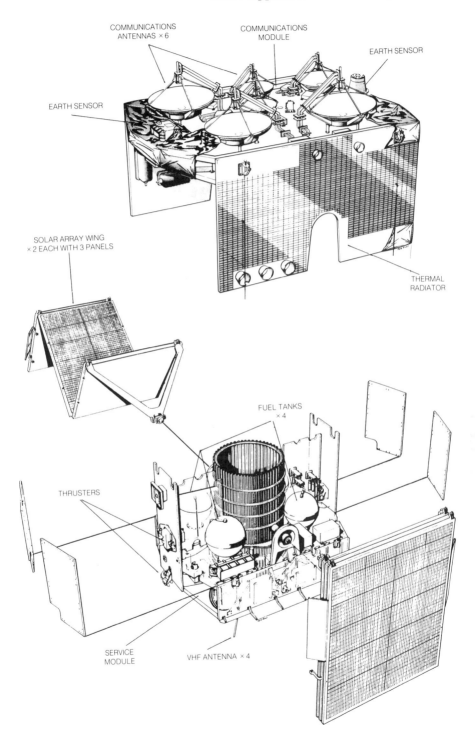

Fig. 12-17. Exploded view of ECS-1.

System. The satellite system is defined by the following general design and mission constraints:

Mission

—12 × 20 watt 11-14 GHz repeaters,
—Eurobeam receiver antenna coverage,
—Eurobeam and triple spot beam transmit antenna coverage,
—A satellite design lifetime of seven years,
—To be launched by Ariane (together with a passenger and also having compatibility with the Delta launcher).

Configuration

—Based on OTS 3-axis stabilized configuration,
—Modular construction (separate payload integration with "Bus" service module concept),
—Sun-tracking solar array,
—Stable Earth-pointing antenna platform,
—North and south faces carrying large power dissipating largely elements to facilitate the use of a passive thermal control subsystem.

An early system level study resulted in the choice of a solid propellant apogee boost motor (ABM) which in turn imposed the constraint of stable spin configuration in transfer orbit and during apogee motor firing.

Recognizing that the communications mission did not require the satellite orbit to be maintained at zero inclination, the ability to operate with limited orbit inclination, but Earth-pointing through body steering, was also incorporated. This also ensures that there is flexibility to accommodate other missions with different orbit control constraints.

The final satellite configuration has the following summary mass and power characteristics.

	ECS first flight unit	ECS second unit and following
Mass at liftoff	1030 kg	1140 kg
Power in transfer orbit	110 Watts	110 Watts
Power at equinox	910 Watts	910 Watts
Power at solstice	802 Watts	802 Watts
Power in eclipse	500 Watts	761 Watts

Structure. The way in which these constraints shaped the satellite system design can be seen in Fig. 12-17. The structural skeleton carries the subsystems and equipment; the skeleton consists of two modules, the service module (SM) and the communications module (CM).

The SM has a central thrust cylinder, expanding into a cone at the lower end, which interfaces with the launcher. A horizontal platform extends from the thrust cylinder to the limits of the launcher envelope. Since this platform carries on its faces most of the principal mass elements of the SM, it is braced by struts down to the launcher interface ring.

The CM has an inverted "U" shape, mounting onto the upper end of the thrust cylinder with additional support stiffening provided by shear panels. The top of the "U" provides a stable platform for the antennas and carries many of the low power repeater elements. A good Earth-pointing performance is ensured by mounting the Earth sensors together with the antennas on this top platform. The high power dissipating elements (the repeater TWTA's) are mounted on the arms of the "U", these being the N/S faces and subject therefore to minimum changes in solar irradiation; 6 TWTA's are mounted on each face. The Sun-tracking solar arrays are mounted on their bearings (BAPTA's) at the centers of the north and south faces and they consist of three panels; these panels are folded flat to the faces during the spin-stabilized phases, power being limited then to one panel on each wing. See Fig. 12-18.

The passively spin-stabilized mode requires the principal inertia axis to lie along the axis of the thrust cylinder.

Thermal Control. It is by means of the thermal control subsystem that a suitable operating temperature environment is provided for the onboard equipment. Certain peripheral equipments (antennas, arrays, thrusters, etc.) are controlled indirectly by the thermal properties of the materials used.

The main thermal control subsystem is of a passive design with some active control for specific units (battery, TWTA, PSU, and the complete RCS). The overall temperature level of the spacecraft is controlled by using multilayer super-insulation blankets over the majority of the external surface, specific

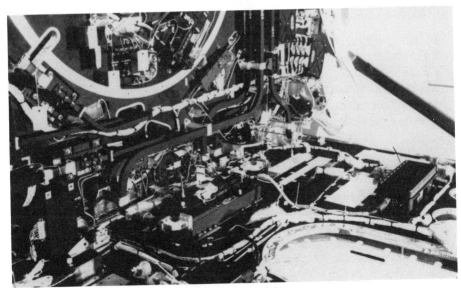

Fig. 12-l8. Internal detail of the communications module.

control radiator areas externally covered by second surface mirrors, and a high emittance coating on all internal surfaces to maximize internal radiative heat exchange. Certain high power dissipation units (TWT, PSU and shunt electronics) are mounted on separate radiators located on the north and south walls of the spacecraft.

Other units like batteries and flow control valves are controlled by heaters switched automatically by thermistors, while the remaining active thermal control elements are operated by ground command.

The thermal subsystem is controlled by telemetry commands given directly via the power subsystem, which contains the conditioning circuitry for a number of temperature sensors. Other sensors located within equipment are conditioned within their own subsystems. In addition there are telemetry channels within the power subsystem giving the switching status of the thermal control subsystem heaters.

All operational functions of the thermal control subsystem are controlled by the power subsystem (electrical integration unit) to which the telecommands are directed.

The Communication Subsystem—Antennas. The ECS primary service payload receives signals from within the European coverage zone and (re)transmits them into four zones: the European, the Atlantic, the Western and the Eastern zones. The ECS multiservice payload receives signals from and (re)transmits them in a zone covering the European continent but narrower than the primary service European zone.

The ECS-1 spacecraft, not carrying the multiservice payload is equipped with two redundant receive Eurobeam antennas, one transmit Eurobeam antenna and three identical spotbeam antennas covering the Atlantic, Western, and Eastern zones. See Fig. 12-19.

The ECS-2 and subsequent spacecraft which also carry the multiservice payload are equipped with the following antennas:

—Primary service payload:
1 Eurobeam receive antenna
1 Eurobeam transmit antenna
3 spotbeam transmit antennas
—Multiservice payload:
1 receive/transmit antenna.

All the communication antennas of the ECS-1 spacecraft are of the front-fed type, derived from the OTS program. They generate a linearly polarized electromagnetic wave and allow frequency re-use by orthogonal polarization discrimination. As the Eurobeam disk has a rather small diameter (32 cm), the signal blockage caused by the feed, the waveguides, and the trusses supporting the pad have a measurable effect on the antenna performances. In particular gain and cross polarization isolation are noticeably affected.

This has led to the development of a new type of antenna that is used from ECS-2 onwards as Eurobeam receive and transmit antennas and as a multiservice receive/transmit antenna. This new antenna is of the double reflector type, known in the literature as Dragone double reflector antenna. Its outstanding performances are the increased gain, increased cross polarization purity and the large bandwidth capability that allows it to be used as a receive/transmit antenna over a frequency range of at least 2 GHz (12.5 GHz to 14.5 GHz). It is expected that due to the extensive use of carbon fiber reinforced plastic the in-orbit thermal deformation, resulting in undesirable beam swing, is reduced with respect to the front-fed antenna design. Since the area of the spotbeam antenna dish is big with respect to the area of the blocking elements, the front-fed design has been maintained for the spotbeam antenna for all five ECS satellites. See Fig. 12-20.

Table 12-5 lists the typical performances of the various antennas of the ECS satellites.

Communication Subsystem—Repeaters. The ECS repeater is designed for the transmission of six 80 MHz broad communications channels in each of two orthogonal polarization directions and one beacon channel. Five of the twelve communications channels are switchable, i.e., each of them can be connected optionally, with two of a total of four transmit antennas.

The repeater configuration is based on the use of redundant receive antennas and receive sections and of non-redundant transmit chains.

From ECS-2 onwards two additional chains are incorporated to provide multiservice capacity. These two chains can be operated in parallel with the primary service channels referred to above, but a maximum of nine channels only may be operated simultaneously. To avoid interference, the primary service channels 4X and 4Y must not be operated when the multiservice channels are switched on. Since all four broadband sections have to be operated simultaneously when primary service channels and multiservice channels are both operated, no redundancy is available for this mode of operation. In the case of a failure of one or two broadband sections the primary service payload will have priority.

The hardware of the primary and multiservice channels is very similar: only in the power section, due to the differences of frequency and also to the absence of adjacent channels, different types of output are used. See Figs. 12-21 and 12-22.

The function (i.e., from ECS-2 onwards this description is valid when the multiservice payload is switched off) is described as follows:

The X and Y polarized uplink signals are received by the redundant broadband receive sections, each consisting of an input bandpass filter, a parametric amplifier for low-noise preamplification, and a broadband receiver for amplification and down conversion of the overall receive band to the i-f frequency range of 800-1300 MHz.

Only one receive section in each polarization direction is operating, the redundant one is switched off via the dc supply voltages.

The branching network combines the outputs of the redundant receive sections and distributes the communications signals to the individual channelized sections.

In each channelized section the communications channel is band limited

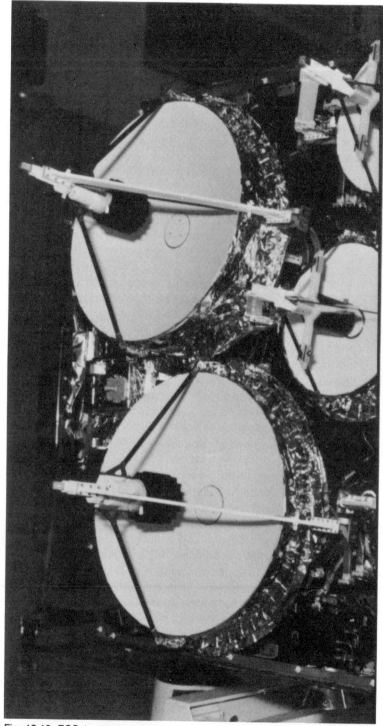

Fig. 12-19. ECS-1 antenna platform.

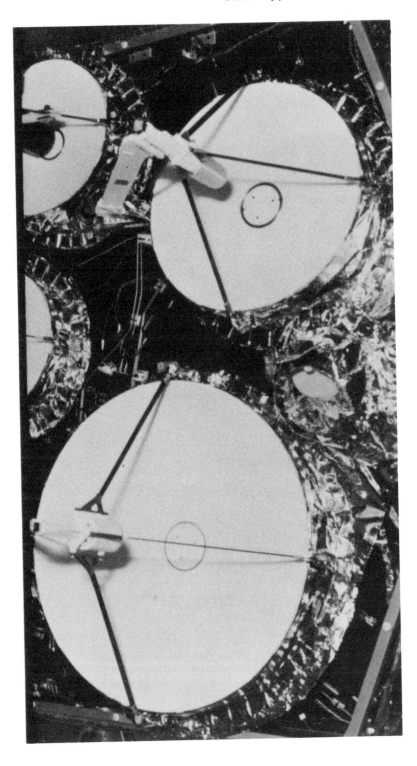

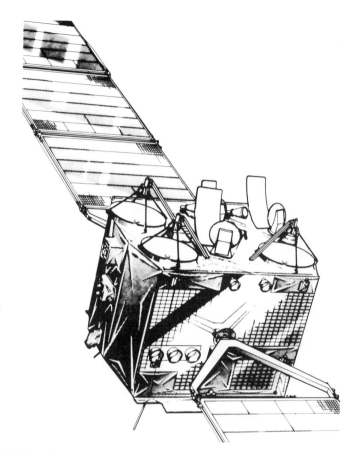

Fig. 12-20. Drawing of ECS-2 antenna configuration.

Table 12-5. ECS Antenna Performance Specifications.

| | | gain/db | | cross |
		edge of coverage	beam centre	polar isol. dB
F1—Eurobeam receive transmit	front fed	23.4 23.4	28.7 28.7	31.5 31.5
F2..5 Eurobeam receive transmit	double refl.	25.0 25.0	29.0 29.0	32.5 32.5
F1..5 Spotbeam transmit	front fed	30.0	34.5	33.0
F2..5 Multiservice (rec/transmit)	double refl.	29.0	32.0	32.5

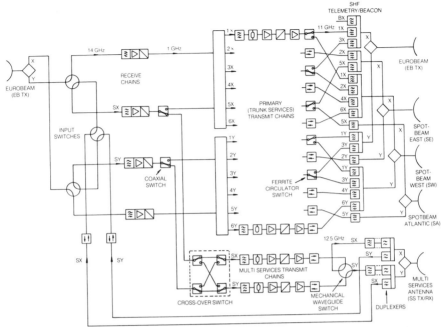

Fig. 12-21. ECS communications subsystem block diagram (including multiservices).

in the i-f range by an i-f filter, amplified by the i-f main amplifier and upconverted to the transmit frequency range by the upconverter.

The i-f main amplifier contains a PIN-attenuator the insertion loss of which is switchable by telecommand allowing gain adjustment and different gain settings in orbit. The control voltage for the PIN attenuator is provided by a control box.

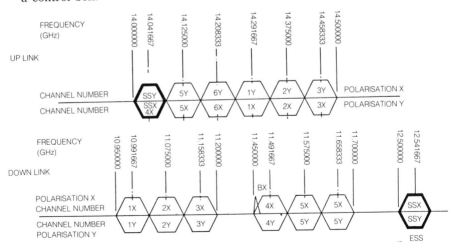

NOTE: CHANNEL SUFFIX REFERS TO DOWNLINK POLARISATION

Fig. 12-22. ECS frequency plan.

In the TWTA the communications signal is amplified to the required transmit power.

The rf switch assemblies connect the outputs of the corresponding TWTA's alternatively to one of two transmit antennas. In the non-switchable channels the switches are replaced by rf isolators.

An rf switch assembly consists of the proper distribution switch and one rf isolator at each output for the suppression of undesired echoes. In order to meet the stringent multipath requirements these isolators are switched together with the distribution switch so that they change their forward direction.

The output multiplexers combine up to four communications channels before feeding them into the ortho-mode transducers of the transmit antennas. The bandpass filters of the multiplexers as well as the output filter in channel 2Y and in channels 5X and 5Y limit the transmit frequency bands and provide the required suppression of out-of-band signals.

The triplexer connected to the Eurobeam transmit antenna includes also an input for the telemetry beacon signal.

The communications signals received in one polarization direction are transmitted in the orthogonal polarization.

In the synthesizer of the carrier supply the carriers for the frequency converters are derived from one master oscillator by frequency multiplication. Thus a coherent down- and up-conversion is achieved. The carrier supply contains two independent redundant synthesizers, which can be activated by telecommand.

The repeater power supply provides the dc supply voltages for the active equipment. It consists of two identical redundant chains activated by telecommand.

The addition of the multiservice channels necessitated only a limited modification to the primary service channels. The redundant receive antenna has been deleted, the connection between the receive antenna and the two redundant broadband sections is now performed by mechanical waveguide switches. They are of the four-part transfer type, allowing simultaneous connection of two inputs to two outputs, the two inputs being the primary and the multiservice signals.

The function of the multiservice channels is identical to the primary service channels, only the output filter is different in view of the relaxed requirements in terms of out-of-band attenuation, which could be granted as there were no "neighbor" channels. The output filter is integrated with the input filter and forms a diplexer that separates receive and transmit paths.

Rf switches at the input and the output of the channelized section enable any combination of up and downlink polarization.

Telemetry Tracking and Command Subsystem. The telemetry, tracking and command subsystem (TTC) performs the following main functions:

—The telemetry function, that is the multiplexing together and transmitting to Earth of the data that establishes the status and the performance of the on-board equipment;

—The tracking function, that is the provision of on-board facilities for:

A. Angular tracking from ground stations using the telemetry carrier,

B. Range measurements from ground stations; the command function, that is the reception, demodulation, decoding, validation and distribution of coded messages sent from the ground stations to control or change the operational status of the satellite and its subsystems.

☐ **The Telemetry Function.** Analog and digital parameters are time multiplexed, digitized, encoded, and serialized into an output message. This message then modulates an rf carrier and is transmitted. During transfer orbit (and for on-station back-up) the signal is transmitted at vhf. During the on-station phase, the signal is transmitted at shf, via the beacon. On the ground, the signal is demodulated, decoded and the data recorded and processed. The execution of commands is verified by the telemetry link.

Each satellite has an identification code built into the telemetry format. There are two formats provided, one for transfer orbit including high bit rate data from the attitude and orbit control subsystem, the second during on-station operation including communications equipment data.

☐ **The Tracking Function.** The tracking is performed at vhf during transfer and drift orbit phases and by using an shf beacon during the on-station phase.

At vhf, angular monopulse tracking is performed by reception at the station of a sufficiently large carrier component in the downlink spectrum.

On station, tracking is performed at shf, data being derived from the shf antennas at the ECC.

Range is obtained using a tone ranging system. During early orbits the uplink (vhf) is modulated by ranging tones. These tones are received by the spacecraft and retransmitted. The range is found from the phase delay between the transmitted uplink tones and the received downlink tones on the ground. During the on-station phase, ranging is performed through one communication channel at shf.

Accurate measurements are performed on the major tones only, ambiguity resolution being performed by sequentially switching the minor tones. The ranging function is switched on and off by ground command.

☐ **The Telecommand Function.** The commands to be transmitted to the satellite are first encoded into a particular form (message) before modulating an rf carrier. Commands are transmitted at vhf. Reception and decoding of commands is normally immediate, although a "time-tagged" or delayed command operation can be used to fire the ABM. Each spacecraft has individual address words for the command message initiation and validation.

There are three different types of command:

—High-level ON/OFF commands providing sufficient power to operate relays.

—Low-level ON/OFF commands providing pulses for logic circuits.

—The memory load commands having a multi-functional role, comprising 16 bits of information and an address of up to 8 parallel bits.

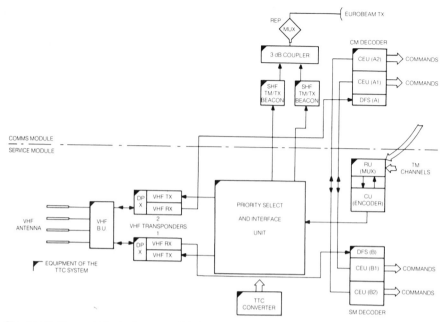

Fig. 12-23. General block diagram of TTC subsystem.

□ **Subsystem Configuration.** The subsystem (Fig. 12-23) comprises the following units:

The *telemetry encoder* is located in the service module and consists of two basic sub-units, namely the central unit (CU) and the remote unit (RU). The CU contains oscillator, counting chains, format generators, subcarrier modulators and interface circuits to the RU. The RU contains multiplexers, ADC's, bit detectors and interface circuits to the CU. The encoder has internal redundancy. 2 TM formats are available, made of 3 bit words arranged in a 12×33 word matrix of a total length of 19.2 secs. The telemetry bit rate is 160 b/s.

The *command decoder* consists of two complete command chains, one in the service module and one located in the payload module. As shown in the system diagram, both units contain a demodulator and frame synchronizer (DFS) as well as command execution units (CEU) providing commands to users in each module.

The *payload module unit* and the service module unit are cross-coupled as shown. A time-tag unit (TTU) is also included in the service module unit for the firing of the apogee boost motor (ABM).

The *priority select and interface unit* (PSIU) performs the functions of switching the various units of the TTC subsystem and conditioning the TM I/P's to both vhf and shf/beacon Tx's. It also provides the current-limiting functions for the ECS vhf transponder receivers.

It also automatically switches on vhf TR Tx No. 1 utilizing a signal from the separation switch at launcher/spacecraft separation. It is not a requirement that simultaneous ranging and command operation be provided by the on-board systems via VHF.

The *redundant vhf transponders* are of the crystal coherent type with diplexer built into the units. Each transponder has a receiving and transmitting element, using a common master crystal-controlled oscillator to preserve Doppler (range rate) information when transponding ranging tones for measurements.

The *VHF antenna branching unit* (ABU) ensures proper phasing and amplitude distribution/combining of signals transmitted to/from the antenna radiators and the ABU ports towards the diplexer.

The *VHF antenna* consists of four whip antennas, which are designed to provide near omni-directional pattern for up and downlink frequencies.

The *TTC converter unit*, operating from the main bus or batteries, provides the supply voltages for the subsystem. The design is modular, each supply being provided by a number of power modules and a majority voted control regulator. Hot redundancy is effected by means of an additional power module/supply rail.

The spacecraft carries *redundant SHF telemetry/beacon transmitters* which during the on-station phase provide the polarization plane reference for the Earth-stations, and additionally transmit telemetry.

Power Subsystem. The power subsystem obtains its electrical power, during the sunlight phase of operation from two independently steerable rigid solar array wings. These wings are mounted on the north and south faces of the satellite and are capable of being aligned to the Sun. Each wing consists of 3 panels. See Fig. 12-24.

The bearing and power transfer assemblies (BAPTA) are used for control of array pointing and transfer of the power into the spacecraft. The pointing signals for the BAPTA control electronics are generated by the solar array mounted Sun sensors (SASS). The power is regulated by the power subsystem to 50 volts $\pm 2\%$ at the user's input. A sequential switching shunt regulator is used to regulate the power.

In eclipse, two nickel-cadmium batteries provide power to the spacecraft via boost discharge regulators. The batteries are sized to allow for ECS-1, limited operation of the communication payload during the longest eclipses while not exceeding the permitted depth of discharge (DOD) of the batteries. ECS-2 and later satellites will have larger batteries and full eclipse capability.

The batteries are also used to complement the spinning satellite array output during transfer orbit.

Battery charge power will be obtained from the outer-most sections of the solar array. The battery charge philosophy is such that the batteries are maintained at or near the fully charged state. In order to achieve the specified long life performance, over-charging of the batteries is avoided, and battery temperature is kept low.

The batteries can be trickle charged and reconditioned during solstice seasons.

A separate power bus (called the inboard loadbus) provides power for the

Fig. 12-24. Solar array wing.

heaters during the on-station sunlight phase. The bus is isolated during the launch and spin phases. Its voltage is 50.25 volts ±5% and is derived from the inboard active array panels.

Protection is provided against short circuit failures on the main bus and the heater bus and against overcharge/discharge and the over-temperature of the batteries.

Non-essential loads can be disconnected when abnormal load conditions occur.

The collection and distribution of electrical power and the protection of the main bus is the function of the electrical integration unit (EIU) which also includes the spacecraft instrumentation conditioning circuits.

An auxiliary power supply (APS) provides power to units within the power subsystem. Units which use the APS are so designed as to ensure that single component failures do not short-circuit the APS outputs.

The battery control unit (BCU) contains the following functions for each battery:

—Cell voltage monitoring including cell reversal protection by level sensing and generation of an output signal which is used for battery disconnection;
—End of charge (EOC) and end of discharge (EOD) level detectors (threshold levels are adjustable by command).

In summary, the power subsystem comprises the following items:

—Solar array,
—BAPTA and associated electronics,
—Shunt regulator unit (SRU),
—Nickel-cadmium batteries,
—Battery control unit (BCU),
—Battery discharge regulators (BDR),
—Electrical integration unit (EIU),
—Auxiliary power supply (APS).

The electrical distribution (EDS) distributes power and signals to communications, AOCS, TTC, and thermal subsystems.

The complete EDS comprises:

—The cable harness, including connectors, between all the SM units and to the interface brackets with the CM harness,
—The grounding and shielding provisions of the wiring,
—The facility for isolating the batteries for ground storage and operations,
—The spacecraft wiring to umbilical and skin test connectors and the spacecraft mounted umbilical and skin connectors.

The following cabling is part of the relevant subsystem:

—Communications subsystem cables,
—Rf cables between the vhf antennas and the ABU,
—The firing lines of the pyrotechnic subsystem,
—The connectors mounted on units, including pyrotechnic devices.

Altitude and Orbit Control Subsystem (AOCS). The primary functions of the AOCS are:

—To provide the attitude determination and control functions required for transfer orbit;
—Following ABM firing, to perform the transition from spin stabilized to three-axis stabilized conditions;
—To control the attitude and orbit of the satellite in its three-axis stabilized configuration.

The transfer orbit functions require the following, the overall objective being to ensure alignment of the apogee boost motor (ABM) for minimum injection error and propulsion loss:

—Processing attitude data from the Earth and Sun elevation sensors (ESS) as the spacecraft spins about the yaw axis,

—Adjustment of the spacecraft orientation through operation of the pitch thrusters on ground command via the spin phase and thruster drive electronics (SPTE),

—And passive nutation damping throughout this phase.

Transition from spin stabilized to three-axis stabilized requires the following:

—Despin using roll and yaw thrusters,

—Sun acquisition using the hydrazine thrusters controlled by processing signals from the Sun acquisition sensors (SAS) to point the roll axis to the Sun, in either positive or negative direction; when this is achieved, the solar arrays are deployed and rotated to acquire the Sun;

—Earth acquisition is performed when the spacecraft-Earth and spacecraft-Sun directions are orthogonal. It is accomplished by rotating the satellite around the roll axis (which is Sun-pointing) until the yaw axis lies in the orbit plane and is Earth-pointing.

—Spin-up of momentum wheels,

—Station acquisition which consists of a series of maneuvers to remove launch vehicle and ABM injection errors and to reach the desired orbit location.

The results of this phase is that on-station communications links will be established within 21 days, with the satellite at the required orbital position.

Finally on station three-axis stabilized control is accomplished by:

—Automatic normal mode operations, using infrared Earth sensor data to provide the required pointing reference,

—Automatic wheel off-loading by thrusters to maintain wheel kinetic momentum in the nominal range,

—Station keeping using hydrazine thrusters under ground command for correcting east/west orbital drift, while maintaining automatic normal mode operations. The ECS satellites are designed to maintain the nominal orbital longitude within $\pm 0.1°$ of the nominal position throughout their lifetime. See Figs. 12-25 through 12-27.

☐ **Subsystem Configuration.** The AOCS subsystem comprises sensors, actuators and the necessary control and drive electronics. Stand-by hardware redundancy is available to provide the reliability for seven years. Redundant units are automatically activated by on-board failure detection and protection circuitry or by ground command.

Sensors

Seven sensors (see Fig. 12-28) are available to ensure attitude control during all phases of the mission:

—Two infrared Earth sensors of high accuracy (IRES). One sensor is based on scanning the horizon with four bolometers, the other based on achieving

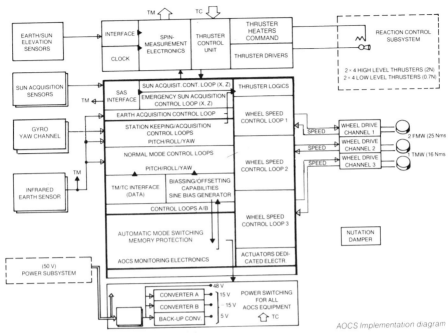

Fig. 12-25. AOCS implementation diagram.

local radiance balance at the horizon. Accuracy is maintained also at Sun and Moon crossings. The internal redundancy of each sensor is rather extensive, allowing graceful degradation in the highly unlikely event of multiple failures. These sensors are used during acquisition, normal mode operation and orbit control;

— Two Sun acquisition sensors (SAS) which together provide 4π steradian field of view, ensuring Sun acquisition from any conceivable initial conditions;

— An Earth and Sun sensor assembly (ESS) comprising two pencil beam Earth sensors and two slit sun sensors; the output of all sensors is available throughout the transfer orbit, when the spacecraft is spin stabilized at 65 rpm; it provides redundant information that enables attitude determination with an accuracy of $< 0.025°$,

— One nutation damper; this reduces spacecraft nutation through passive liquid damping; the damper is tuned to recover from an initial nutation angle of $3°$ with a time constant of 30 minutes. Residual nutation is less than 0.02 degrees;

— One gyropackage (GYP) comprising two rate integrating gyros in standby redundancy, to measure yaw rates and relative angles. The gyros are operated in a well controlled thermal environment of $70°C$.

Actuators

Apart from the nutation damper, two other types of actuators are utilized in the AOCS:

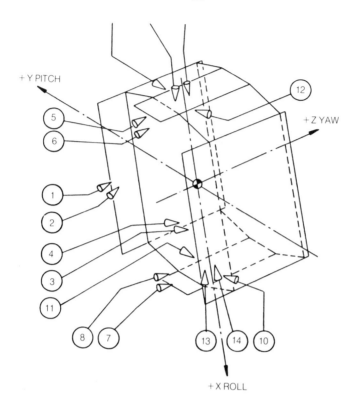

SPACECRAFT MANOEUVRES		THRUSTER
SPIN AXIS RE-ORIENTATION		5 + 7 OR 6 + 8
ATTITUDE ACQUISITION	ROLL:	1 + 3 OR 2 + 4
	PITCH:	5 + 7 OR 6 + 8
	YAW:	9 + 11 OR 10 + 12
STATION ACQUISITION AND/OR KEEPING	EAST/WEST	13 + 15 OR 14 + 16
	NORTH/SOUTH	9 + 11 OR 10 + 12
NORMAL MODE OPERATION		1 + 3 + 9 + 11 OR 2 + 4 + 10 − 12
E + W STATION KEEPING		13 + 15 OR 14 + 16
MOMENTUM WHEEL OFF-LOADING		5 + 7 OR 6 + 8

Fig. 12-26. Thrusters location and functions.

Fig. 12-27. Example of a ECS thruster on the side of the satellite.

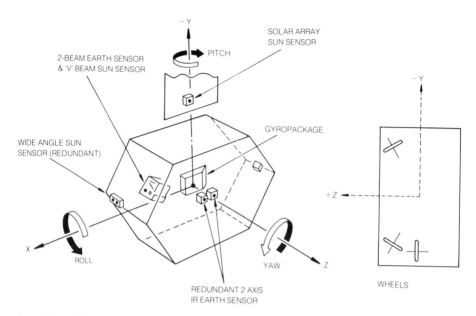

Fig. 12-28. AOCS sensors and wheel locations.

—Three momentum wheels arranged in a triangle in the YZ plane; two of these are fixed momentum wheels (FMW), of 25 Nms skewed by ± 35 °, with respect to the pitch axis; the third wheel is the transverse momentum wheel (TMW) of 20 Nms which can run either clockwise or counterclockwise to provide redundancy; each wheel has its own commutation electronics; for the TMW the direction of rotation is selected through the control law electronics (CLE);

—Hydrazine thrusters, eight of 2 N and eight of 0.75 N level; they are grouped in prime and redundant branches, supplied by four tanks arranged in two pairs; four latching valves provide full cross-strapping, as well as fuel management capabilities.

For the transfer orbit only the yaw and pitch thrusters are required, to perform spin axis reorientation and, as a contingency, to provide a perigee raising capability. For normal mode and station keeping, thrusters are required to operate as commanded by automatic loops in the CLE.

Control and drive electronics

The control and drive electronics utilize conventional analog and digital technology. They are arranged in four units for improved flexibility and reduced electromagnetic interference:

—Power supply electronics (PSE), supporting all other AOCS electronic units with the necessary power conditioning and switching functions; it contains the automatic failure detection circuits for the safe operation of the various power buses, as well as the automatic failure recovery switching logic.

It also provides for a safe Sun seeking mode in case Earth pointing cannot be re-established through the safety logic.

—Spin phase and thruster driver electronics (SPTE) which provide the capability for ground command of all thrusters during transfer orbit phase, for station keeping, and on any occasion where functions of the automatic control systems in the CLE need to be overridden.

—Control law electronics (CLE) containing all the automatic control logic and attitude monitoring electronics including fail safe detection circuitry, i.e., the station acquisition control loops, attitude acquisition control loops, normal mode control loops, wheel speed loop electronics, emergency Sun re-acquisition loops, telecommand centralized electronics, automatic monitoring electronics, memory protection electronics, and station keeping loops, utilizing microprocessor technology.

—The automatic failure recovery mode (ARM) is a special feature that is initiated through coordinated action of CLE and PSE, which has the capability to recover from most on-board failures and ground segment errors without disrupting the communication links;

—Wheel drive electronics (WDE) containing three separate drive amplifiers operating in pulse width modulation for wheel spin up and fine control at any wheel torque level.

Reaction Control Subsystem (RCS). The reaction control system consists of 16 hydrazine thrusters, eight of 2 N nominal thrust and eight of 0.5 N nominal thrust. They are arranged in prime and redundant branches and are supplied with mono-propellant hydrazine from four tanks arranged in two pairs. Cross strapping and fuel management is provided by means of four latching valves. The tanks are pressurized by compressed nitrogen separated from the hydrazine by an elastic diaphragm within the tanks. See Fig. 12-29.

For the transfer orbit only the yaw and pitch axis thrusters are required to provide spin up, axis reorientation and, in emergency, to raise the perigee.

During normal on-station mode of operation, thrusters are required to operate as commanded by the automatic loops in the control law electronics circuitry.

☐ **Operation.** Thrust is generated by the reaction of hydrazine with a catalyst, which causes the hydrazine to disassociate. The disassociation is exothermic, that is, it generates heat expanding the gaseous products and expelling them at high velocity from the thrusters.

Control of the thrust is by means of flow control valves that enable the thrusters to be operated in a pulsed mode for normal "on-station" orientation of the spacecraft, and in continuous mode for spin up, spin down and east/west and north/south station keeping.

Pyrotechnic Subsystem. The purpose of the pyrotechnic subsystem is to generate current pulses on telecommand and route them to the satellite pyrotechnic devices. The subsystem also provides arming circuits so that the

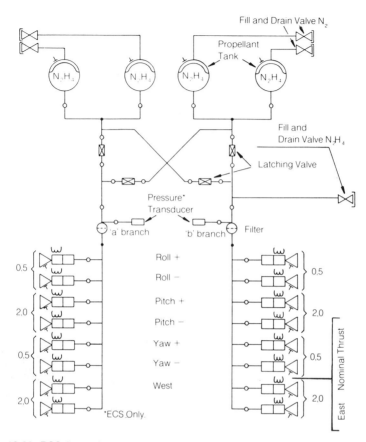

Fig. 12-29. RCS flow scheme.

devices can be isolated individually from the power supply until immediately before the required operation time. At launch, the entire subsystem is isolated from the power supply by "separation" switches which close only when the spacecraft separates from the launch vehicle.

The pyrotechnic subsystem is used to initiate the following operations:

☐ **ABM Ignition.** The AMB is fired when the spacecraft is near the chosen apogee in the transfer orbit. Because of the critical nature of this operation and the need for exact timing, the command to the pyrotechnic subsystem originates from a time tag unit situated in the TTC subsystem and preset by telecommand. Should a failure occur in this unit, a back-up is provided by direct command to fire the ABM.

Before the ABM can be fired, the firing circuit is connected to the power supply by "arming" telecommands.

☐ **Solar Array Deployment.** The solar arrays are retained folded against the spacecraft throughout launch and in the transfer orbits. After Sun acquisition by the AOCS, the arrays are released by severing the restraining cables with pyrotechnic guillotines. The two guillotines are operated sequentially. The circuits involved are connected to the power supply by "arming" telecommands. See Fig. 12-30.

☐ **BAPTA Release.** During launch and transfer orbit phases the BAPTAs are caged by a mechanism that protects their bearings from possible damage under vibration. They are released by ground command after array deployment and placed in the Sun acquisition mode to initiate rotation of the shafts so that the arrays face the sun. The relevant firing circuits are connected to the power supply by the arming commands that arm both the array release and BAPTA firing circuits.

☐ **Subsystem Configuration.**

The subsystem comprises the following:

—arming circuitry
—firing circuitry
—pyrotechnic devices

The design of the pyrotechnic firing circuits is such that each of the five units requiring pyrotechnics, i.e., the ABM, the two array wings and the two

Fig. 12-30. Solar array deployment test at Matra, Toulouse, France.

Fig. 12-31. MAGE II apogee boost motor on the integration stand.

BAPTAs has two squibs, each connected by a completely separate circuit to the battery, thus providing full redundancy.

Apogee Boost Motor Subsystem. The function of the apogee boost motor (ABM) is to accurately accelerate the spacecraft when it is near to the designated apogee of its transfer orbit to the velocity required to inject it into the desired quasi-stationary orbit.

The spacecraft baseline design uses the European MAGE II ABM but is capable of accommodating the Thiokol STAR 30B motor. The design of both motors permits the propellant load to be varied at the time of manufacture to provide different velocity increments or accommodate changes in space-craft mass.

For MAGE II, the ABM propellant burn is initiated from an ignitor chamber located in the center of the motor forward dome. The ignitor chamber is fitted with a small charge of propellant that is in turn ignited by a pair of explosive transfer lines that are, relatively, slow burning cord-type charges stretching from the safe and arm unit initiator squibs to the ignitor charge. See Fig. 12-31.

The ignition chain described is kept in a safe condition by a metal barrier disc mechanically isolating the squibs from the explosive transfer lines and by a switch breaking the electrical lines that power the squib firing lines, all located in the safe and arm unit.

An electric motor rotating the barrier plate arms the remotely located S&A unit, allowing the squib gases to impinge on the explosive transfer lines when the squibs are fired from the pyro relay unit (which has its own arming relay incorporated.)

The S&A unit is rotated to arm it late in the count-down procedure prior to launch and arming of the pyro relay unit is carried out by telecommand just prior to the ABM burn.

Initiation of the ignition chain finally results in hot gases from the ignitor chamber impinging on the main propellant grain to produce a controlled explosive burn of the main charge, the combustion products of which are expelled through a single axisymmetric carbon/carbon divergent nozzle, producing an essentially constant thrust for about 40 to 60 seconds.

The temperature of the ABM case during this burn and the heat "soak-back" after the burn is monitored via two temperature sensors located on the case outer-face. The fit of these temperature profiles to known previous performance profiles can help verify the correct performance of the ABM.

Appendix

Presented in the following pages are tables, charts and formulas that may be useful to the reader.

Worldwide Successful Space Launches

Year	USSR	US†	France	Australia	China	Japan	UK	ESA**	India	Total
1957	2									2
1958	1	7								8
1959	3	11								14
1960	3	16								19
1961	6	29								35
1962	20	52								72
1963	17	38								55
1964	30	57								87
1965	48	63	1							112
1966	44	73	1							118
1967	66	58*	2	1						127
1968	74	45								119
1969	70	40								110
1970	81	29*	2		1	1				114
1971	83	32*	1		1	2	1			120
1972	74	31*				1				106
1973	86	23								109
1974	81	24*				1				106
1975	89	28*	3		3	2				125
1976	99	26			2	1				128
1977	98	24				2				124
1978	88	32			1	3				124
1979	87	16				2		1		106
1980	89	13				2			1	105
1981	98	18			1	3		2	1	123
1982	101	18			1	1				121
1983	98	23‡			3	3		1‡		127
1984	97	22			1	3		4	1	129
1985	98	17			1	2		3		121
Total	**1,831**	**865**	**10**	**1**	**15**	**29**	**1**	**11**	**3**	**2,766**

* Italy has launched eight spacecraft from its San Marco platform. U.S. Scout rockets were used for these launches. so NASA includes them in the U.S. launch total.

** European Space Agency

† Does not include payloads launched from space shuttle.

‡ ESA launch from WSMC used U.S. Delta 3914 and is included in U.S. total

Table based on information reported by the Science Policy Research Division, Library of Congress

Manned Space Flight Records

	Total Flights	Earth Missions	Lunar Missions	Total Crew-Hours
United States	54	45	9	41,460
USSR	58	58	–	98,385
Total	112	103	9	139,845

First manned flight: Vostok 1
First EVA: Voskhod 2
Longest mission: Soyuz T-5/T-7 (211 days, 9 hrs, 05 mn)
First rendezvous mission: Gemini 7 and 8
First docking flight: Gemini 8
First lunar orbit: Apollo 8
First crew exchange: Soyuz 4 and 5
First manned lunar landing: Apollo 11
First triple launch/manned mission: Soyuz 6, 7, 8

Heaviest manned spacecraft: Skylab 1 (164,597 lb)
First use of reusable manned spacecraft: STS 1
First modular space station assembly: Salyut 7/Kosmos 1443
First space station repair: Solyut 7/ Soyuz T-13
First retrieval, repair of satellite: STS 41-C
First space construction: STS 61-B

Payloads In Orbit*

Australia	3	Italy	1
Brazil	1	Japan	28
Canada	14	Mexico	2
ESA	16	NATO	6
France	13	PRC	3
France/FRG	2	Saudi Arabia	2
FRG	5	Spain	1
India	7	UK	9
Indonesia	3	US	516
ITSO	35	USSR	906
Subtotal	99	Subtotal	1,474
Total			1,573

* December 31, 1985

STS Payload Deployment Record 1982–1985

Year	Successes*	Failures	Total
1982	2	0	2
1983	4	1**	5
1984	8	2†	10
1985	18	1‡	19
Total	32	4	36

*Includes satellites, platforms, experiments

**IUS failure cast Tracking and Data Relay Satellite into incorrect orbit; spacecraft subsequently boosted to geosynchronous orbit using its own thrusters

†Perigee kick motor failures. Westar VI and Palapa B returned to earth on later STS mission

‡Faulty switch prevented perigee kick motor firing; Leasat 3 repaired on orbit on later STS mission and successfully boosted into geosynchronous orbit

WIRE AND WINDING DATA

Sizes, Areas, Resistance & Current Capacities for Synthetic Film Insulated Wire †

WIRE SIZE AWG	WIRE AREA (MAX) * (CIRCULAR MILS)	
	HEAVY	TRIPLE
8	18,010	18,360
9	14,350	14,670
10	11,470	11,750
11	9,158	9,390
12	7,310	7,517
13	5,852	6,022
14	4,679	4,830
15	3,758	3,894
16	3,003	3,114
17	2,421	2,520
18	1,936	2,025
19	1,560	1,632
20	1,246	1,310
21	1,005	1,063
22	807	853
23	650	692
24	524	562
25	424	458
26	342	369
27	272	296
28	219	240
29	180	199
30	144	161
31	117	132
32	96.0	110
33	77.4	90.2
34	60.8	70.6
35	49.0	57.8
36	39.7	47.6
37	32.5	38.4
38	26.0	31.4
39	20.2	25.0
40	16.0	19.4
41	13.0	16.0
42	10.2	13.0
43	8.4	10.2
44	7.3	9.0

*Areas are for maximum wire area plus maximum insulation buildup.
**Based on 1,000 cir. mils/amp., current capacity will vary according to the geometry of the unit and may range from 750 to 1200 cir. mils/amp.
†Includes Formvar and Poly-Thermaleze types.

QUAD	OHMS/1000'	CURRENT (MA) CAPACITY(**)
18,960	.6281	16,510
15,200	.7925	13,090
12,230	.9987	10,380
9,821	1.261	8,226
7,885	1.588	6,529
6,336	2.001	5,184
5,112	2.524	4,109
4,147	3.181	3,260
3,329	4.020	2,581
2,704	5.054	2,052
2,190	6.386	1,624
1,781	8.046	1,289
1,436	10.13	1,024
1,170	12.77	812.3
949	16.20	640.1
778	20.30	510.8
635	25.67	404.0
520	32.37	320.4
424	41.02	252.8
342	51.44	201.6
276	65.31	158.8
231	81.21	127.7
188	103.7	100.0
154	130.9	79.21
128	162.0	64.00
104	205.7	50.41
82.8	261.3	39.69
67.2	330.7	31.36
54.8	414.8	25.00
44.9	512.1	20.25
36.0	648.2	16.00
28.1	846.6	12.25
22.1	1079.6	9.61
	1323.	7.85
	1659.	6.25
	2143.	4.84
	2593.	4.00

CONVERSION FACTORS

MULTIPLY	BY	TO OBTAIN
WEIGHT		
Pounds	453.59	grams
Pounds	0.45359	kilograms
Grams	0.0022046	pounds
Kilograms	2.2046	pounds
LENGTH		
Feet	30.480	centimeters
Inches	2.5400	centimeters
Centimeters	0.032808	feet
Centimeters	0.39370	inches
Inches	0.025400	meters
Meters	39.370	inches
AREA		
Square feet	929.04	square centimeters
Square inches	6.4516	square centimeters
Square centimeters	1.0764×10^{-3}	square feet
Square centimeters	0.15500	square inches
Square inches	6.4516×10^{-4}	square meters
Square meters	1.5500×10^{3}	square inches
Square centimeters	10^{-4}	square meters
Square meters	10^{4}	square centimeters
SINUSOIDAL WAVEFORM		
Peak current or voltage	0.70711	rms current or voltage
Peak current or voltage	0.63662	average current or voltage
Rms current or voltage	1.4142	peak current or voltage
Rms current or voltage	0.90032	average current or voltage
Average current or voltage	1.5708	peak current or voltage
Average current or voltage	1.1107	rms current or voltage
MAGNETIC INDUCTION, B		
Gausses	6.4516	lines per square inch
Gausses	6.4516×10^{-8}	webers per square inch
Gausses	10^{-4}	webers per square meter (teslas)
Lines per square inch	0.15500	gausses
Lines per square inch	1.5500×10^{-5}	webers per square meter (teslas)
Lines per square inch	10^{-8}	webers per square inch
Webers per square inch	1.5500×10^{7}	gausses
Webers per square inch	10^{8}	lines per square inch
Webers per square inch	1550	webers per square meter (teslas)

MULTIPLY	BY	TO OBTAIN
MAGNETIZING FORCE, H		
Oersteds	2.0213	ampere-turns per inch
Oersteds	0.79577	ampere-turns per centimeter
Oersteds	79.577	ampere-turns per meter
Ampere-turns per centimeter	1.2566	oersteds
Ampere-turns per centimeter	2.5400	ampere-turns per inch
Ampere-turns per centimeter	100.00	ampere-turns per meter
Ampere-turns per inch	0.49474	oersteds
Ampere-turns per inch	0.39370	ampere-turns per centimeter
Ampere-turns per inch	39.370	ampere-turns per meter
Ampere-turns per meter	0.012566	oersteds
Ampere-turns per meter	10^{-2}	ampere-turns per centimeter
Ampere-turns per meter	0.025400	ampere-turns per inch
PERMEABILITY		
Gausses per oersted	3.1918	lines per ampere-turn inch
Gausses per oersted	3.1918×10^{-8}	webers per ampere-turn inch
Gausses per oersted	1.2566×10^{-6}	webers per ampere-turn meter
Webers per ampere-turn meter	7.9577×10^{5}	gausses per oersted
Webers per ampere-turn meter	2.5400×10^{6}	lines per ampere-turn inch
Webers per ampere-turn meter	0.025400	webers per ampere-turn inch
Webers per ampere-turn inch	3.1330×10^{7}	gausses per oersted
Webers per ampere-turn inch	10^{8}	lines per ampere-turn inch
Webers per ampere-turn inch	39.370	webers per ampere-turn meter
Lines per ampere-turn inch	0.31330	gausses per oersted
Lines per ampere-turn inch	39.370×10^{-8}	webers per ampere-turn meter
Lines per ampere-turn inch	10^{-8}	webers per ampere-turn inch

Temperature Conversion

Given	Celsius	Fahrenheit	Kelvin	Reaumur	Rankine
Cels.	—	$\left(\dfrac{9}{5}C\right) + 32$	$C + 273.16$	$\dfrac{4}{5}C$	$1.8\,(C + 273.16)$
Fahr.	$\dfrac{5}{9}(F - 32)$	—	$\left[\dfrac{5}{9}(F - 32)\right] + 273.16$	$\dfrac{4}{9}(F - 32)$	$F + 459.7$
Kelvin	$K - 273.16$	$\left[\dfrac{9}{5}(K - 273.16)\right] + 32$	—	$\dfrac{4}{5}(K - 273.16)$	$K \times 1.8$
Reau.	$Re \times \dfrac{5}{4}$	$\left(\dfrac{9}{4}Re\right) + 32$	$\left(\dfrac{5}{4}Re\right) + 273.16$	—	$\left(\dfrac{9}{4}Re\right) + 491.7$
Rank.	$\dfrac{Ra}{1.8} - 273.16$	$Ra - 459.7$	$\dfrac{Ra}{1.8}$	$\dfrac{4}{9}(Ra - 491.7)$	—

Five major temperature scales are in use at present. They are: Fahrenheit, Celsius, Kelvin (Absolute), Rankine, and Reaumur. The interrelationship among the scales is shown here.

Bibliography

1. Meltzer, S.A. *Designing for Reliability.* IRE Transactions on Reliability and Quality Control, PGRQC-8, September 1956.
2. Bazovsky, Igor *Reliability Theory & Practice.* Prentice-Hall, 1961.
3. Koelle, H.H. *Handbook of Astronautical Engineering.* McGraw-Hill Company, Inc., 1961.
4. James, D.C. et al, *Redundancy and the Detection of First Failures.* IRE Transactions on Reliability and Quality Control, Vol.RQC-11, No. 3, Oct. 1962.
5. Farrell, Edward J. *Improving the Reliability of Digital Devices with Redundancy: an Application of Decision Theory.* IRE Transactions on Reliability and Quality Control, Vol.RQC-11, No. 1, May 1962.
6. Klein, W.R. and Lehr, S.N. *Reliability of Solar Array.* IRE Transactions on Reliability and Quality Control,Vol.RQC-11, No. 3, pp 71-80, October 1962.
7. Fischell, Robert E. and Mobley, Frederick F. *A System For Passive Gravity-Gradient Stabilization of Earth Satellites.* AIAA Guidance and Control Conference, August 12-14, 1963.
8. Balakrishnan, A.V. *Space Communications.* McGraw-Hill Book Company, Inc., 1963.
9. *Progress in Astronautics and Aeronautics.* Vol. 11, Power Systems for Space Flight, 1963.
10. Thomson, William Tyrrel *Introduction to Space Dynamics.* New York, John Wiley and Sons, Inc., 1963.

11. Leinwoll, Stanley *Space Communications.* John F. Rider Publisher, Inc., 1964.

12. Blasingame, Benjamin P. *Astronautics.* McGraw-Hill Book Company, 1964.

13. Mueller, G.E. and Spangler, E.R. *Communications Satellites.* Wiley, 1964.

14. Geyger, W.A. *Non-linear Magnetic Control Devices.* New York, McGraw-Hill Inc., 1964.

15. *Multiple Access To a Communication Satellite With a Hard-limiting Repeater.* Report R-108, Institute for Defense Analyses, Research and Engineering Support Division, 1965.

16. Hilton, W.F. *Manned Satellites.* Harper & Row, Publishers, 1965.

17. Haviland and House, C.M. *Handbook of Satellite and Space Vehicles.* Van Nostrand, 1965.

18. *Progress in Astronautics and Aeronautics.* Vol. 16, Space Power Systems Engineering, 1966.

19. Stiltz, Harry L. *Aerospace Telemetry.* Prentice-Hall, 1966.

20. *Advance Study of an Applications Technology Satellite (ATS-4) Mission.* Final Report, by Lockheed Missiles & Space Company, 1966.

21. Dorrell, Russel and Cooke, Patrick *Tips to Experiments.* Quality Assurance Branch, Test and Evaluation Division, NASA Goddard Space Flight Center, Greenbelt, Maryland, December 1966.

22. *Electrical Power Generation System for Space Applications.* NASA-SP-79, P 18, 1966.

23. Corliss, William R. *Scientific Satellites.* NASA SP-133, 1967.

24. Baker, R.M.L Jr., and Makemson, M.W. *An Introduction to Astrodynamics.* Academic Press, 1967.

25. Baker, J.K. et al. *Power Conditioning Reliability Improvement through Standby Redundancy and Automatic Failure Detection.* IEEE Intersociety Energy Conversion Engineering Conference, pp 587-593, Aug. 1967.

26. Jenson, K.J. *Synchronization and Failure Isolation for Redundancy Low Input Voltage Converters.* EASTCON Convention Record 1967.

27. Hatcher, Norman M. *A Survey of Attitude Sensors for Spacecraft.* NASA-SP-145, 1967.

28. Bauer, Paul *Batteries for Space Power Systems.* NASA-SP-172, 1968.

29. Brooman and McCallum *The Thermal Properties and Behaviour of Nickel-Cadmium and Silver-Zinc Cells and Their Components.* Battelle Memorial Institute, AFAPL-TR-68-41, June 1968.

30. Clarke, Arthur C. *The Promise of Space.* New York, Harper & Row, 1968.

31. *Spacecraft Earth Horizon Sensors.* NASA-SP-8033, December 1969.

32. Mantell, C.L. *Batteries and Energy Systems.* McGraw-Hill Book Company, 1970.

33. Voss, J.M. and Gray, J.G. *A Regulated Solar Array Model—A Tool for Power System Analysis.* IEEE Power Conditioning Specialists Conference Record, April 20-21, 1970.

34. *Spacecraft Star Trackers.* NASA-SR-8026, July 1970.

35. Landsman, E.E. *Modular Converters for Space Power Systems.* IEEE Power Conditioning Specialists Conference Record, pp 87-99, April 1970.

36. Bazin, A. *Etude du Circuit de Controle Multiphase de la Commande de n etages de puissance en parallel.* Contract ESTEC No. 830/69/HP, 1970.

37. Paulkevich, John, and Ford, Floyd E. *Coulometer and Third Electrode Battery Charging Circuit.* NASA-Case-GSC-10487-1, US-Patent-3541422, NASA, Goddard Space Flight Center, Greenbelt, Maryland, November 1970.

38. Dumont, A. et al. *Development of an Advanced Control Circuit for Satellite Power Systems.* ESRO-CR-15, March 1970.

39. *The ATS-F and -G Data Book.* NASA Goddard Space Flight Center, 1971.

40. Bate, R.R., Mueller, D.D. and White, J.E. *Fundamentals of Astrodynamics.* Dover, 1971.

41. *Design of Liquid Propellant Rocket Engines.* NASA-SP-125, 1971.

42. Wick, H.M. Jr. *A Design for Thick-film Microcircuit dc to dc Converter Electronics.* IEEE Transactions On Aerospace and Electronic Systems, pp 528-531, Vol.AES.7, No. 3, May 1971.

43. Hnatek, F.R. *Design of Solid State Power Supplies.* Van Vostrand Reinhold Company, New York, 1971.

44. Frohlich, H. and Muller, W. *Modularised Power System Concept for General Application in Satellite at Power Ranges up to 5 KW.* ESRO-SP-84, pp 117-129, July 1972.

45. Boehringer, A. and Haussmann, J. *Dynamic Behaviour of Power Conditioning Systems for Satellites with a Maximum Power Point Tracking System.* ESRO-SP-84, pp 21-28, 1972.

46. Capel, A. and O'Sullivan, D.M. *A Sequenced PWM Controlled Power Conditioning Unit for a Regulated Bus Satellite Power System.* IEEE Power Processing and Electronics Specialists Conference Record, pp 48-58, 1972.

47. O'Sullivan, D.M. *The Multiphase PWM Shunt.* ESRO-SP-84, pp 135-147, April 1972.

48. O'Sullivan, D.M. and Capel, A. *Modularised Power Conditioning Units for High Power Satellite Applications.* IEEE Intersociety Energy Conversion Engineering Conference Record, pp 639-650, Sept. 1972.

49. Lilli, V.R. and Schoefeld, A.D. *ASDTIC - Duty Cycle Control for Power Converters.* IEEE 1972 Power Processing and Electronics Specialists Conference, pp 96-102, May 1972.

50. Polcyn, Kenneth A. *An Educator's Guide To Communication Satellite Technology.* Information Center on Instructional Technology, 1973.

51. *U.S. Space Launch Systems (U).* 1 March 1973, Report No NSSA-R-20-72-2.

52. Wolf, M. *Potential Improvements in Efficiency and Cost of Solar Cells.* IEEE Photovoltaic Specialists Conference, November 13-15, 1973, Palo Alto, California.

53. Dee. S. *Recent Trends in Development of Solar Cells and Economics of Solar Energy Conversion.* Presented at the International Solar Energy

Conference, "The Sun in the Services of Mankind", Paris, France, July 1973.

54. Treble, F.C. *Solar Arrays for the next generation of Communication Satellites.* JBIS, Vol. 26, No. 8, pp 449-465, Aug. 1973.

55. Call, R.A. *An Induced Junction Photovoltaic Cell.* IEEE Photovoltaic Specialists Conference Record, pp 64-68, 1973.

56. Sater, B.L. et al. *The Multijunction Edge Illuminated Solar Cells.* IEEE Photovoltaic Specialists Conference Record, pp 188-193, 1973.

57. Chetty, P.R.K. et al. *Status of Solar Cells for Space Applications.* Proceedings of the 5th meeting of All India Solar Energy Conversion Conference, IIT., Madras, November 1973.

58. Ashfold, E. and Lechte, H. *Satellite System Considerations in the Selection of Solar Array Configuration Generators.* Presented at the International Solar Energy Conference, "The Sun in the Service of Mankind", Paris, France, July 1973.

59. Sander, William August III *Static DC to DC Power Conditioning Active Ripple Filter, 1 MHz DC to DC Conversion, and Non-linear Analysis.* Ph.D. Dissertation, Dept. of Electrical Engineering, Duke University, 1973.

60. Cardwell, G.I. and Neel, W.O. III *Bilateral Power Conditioner.* IEEE Power Processing and Electronics Specialists Conference Record, pp 214-221, June 1973.

61. Wester, G.W. *Stability analysis and Compensation of a Boost Regulator with two loop Control.* IEEE Power Electronics Specialists Conference Record, 1974.

62. Capel, A. et al. *Stability Analysis of a PWM Controlled DC/DC Regulator with DC and AC Feedback Loops.* ESRO-SP-103, pp 233-243, September 1974.

63. Denzinger, W. *Square-wave Generation and Distribution.* Spacecraft Power Conditioning Electronics Conference, ESRO-SP-103, pp 65-70, September 1974.

64. Hehnew, R. *AC Power Systems in the Kilowatt Range.* Spacecraft Power Conditioning Electronics Conference, ESRO-SP-103, pp 55-62, September 1974.

65. Marsh, M.J. *Advances in Methods of Construction including Thick-Film Multichip Integration for Future Spacecraft Power System.* ESRO-SP-103, pp 315-323, September 1974.

66. Gohrbandt, B. *Modular DC Power System for the 0.5 to 5 KW Range.* ESRO-SP-103, pp 77-82, September 1974.

67. Dennies, Norman G. *Insight into Standby Redundancy via Unreliability.* IEEE Transactions on Reliability, Vol-R-23, No. 5, December 1974, pp 305-313.

68. Young, R.W. *Power Subsystem Configuration for Geosynchronous Application Satellites.* Proceedings of Spacecraft Power Conditioning Electronics Seminar, 1974, ESA-SP-103, pp 13-18.

69. Lindeman, Gerald A. *Regulation and Control of a Multi-Kilowatt Space-*

craft Power System. Proceedings of IECEC, 1974, pp 232-241.

70. Rusta, D. and Sternberg, R. *Power Subsystem Simulation Studies for the Timation IIIA Spacecraft.* IECEC 1974.

71. Corbett, R.F., Glass, M.C. and Matsu, R.G. *Development of a Power Module using Third Electrode Battery Charge Control.* 9th Intersociety Energy Conversion Engineering Conference Record, Aug. 1974.

72. Lechte, H. *Monitoring, Control and Protection Techniques for Storage Batteries of Application Satellites.* ESRO-SP-103, pp 285-297, September 1974.

73. Hogsholm, A. *Design Aspects of the PCU for GEOS power System.* ESRO-SP-103, pp 83-92, September 1974.

74. Glaser, P.E., Maynard, O.E., Mackovcisk, J., Jr. and Ralph, E.L. *Feasibility Study of a Satellite Solar Power Station.* NASA-CR-2357, 1974.

75. *CTS Reference Book.* NASA TM X-71824, 1975.

76. *Low Cost Modular Spacecraft Description* NASA Goddard Space Flight Center, X-700-75-140, 1975.

77. *Nickel-Cadmium Battery - Application Engineering Handbook.* Second Edition, General Electric, 1975.

78. Von Braun, Werner and Ordway, Frederick I. III, *History of Rocketry & Space Travel.* 1975.

79. *Solar Array Status.* Internal Report, Aerospatiale, Division Systems Balistiques et spatiales, Establishments de Connes, June 1975.

80. Nagler, R.G. and Schlue, John W. *EODAP Programmatic Assessments Summary, Satellite and Sensor Capabilities, Applications and Science Opportunities.* Jet propulsion laboratories, Pasadena, California, October 1975.

81. Capel, A., et al. *State Variable Stability Analysis of Multi-loop PWM Controlled DC/DC Regulators in Light Heavy Mode.* IEEE Power Electronics Specialists Conference Record, 1975.

82. Adcole Corp, *Sun Angle Sensor Systems Short Form Catalog.* February 1975.

83. Green, M.A. *Resistivity Dependence of Silicon Solar Cell Efficiency and its Enhancement using a Heavily Doped Back-Contact Region.* IEEE Transactions on ED, Vol. 23, No. 1, pp 11-16, January 1976.

84. Smiths, F.M. *History of Silicon Solar Cells.* IEEE Transactions on Electron Devices, Vol. ED. 23, No. 7, pp 640-643, July 1976.

85. Restrepo, F., and Backus, C.E. *On Black Solar Cells for the Tetrahedral Texturing of a Silicon Surface.* IEEE Transactions on Electron Devices, Vol. 23, No. 10, pp 1195-1197, October 1976.

86. Du Bow, Joel *From Photons to Kilowatts? Can Solar Technology Deliver?* Electronics, Vol. 49, No. 23, pp 86-90, November 1976.

87. Costogue, E.N., and Lindena, S. *Comparison of Candidate Solar Array Maximum Power Utilization Approaches.* Proceedings of 11th Intersociety Energy Conversion Engineering Conference, pp 1449-1456, 1976.

88. *Constraints imposed on passengers flown on Ariane Test Flights.* AR (75)03,

Issue (3), October 1976, European Space Agency.

89. Mennie, Don *Batteries Today and Tomorrow*. IEEE Spectrum, Vol. 13, No. 3, pp 36-41, March 1976.

90. Dryden, M.H. *Design for Reliability*. Mullard technical Communication, No.130, April 1976, pp 395-432.

91. Salim, Abbas A. *A Simplified Minimum Power Dissipation Approach to Regulate the Solar Array Output Power in a Satellite Power Subsystem.* Proceedings of IECEC, 1976, pp 1437-1442.

92. *Solar Cell Array Design Handbook*. Volumes 1 & 2, JPL SP 43-38, 1976.

93. *Outlook For Space*. NASA SP-386, 1976.

94. Peterson, W.A. *A Frequency Stabilized Free Running DC-to-DC Converter Circuit employing Pulse Width Control Regulation.* IEEE Power Electronics Specialists Conference Record, pp 200-205, June 1976.

95. Kaplan, Marshal H. *Modern Spacecraft Dynamics and Control.* New York, John Wiley and Sons Inc., 1976.

96. *A Forecast of Space Technology*. NASA-SP-387, 1976.

97. Hussey, W. John *The TIROS-N Polar Orbiting Environmental Satellite System.* PB 280 743, National Oceanic and Atmospheric Administration, 1977.

98. Rosen, Harold A., and Jones, Richard *STS-Optimised Satellite Concept.* Astronautics and Aeronautics, pp 48-53, June 1977.

99. Putois, F. *Feasibility of Silver - Hydrogen Rechargeable Cells.* Third ESTEC Spacecraft Power Conditioning Seminar, ESA-SP-126, pp 21-23, September 1977.

100. Muller, W., et al., *Square-wave AC Power Generation and Distribution of High Power Spacecraft.* Third ESTEC Spacecraft Power Conditioning Seminar, ESA-SP-126, September 1977.

101. Marsh, M.J. *Thick-film Development for Spacecraft.* Third ESTEC Spacecraft Power Conditioning Seminar, ESA-SP-126, September 1977.

102. Genzel, J.P. *Management of Heat Generated Onboard Spacecraft.* Proceedings of the Third ESTEC Spacecraft Power Conditioning Seminar held at Noordwijk, The Netherlands, ESA-SP-126, pp 33-38, 1977.

103. Web, A.D. *Designing Electrical Interfaces.* Proceedings of Spacecraft Power Conditioning Seminar, 21-23 Sept. 1977, ESA-SP-126, pp 39-48.

104. *Global Satellite Communications.* Scientific American, Vol. 236, No. 2, February 1977.

105. Lindmayer, J. and Wrigley, C.Y. *Ultrathin Silicon Solar Cells.* IEEE Photovoltaic Specialists Conference 1978, pp 450-453.

106. O'Sullivan, D. and Weinberg, A. *Developments in Modular Spacecraft Power Conditioning for Applications Satellites.* Proceedings of IECEC, 1978, pp 28-36, Paper No. 789005.

107. Chetty, P.R.K. *Spacecraft Power Systems - Some New Techniques for Performance Improvement.* Ph.D. Thesis, Indian Institute of Science, Bangalore, India, 1978.

108. Scott, W.R. and Rusta, D.W. *Sealed cell Nickel-Cadmium Battery Application Manual.* NASA Reference Publication 1052, December 1979.

109. Chetty, P.R.K., and Sivaprasad, N.V. *An Improved Relay Driver Circuit.* Electronic Engineering, U.K., August 1979.
110. *Solar Power Satellites: Microwaves Deliver the Power.* Spectrum, 1979, pp. 36-42.
111. Jones, W.L., et al., *SEASAT Scatterometer: Results of the Gulf of Alaska Workshop.* Science 204, pp 1413-1415, 1979.
112. Murrel, James W. *The NASA Multimission Spacecraft Modular Attitude Control System.* presented at Annual Rocky Mountain Guidance and Control Conference, February 24-28, 1979, Keystone, Colorado.
113. ITHACO Inc., IPS 0006, *Conical Earth Sensor.* July 1979.
114. *Sealed Nickel-Cadmium Battery Applications Manual*, by Willard R. Scott and Douglas W. Rusta, NASA Reference Publication 1052, December 1979.
115. *30/20 GHz Communications System.* 1-1-MO-1-T1, Motorola Inc., Prepared for NASA Lewis Research Center, 1981.
116. *Ariane into Space.* Spaceflight, Vol. 23, October 8, 1981.
117. *Electric Power from Orbit: A Critique of a Satellite Power System.* National Academy press, 1981.
118. *NASA GSFC Battery Workshop Proceeding*, 1981.
119. *India's SLV-3 Launch Vehicle.* Spaceflight, Vol. 24, February 2, 1982.
120. Betz, Frederick E. *Real and Potential Nickel Hydrogen Superiority.* NASA GSFC Battery Workshop Proceedings, pp 416-429, 1982.
121. Dunlop, J.D. *Ni-H$_2$ Batteries for Communication Satellites.* NASA GSFC Battery Workshop Proceedings, pp 389-416, 1982.
122. *NASA GSFC Battery Workshop Proceedings.* 1982.
123. *Solar Cell Radiation Handbook, Third Edition.* Jet Propulsion Laboratory, Publication 82-69, 1982.
124. Slifer, L.W. *Comparative Values of Advanced Space Solar Cells.* NASA Technical Memorandum 84951, September 1982.
125. Kunzi, K.F., Paril, S. and Rott, H. "Snow Cover Parameters Retrieved From Nimbus-7 SMMR Data", *IEEE Transactions on Geoscience & Remote Sensing.* GE-20, pp 452-467, 1982.
126. Stadnick, Steven J. "Ni-H$_2$/Ni-Cd Battery Trade Studies" *NASA GSFC Battery Workshop Proceedings.* pp 430-438, 1982.
127. *Aviation Week and Space Technology.* 5 April 1982, p 57.
128. *Aviation Week and Space Technology.* 19 April 1982, p 24.
129. *Spaceflight, Vol. 24.* December 12, 1982, p 472.
130. Graf, Rudolf F. *Electronic Data Book, 3rd Edition.* TAB Books Inc., 1983.
131. Schwenk, Francis C. *Summary Assessment of the Satellite Power System.* American Institute of Aeronautics and Astronautics, Journal of Energy, Vol. 7, No. 3, May-June 1983.
132. *NASA GSFC Battery Workshop Proceeding.* 1983.
133. Morgan, Walter L. "INSAT", *Satellite Communications.* August 1984.
134. "Broad Spectrum of Business Involved in Space Commercialization", *Aviation Week & Space Technology.* June 25, 1984, pp 62-63.
135. "Unique Products, New Technology Spawn Space Business", *Aviation*

Week & Space Technology. June 25, 1984, pp 41-51.

136. "Medicine Sales Forecast at $1 Billion", *Aviation Week & Space Technology.* June 25, 1984, pp 52-56.

137. *NASA GSFC Battery Workshop Proceeding.* 1984.

138. Matsunda, S., et al., *Development of AlGaAs/GaAs Solar Cells with Space Qualifications.* IEEE Photovoltaic Specialists Conference 1984.

139. Ralph, E.L. *Photovoltaics - 10 years After Cherry Hill.* 17th IEEE Photovoltaic Specialists Conference Record, May 1984.

140. Knechtli, R.C., Loo, R.Y., and Kamath, G.S. *IEEE Transactions on Electron Devices.* Vol. 31, 1984.

141. *Evaluation Program for Secondary Spacecraft Cells - 20th Annual Report of Cycle Life Test.* Dept of Navy, Crane, Indiana, January 1984.

142. Ni-H_2 *System Current Status.* Eagle-Picher Industries Inc. September 1984.

143. *Satellite Communications.* July 1984, pp 36-37.

144. *Satellite Communications.* August 1984, p 48.

145. *Space.* December 1984, Vol. 1, No. 3, Shearson Lehman - American Express.

146. *What's the Payoff?* Commercial Space, Summer 1985.

147. Ni-H_2 *System Technology Update.* Eagle-Picher Industries Inc., May 1985.

148. Scott-Monck, L. and Stella, P *Current Status of Advanced Solar Array Technology.* IECEC 1985.

149. *NASA GSFC Battery Workshop Proceeding.* 1985.

150. Phenneger, M.C., Singhal, S.P. and Lee, T.H. *Infrared Horizon Sensor Modeling for Attitude Determination and Control: Analysis and Mission Experience.* NASA Technical Memorandum 86181, 1985.

151. Chetty, P.R.K. *Switch-Mode Power Supply Design.* TAB Books Inc., 1986

152. Houghton, John T. *The Role of Satellites in Meteorology and the Contribution of Europe to Space Meteorology.* European Space Directory, 1986.

153. "Chinese Prepare Long March 3 for Launch", *Aviation Week & Space Technology.* September 15, 1986.

154. Martin, Graham J. *Gyroscopes May Cease Spinning.* IEEE Spectrum, February 1986, pp 48-53.

155. Lillington, D.R., and Kukulka, J.R. *Further Advances in Silicon Solar Cell Technology for Space Application.* IECEC 1986, pp 1446-1451.

156. Kamath, G.S. *Advanced Solar Cells for Space Application.* IECEC 1986, pp 1423-1426.

157. Ralph, E.L. *Photovoltaic Technology Assessment.* IEEE Aerospace and Electronic Systems Magazine, October 1986, pp 2-7.

158. Billerbeck, W.J. *High Power Needs for Commercial Satellites.* IEEE Aerospace and Electronic Systems Magazine, October 1986, pp 20-28.

159. Wertz, James R. *Spacecraft Attitude Determination and Control.* D. Reidel Publishing Co.

160. *Development of Flexible, Fold-out Solar Array.* G. Barkats, SNIAS, BP

52-06322 CANNES LA BOCCA, France and J. FREMY, SAT, 41 Rue de Centegral -75013, Paris, France.

161. *Silicon Photovoltaic Cell Solar Generator.* Societe Anonyme de Telecommunications Catalogue, 41, Rue Cantegral, Paris-13.

162. Babaa, I.M.H., et al., *DC to DC Converter Regulated by a Constant Frequency Duty Cycle Generator.* Bell Telephone Laboratories, Winston Salem, N.C.

163. *Symphonie Power System Report.* CNES (French Space Agency), France.

164. *Reliability Stress and Failure Rate Data for Electronic Equipment.* MIL-HDBK-217A.

165. *Gallium Arsenide Solar Cells Catalog.* Spectrolab Inc., a subsidiary of Hughes Aircraft Co.

166. *Design Manual Featuring Tape Wound Cores.* TWC-300R, Magnetics, USA.

167. *Space Flight Actuators Catalog.* Schaeffer Magnetics, Inc., USA.

168. *Biaxial Drive System, Type 55, Catalog.* Schaeffer Magnetics, Inc., USA

169. *A New Generation of Solar Array Drive and Power Transfer Assembly Catalog.* SPAR, Canada.

170. Bate, Roger R., Mueller, Donald D., and White, Jerry E. *Fundamentals of Astrodynamics.* Dover Publications, Inc., USA.

171. *Magnetic Moment Compensator.* IPS 4-4/79, Ithaco, Inc., USA.

172. *Spinner Magnetic Attitude Control System.* IPS 11-8/80, Ithaco, Inc., USA.

173. *Magnetic Acquisition/Despin System.* IPS 0009, 7/79, Ithaco, Inc., USA.

174. *Three Axis Magnetic Attitude Control System.* IPS 7-7/79, Ithaco, Inc., USA.

175. *Scanwheel.* IPS 0005, 6/83, Ithaco, Inc., USA.

176. *Torqrod.* IPS 0003, 4/79, Ithaco, Inc., USA.

177. *Boresight Limb Sensor.* IPS 0012, 5/83, Ithaco, Inc. USA.

178. *Conical Earth Sensor.* IPS-6, 12/83, Ithaco, Inc., USA.

179. *Steerable Horizon Crossing Indicator.* IPS 1-4/79, Ithaco, Inc., USA.

180. *Horizon Crossing Indicator.* IPS 0002, 6/81, Ithaco, Inc., USA.

181. *Components for Space Applications, Momentum Wheels for Satellite Stabilization, Type DRALLRAD; Reaction Wheels for Satellite Attitude Control, Type RSR; Magnetic Bearing Reaction Wheels, Type MRR; Catalog.* TELDIX HEIDELBERG, Fed. Rep. of Germany.

182. *DRALLRAD + WDE, Ball Bearing Momentum & Reaction Wheels + Wheel Drive Electronics, Catalog.* TELDIX HEIDELBERG, Fed. Rep. of Germany.

183. *Solid State Star Scanner CS-207.* Data Sheet, Ball Aerospace Systems Division, USA.

184. *Standard Star Tracker.* Data sheet, Ball Aerospace Systems Division, USA.

185. *Applications for Landsat Imagery.* EOSAT Catalog, Earth Observation Satellite Company, USA.

186. *METEOSAT Europe's Weather Satellite System.* Brochure, European Space Agency.

187. *Oceanography from Space - A Research Strategy for the Decade 1985-1995.* Joint Oceanographic Institutions Inc., USA.

188. *Advanced TIROS-N (ATN) NOAA-E.* Brochure, NASA Goddard Space Flight Center, Greenbelt, Maryland, USA.

189. *The Soviet Space Program Revisited.* TRW Space Log.

190. *NASA ACTS - The Next Generation of Space Communications.* Brochure, NASA GSFC, Greenbelt, Maryland, USA.

191. *Ariane.* Brochure, Arianespace, France.

192. *Hubble Space Telescope.* Brochure, Lockheed Missiles & Space Company, Inc.

193. Seifert, Howard S. and Brown, Kenneth *Ballistic Missile and Space Vehicle Systems.* John Wiley & Sons, Inc.

194. *MMS External I/F Specifications and Users Guide.* S-700-11.

195. *Space Transportation System (STS).* TRW Space Log, Volume 19.

196. *Fleet Satellite Communications (FLTSATCOM).* TRW Space Log, Volume 19.

197. *Maritime Communications Satellite (MARECS-1).* TRW Space Log, Volume 19.

198. *TRW Space Log.* Twenty-Fifth Anniversary of Space Exploration, 1957-1982, Volume 19.

199. *TRW Space Log.* 1984-1985, Volume 21.

200. Gordon, Gary *Notes on Satellite Structu es.* COMSAT.

201. *TIROS-N/NOAA Weather Satellite Ser. .* Brochure, RCA News.

202. *DSCS - A Generation of High Performance, Reliable, Defense Communications Satellites.* Brochure, General Electric.

203. *Metallic Materials & Elements for Aerospace Vehicle Structures.* MIL-HDBK-5C.

Index

418

Edited by Roland Phelps